Mathematical Evolutions

©2002 by
The Mathematical Association of America (Incorporated)
Library of Congress Catalog Card Number 2001095774

ISBN 0-88385-536-4

Printed in the United States of America

Current printing (last digit):
10 9 8 7 6 5 4 3 2 1

Mathematical Evolutions

Edited by

Abe Shenitzer and John Stillwell

Published and Distributed by
The Mathematical Association of America

510.1
M423

SPECTRUM SERIES

Published by

THE MATHEMATICAL ASSOCIATION OF AMERICA

Committee on Publications
William Watkins, *Chair*

Spectrum Editorial Board
Gerald L. Alexanderson, *Editor*

Robert Beezer
William Dunham
Michael Filaseta
William J. Firey
Erica Flapan
Dan Kalman
Eleanor Lang Kendrick
Ellen Maycock
Russell L. Merris

Jeffrey L. Nunemacher
Jean Pedersen
J. D. Phillips
Jennifer J. Quinn
Marvin Schaefer
Harvey J. Schmidt, Jr.
Sanford L. Segal
Franklin Sheehan
Francis Edward Su

John E. Wetzel

The Spectrum Series of the Mathematical Association of America was so named to reflect its purpose: to publish a broad range of books including biographies, accessible expositions of old or new mathematical ideas, reprints and revisions of excellent out-of-print books, popular works, and other monographs of high interest that will appeal to a broad range of readers, including students and teachers of mathematics, mathematical amateurs, and researchers.

All the Math that's Fit to Print, by Keith Devlin

Circles: A Mathematical View, by Dan Pedoe

Complex Numbers and Geometry, by Liang-shin Hahn

Cryptology, by Albrecht Beutelspacher

Five Hundred Mathematical Challenges, Edward J. Barbeau, Murray S. Klamkin, and William O. J. Moser

From Zero to Infinity, by Constance Reid

The Golden Section, by Hans Walser. Translated from the original German by Peter Hilton, with the assistance of Jean Pedersen

I Want to Be a Mathematician, by Paul R. Halmos

Journey into Geometries, by Marta Sved

JULIA: a life in mathematics, by Constance Reid

The Lighter Side of Mathematics: Proceedings of the Eugène Strens Memorial Conference on Recreational Mathematics & Its History, edited by Richard K. Guy and Robert E. Woodrow

Lure of the Integers, by Joe Roberts

Magic Tricks, Card Shuffling, and Dynamic Computer Memories: The Mathematics of the Perfect Shuffle, by S. Brent Morris

The Math Chat Book, by Frank Morgan
Mathematical Carnival, by Martin Gardner
Mathematical Circus, by Martin Gardner
Mathematical Cranks, by Underwood Dudley
Mathematical Evolutions edited by Abe Shenitzer and John Stillwell
Mathematical Fallacies, Flaws, and Flimflam, by Edward J. Barbeau
Mathematical Magic Show, by Martin Gardner
Mathematical Reminiscences, by Howard Eves
Mathematics: Queen and Servant of Science, by E. T. Bell
Memorabilia Mathematica, by Robert Edouard Moritz
New Mathematical Diversions, by Martin Gardner
Non-Euclidean Geometry, by H. S. M. Coxeter
Numerical Methods That Work, by Forman Acton
Numerology or What Pythagoras Wrought, by Underwood Dudley
Out of the Mouths of Mathematicians, by Rosemary Schmalz
Penrose Tiles to Trapdoor Ciphers ... and the Return of Dr. Matrix, by Martin Gardner
Polyominoes, by George Martin
Power Play, by Edward J. Barbeau
The Random Walks of George Pólya, by Gerald L. Alexanderson
The Search for E.T. Bell, also known as John Taine, by Constance Reid
Shaping Space, edited by Marjorie Senechal and George Fleck
Student Research Projects in Calculus, By Marcus Cohen, Arthur Knoebel, Edward D. Gaughan, Douglas S. Kurtz, and David Pengelley
Symmetry, by Hans Walser. Translated from the original German by Peter Hilton, with the assistance of Jean Pedersen
The Trisectors, by Underwood Dudley
Twenty Years Before the Blackboard, by Michael Stueben with Diane Sandford
The Words of Mathematics, by Steven Schwartzman

MAA Service Center
P.O. Box 91112
Washington, DC 20090-1112
800-331-1622   FAX 301-206-9789

INTRODUCTION

Back in 1993 John Ewing, then editor of the *Monthly*, asked me (Abe Shenitzer) to edit a column of expository articles for the *AMM*. They were to shed light on the meaning and significance of important mathematical ideas. This was the genesis of "The Evolution of ..." column that was inaugurated in January 1994. I could not have done the required work without the assistance of a number of people, namely John Ewing, Roger Horn, the present editor of the *Monthly*, John Stillwell, now the coeditor of the column, and last but certainly not least, my friend and former colleague at York University, Hardy Grant, who has advised me on many occasions and has turned many an awkward piece of prose into a piece of prose.

The first column article was preceded by the following introduction:

> An English major may or may not be a novelist or a poet, but would undoubtedly be expected to be able to evaluate a novel or a poem. The term "English major" implies some historical, philosophical, and evaluative training and competence. It is sad but true that the term "mathematician" does not imply corresponding training and competence.
>
> Integration of the narrowly mathematical aspects of our discipline with its historical, philosophical, and critical aspects is bound to make it more meaningful not only to those who identify themselves as mathematicians but also to those who have no more than a tangential interest in the subject.
>
> To promote such integration, and thus encourage an approach to mathematics that emphasizes its meaning and significance, the *Monthly* will publish in alternate issues articles ... under the generic title "The Evolution of.... The core of such an article will be an account of important mainstream mathematics.... (*AMM* Jan. 1994, p. 66.)

Less than two years later I wrote the following COMMENT:

Specialization is the price we pay for creative achievements. This is true, in particular, in mathematics, where many productive mathematicians know little about mathematical ideas outside their specialties and even less about their evolution and role. When teaching Ph.D. students, our specialist-mathematicians may be extremely effective in helping novice researchers to produce their first paper, but as teachers of less exalted categories of students they are likely to be hampered by their mathematical narrowness. The articles in the column "The Evolution of..." may contribute to the enlargement of their mathematical horizons. The column's underlying ideology is perhaps best described by Weyl's idea of "... intermingling of philosophical, mathematical and physical thought...." More specifically:

> 1. The articles in the column are to promote an informed appreciation of mathematical ideas and issues of the last 300 years with emphasis on developments since 1800. (Obviously, in some cases it may be necessary to go

back thousands of years.) A prized quality is accessibility; the articles are not meant for experts.

2. The ideas and issues presented in the column are to be "classical", that is, of established weight and significance, with special attention paid to ideas and issues that overlap different domains of mathematics or overlap mathematics and other disciplines, such as physics or philosophy. Of course, there can be purely mathematical reasons for mathematical developments. Such reasons should be adduced so as to avoid producing instances of chronicles, sequences of disconnected facts.

3. All articles are bound to make some use of the history of mathematics, but an article acceptable to a history journal is probably "too learned" for this column, which aims at the broad audience of mathematicians with a fair amount of technical knowledge and a curiosity about different areas of mathematics, mathematicians who insist on being persuaded that an idea stamped "important" is indeed important. As editor, I prize that last quality so much that I would gladly replace the excessively open-ended, and thus potentially misleading title "The Evolution of..." with the clumsy, but informative, title "What is it? What good is it (in the broadest sense of the term)? How did it get to be what it is?"

4. I hope that the articles in the column will help to eliminate from the classroom presentations that begin with a list of axioms, or rely on some other unmotivated approaches. (*AMM* Oct. 1995, p. 674.)

Today, when writing an introduction for a selection of column articles to be published in the MAA Spectrum Series edited by Professor Gerald Alexanderson, we (that is, my friend and now official coeditor of the column John Stillwell and I) would like to add a few more down-to-earth remarks directed at prospective readers of this collection

I. The essays in this collection cover a great deal of mathematical ground and have ideological, rather than technical, unity.

II The essays follow the genetic approach—"from a problem to its solution"— and in this sense are only partly historical. The drama is not the drama of the lives of the creators of mathematics but the drama of evolving ideas.

We hope that this collection of essays lives up to the motto of a certain publishing house one of us knew as a child: "To teach without boring [and] to entertain without lying".

Abe Shenitzer and John Stillwell
Coeditors

ACKNOWLEDGMENTS

The editors wish to acknowledge the contributions of Don Albers for initiating the idea of assembling this collection of essays and for facilitating its implementation in a variety of ways; John Ewing, who got the column going in the *Monthly* and was tremendously helpful during the initial stages of its existence; Roger Horn, for his editorial assistance, and Gerald Alexanderson, for bringing the book into existence. Last but not least, we wish to thank Elaine Pedreira for making sure that everything went as smoothly as possible, and Beverly Ruedi for her splendid technical work.

A discussion of the nature of any intellectual effort is difficult per se—at any rate, more difficult than the mere exercise of that particular intellectual effort. It is harder to understand the mechanism of an airplane, and the theories of the forces which lift and propel it, than merely to ride in it, to be elevated and transported in it—or even to steer it.

John von Neumann, from his
essay "The Mathematician"

TABLE OF CONTENTS

Introduction — iii

Acknowledgments — v

General
    Mathematics in the 20th Century—Sir Michael Atiyah — 1

Analysis
    Function—N. Luzin — 17
    Two Letters from N. N. Luzin to M. Ya. Vygodskiĭ — 35
    Riemann's Dissertation and Its Effect on the Evolution of Mathematics
        —Detlef Laugwitz — 55
    The Evolution of Integration—A. Shenitzer and J. Steprāns — 63
    On the Calculus of Variations and Its Major Influences on the
        Mathematics of the First Half of Our Century—Erwin Kreyszig — 71

Algebra and Number Theory
    The Evolution of Literal Algebra—I. G. Bashmakova and G. S. Smirnova — 83
    The Evolution of Algebra 1800–1870—I. G. Bashmakova and
        A. N. Rudakov — 99
    What Are Algebraic Integers and What Are They For?—John Stillwell — 105
    The Genesis of the Abstract Ring Concept—Israel Kleiner — 111
    Field Theory from Equations to Axiomatization—Israel Kleiner — 119
    Elliptic Curves—John Stillwell — 133
    Modular Miracles—John Stillwell — 141
    Topology and Abstract Algebra as Two Roads of Mathematical
        Comprehension—Hermann Weyl — 149

Geometry and Topology
    Symmetry—Jacques Tits — 163
    Exceptional Objects—John Stillwell — 171
    The Problem of Squarable Lunes—M. M. Postnikov — 181
    How Hyperbolic Geometry Became Respectable—Abe Shenitzer — 189
    Does Mathematics Distinguish Certain Dimensions of Space?
        —Zdzisław Pogoda and Leszek M. Sokołowski — 195
    Glimpses of Algebraic Geometry—I. G. Bashmakova and E. I. Slavutin — 213

Logic and Foundations
    Four Significant Axiomatic Systems and Some of the Issues Associated
        with Them—Stefan Mykytiuk and Abe Shenitzer — 219
    Foundations of Mathematics in the Twentieth Century
        —Victor W. Marek and Jan Mycielski — 225

Applications
    The Development of Rigor in Mathematical Probability
        (1900–1950)—Joseph L. Doob — 247
    The Evolution of Methods of Convex Optimization—V. M. Tikhomirov — 259

Miscellaneous

    A Few Expository Mini-Essays—Abe Shenitzer   267

    On the Emotional Assumptions without Which One Could Not
        Effectively Investigate the Laws of Nature—Vl. P. Vizgin   275

    The Significance of Mathematics: The Mathematician's Share in the
        General Human Condition—Wilhelm Magnus   283

Biographical Sketches of the Contributors   293

Index   301

Mathematics in the 20th Century[1]

Michael Atiyah

If you talk about the end of one century and the beginning of the next you have two choices, both of them difficult. One is to survey the mathematics over the past hundred years; the other is to predict the mathematics of the next hundred years. I have chosen the more difficult task. Everybody can predict and we will not be around to find out whether we were wrong. But giving an impression of the past is something that everybody can disagree with.

All I can do is give you a personal view. It is impossible to cover everything, and in particular I will leave out significant parts of the story, partly because I am not an expert, and partly because they are covered elsewhere. I will say nothing, for example, about the great events in the area between logic and computing associated with the names of people like Hilbert, Gödel, and Turing. Nor will I say much about the applications of mathematics, except in fundamental physics, because they are so numerous and they need such special treatment. Each would require a lecture to itself. Moreover, there is no point in trying to give just a list of theorems or even a list of famous mathematicians over the last hundred years. That would be rather a dull exercise. So instead I am going to try and pick out some themes that I think run across the board in many ways and underline what has happened.

Let me first make a general remark. Centuries are crude numbers. We do not really believe that after a hundred years something suddenly stops and starts again. So when I describe the mathematics of the 20th century, I am going to be rather cavalier about dates. If something started in the 1890s and moved into the 1900s, I shall ignore such detail. I will behave like an astronomer and work in rather approximate numbers. In fact, many things started in the 19th century and only came to fruition in the 20th century.

One of the difficulties of this exercise is that it is very hard to put oneself back in the position of what it was like in 1900 to be a mathematician, because so much of the mathematics of the last century has been absorbed by our culture, by us. It is very hard to imagine a time when people did not think in our terms. In fact, if you make a really important discovery in mathematics you will get omitted altogether! You simply get absorbed into the background. So going back, you have to try to imagine what it was like in a different era when people did not think in our way.

1. LOCAL TO GLOBAL. I am going to start by listing some themes and talking around them. My first theme is broadly under what you might call the passage from the local to the global. In the classical period people on the whole would have studied things on a small scale, in local coordinates and so on. In this century,

---

[1]This article is based on a transcript of a recording of the author's Fields Lecture at the World Mathematical Year 2000 Symposium, Toronto, June 7–9, 2000.

the emphasis has shifted to try and understand the global, large-scale behavior. And because global behavior is more difficult to understand, much of it is done qualitatively, and topological ideas become very important. It was Poincaré who both made the pioneering steps in topology and forecast that topology would be an important ingredient in 20th-century mathematics. Incidentally, Hilbert, who made his famous list of problems, did not. Topology hardly figured in his list of problems. But for Poincaré it was quite clear that it would be an important factor.

Let me try to list a few of the areas and you can see what I have in mind. Consider, for example, complex analysis ("function theory," as it was called), which was at the center of mathematics in the 19th century, the work of great figures like Weierstrass. For them, a function was a function of one complex variable and for Weierstrass a function was a power series, something you could lay your hands on, write down, and describe explicitly; or a formula. Functions were formulas: they were explicit things. But then the work of Abel, Riemann, and subsequent people moved us away, so that functions became defined not just by explicit formulas but more by their global properties: by where their singularities were, where their domains of definition were, where they took their values. These global properties were the distinguishing characteristic feature of the function. The local expansion was only one way of looking at it.

A similar sort of story occurs with differential equations. Originally, to solve a differential equation people would have looked for an explicit local solution: something you could write down and lay your hands on. As things evolved, solutions became implicit. You could not necessarily describe them in nice formulas. The singularities of the solution were the things that really determined its global properties. This is very much similar in spirit, but different in detail, to what happened in complex analysis.

In differential geometry, the classical work of Gauss and others would have described small pieces of space, small bits of curvature and the local equations that describe local geometry. The shift from there to the large scale is a rather natural one, where you want to understand the global overall picture of curved surfaces and the topology that goes with them. When you move from the small to the large, the topological features become the ones that are most significant.

Although it does not apparently fit into the same framework, number theory shared a similar development. Number theorists distinguish what they call the "local theory," where they talk about a single prime, one prime at a time, or a finite set of primes, and the "global theory," where you consider all primes simultaneously. This analogy between primes and points, between the local and global, has had an important effect in the development of number theory, and the ideas that have taken place in topology have had their impact on number theory.

In physics, of course, classical physics is concerned with the local story, where you write down the differential equation that governs the small-scale behavior; and then you have to study the large-scale behavior of a physical system. All physics is concerned really with predicting what will happen when you go from a small scale, where you understand what is happening, to a large scale, and follow through to the conclusions.

2. INCREASE IN DIMENSIONS. My second theme is different. It is what I call the increase in dimensions. Again, we start with the classical theory of complex variables: classical complex variable theory was primarily the theory of one complex variable studied in detail, with great refinement. The shift to two or more variables fundamentally took place in this century, and in that area new phenomena appear. Not everything is just the same as in one variable. There are quite new features, and the theory of n variables has become more and more dominant, one of the major success stories of this century.

Again, differential geometers in the past would have studied primarily curves and surfaces. We now study the geometry of n-dimensional manifolds, and you have to think carefully to realize that this was a major shift. In the early days, curves and surfaces were things you could really see in space. Higher dimensions were slightly fictitious, things that you could imagine mathematically, but perhaps you did not take them seriously. The idea that you took these things seriously and studied them to an equal degree is really a product of the 20th century. Also, it would not have been nearly so obvious to our 19th-century predecessors to think of increasing the number of functions, to study not only one function but several functions, or vector-valued functions. So we have seen an increase in the number both of independent and dependent variables.

Linear algebra was always concerned with more variables, but there the increase in dimension was to be more drastic. It went from finite dimensions to infinite dimensions, from linear space to Hilbert space, with an infinite number of variables. There was, of course, analysis involved. After functions of many variables, you can have functions of functions, functionals. These are functions on the space of functions. They all have essentially infinitely many variables, and that is what we call the calculus of variations. A similar story was developing with general (non-linear) functions, an old subject, but one that really was coming into prominence in the 20th century. So that is my second theme.

3. COMMUTATIVE TO NON-COMMUTATIVE. A third theme is the shift from commutative to non-commutative. This is perhaps one of the most characteristic features of mathematics, particularly, algebra, in the 20th century. The non-commutative aspect of algebra has been extremely prominent, and, of course, its roots are in the 19th century. It has diverse roots. Hamilton's work on quaternions was probably the single biggest surprise and had a major impact, motivated in fact by ideas having to do with physics. There was the work of Grassmann on exterior algebras—another algebraic system that has now been absorbed in our theory of differential forms. Of course, the work of Cayley on matrices, based on linear algebra, and that of Galois, based on group theory, were other highlights.

All these are different ways or strands that form the basis of the introduction of non-commutative multiplication into algebra, which is the bread and butter of 20th-century algebraic machinery. We do not think anything of it, but in the 19th century all these foregoing examples were, in their different ways, tremendous breakthroughs. Of course, the applications of these ideas came quite surprisingly in different directions. The applications of matrices and non-commutative multiplication in physics came with quantum theory. The Heisenberg commutation relations are a most important example of a significant application of non-commutative algebra in physics, subsequently extended by von Neumann into his theory of algebras of operators.

Group theory has also been a dominant feature of the 20th century and I shall return to this later.

4. LINEAR TO NON-LINEAR. My next theme is the passage from the linear to the non-linear. Large parts of classical mathematics are either fundamentally linear or, if not exactly linear, approximately linear, studied by some sort of perturbation expansion. The really non-linear phenomena are much harder, and have only been seriously tackled in this century.

The story starts off with geometry: Euclidean geometry, geometry of the plane, of space, of straight lines, everything linear; and then through various stages of non-Euclidean geometry to Riemann's more general geometry, where things are fundamentally non-linear. In differential equations, the serious study of non-linear phenomena has thrown up a whole range of new phenomena that you do not see in the classical treatments. I might just pick out two here, solitons and chaos, two very different aspects of the theory of differential equations that have become extremely prominent and popular in this century. They represent alternative extremes. Solitons represent unexpectedly organized behavior of non-linear differential equations, and chaos represents unexpectedly disorganized behavior. Both of them are present in different regimes, and are interesting and important, but they are fundamentally non-linear phenomena. Again, you can trace back the early history of some of the work on solitons into the last part of the 19th century, but only very slightly.

In physics, of course, Maxwell's equations, the fundamental equations of electromagnetism, are linear partial differential equations. Their counterparts, the famous Yang-Mills equations, are non-linear equations that are supposed to govern the forces involved in the structure of matter. The equations are non-linear, because the Yang-Mills equations are essentially matrix versions of Maxwell's equations, and the fact that matrices do not commute is what produces the non-linear term in the equations. So here we see an interesting link between non-linearity and non-commutativity. Non-commutativity does produce non-linearity of a particular kind, and this is particularly interesting and important.

5. GEOMETRY VERSUS ALGEBRA. So far I have picked out a few general themes. I want now to talk about a dichotomy in mathematics that has been with us all the time, oscillating backwards and forwards, and gives me a chance to make some philosophical speculations or remarks. I refer to the dichotomy between geometry and algebra. Geometry and algebra are the two formal pillars of mathematics, and both are very ancient. Geometry goes back to the Greeks and before; algebra goes back to the Arabs and the Indians, so they have both been fundamental to mathematics, but they have had an uneasy relationship.

Let me start with the history of the subject. Euclidean geometry is the prime example of a mathematical theory, and it was firmly geometrical until the introduction by Descartes of algebraic coordinates in what we now call the Cartesian plane. That was an attempt to reduce geometrical thinking to algebraic manipulation. This was, of course, a big breakthrough or a big attack on geometry from the side of the algebraists. If you compare in analysis the work of Newton and Leibniz, they belong to different traditions: Newton was fundamentally a geometer, Leibniz was fundamentally an algebraist, and there were good, profound reasons for that. For

Newton, geometry, or the calculus as he developed it, was the mathematical attempt to describe the laws of nature. He was concerned with physics in a broad sense, and physics took place in the world of geometry. If you wanted to understand how things worked, you thought in terms of the physical world, you thought in terms of geometrical pictures. When he developed the calculus, he wanted to develop a form of it that would be as close as possible to the physical context behind it. He therefore used geometrical arguments, because that was keeping close to the meaning. Leibniz, on the other hand, had the aim, the ambitious aim, of formalizing the whole of mathematics, turning it into a big algebraic machine. This was totally opposed to the Newtonian approach. They also used very different notations. As we know, in the big controversy between Newton and Leibniz, Leibniz's notation won out. We have followed his way of writing derivatives. Newton's spirit is still there, but it got buried for a long time.

By the end of the 19th century, a hundred years ago, the two major figures were Poincaré and Hilbert. I have mentioned them already, and they are, very crudely speaking, disciples of Newton and Leibniz respectively. Poincaré's thought was more in the spirit of geometry, topology, using those ideas as a fundamental insight. Hilbert was more a formalist; he wanted to axiomatize, formalize, and give rigorous, formal, presentations. They clearly belong to different traditions, though any great mathematician cannot be easily categorized.

When preparing this talk, I thought I should put down some further names from our present generation who represent the continuation of these traditions. It is very difficult to talk about living people—whom to put on the list? I then thought to myself: who would mind being put on either side of such a famous list? I have, therefore, chosen two names: Arnol'd as the inheritor of the Poincaré-Newton tradition, and Bourbaki as, I think, the most famous disciple of David Hilbert. Arnol'd makes no bones about the fact that his view of mechanics, in fact, of physics, is that it is fundamentally geometrical, going back to Newton; everything in between, with the exception of a few people like Riemann, who was a bit of a digression, was a mistake. Bourbaki tried to carry on the formal program of Hilbert of axiomatizing and formalizing mathematics to a remarkable extent, with some success. Each point of view has its merits, but there is tension between them.

Let me try to explain my own view of the difference between geometry and algebra. Geometry is, of course, about space; of that there is no question. If I look out at the audience in this room I can see a lot, in one single second or microsecond I can take in a vast amount of information and that is, of course, not an accident. Our brains have been constructed in such a way that they are extremely concerned with vision. Vision, I understand from friends who work in neurophysiology, uses up something like 80 or 90 percent of the cortex of the brain. There are about 17 different centers in the brain, each of which is specialized in a different part of the process of vision: some parts are concerned with vertical, some parts with horizontal, some parts with color, perspective, finally some parts are concerned with meaning and interpretation. Understanding, and making sense of, the world that we see is a very important part of our evolution. Therefore spatial intuition or spatial perception is an enormously powerful tool, and that is why geometry is actually such a powerful part of mathematics—not only for things that are obviously geometrical, but even for things that are not. We try to put them into geometrical form because that enables us to use our intuition. Our intuition is our most powerful tool. That is quite clear if you try to explain a piece of mathematics to a student or a colleague. You have a long, difficult argument and finally the student understands. What does the student say? The student says, "I see!" Seeing is synonymous with understanding, and we use the word "perception" to mean both things as well. At least this is true of the English language. It would be interesting to compare this with other languages. I think it is very fundamental that the human mind has evolved with this enormous capacity to absorb a vast amount of information by instantaneous visual action, and mathematics takes that and perfects it.

Algebra, on the other hand (and you may not have thought about it like this), is concerned essentially with time. Whatever kind of algebra you are doing, a sequence of operations is performed one after the other, and "one after the other" means you have got to have time. In a static universe you cannot imagine algebra, but geometry is essentially static. I can just sit here and see, and nothing may change, but I can still see. Algebra, however, is concerned with time, because you have operations that are performed sequentially and, when I say "algebra," I do not just mean modern algebra. Any algorithm, any process for calculation, is a sequence of steps performed one after the other, the modern computer makes that quite clear. The modern computer takes its information in a stream of zeros and ones and gives the answer.

Algebra is concerned with manipulation in *time*, and geometry is concerned with *space*. These are two orthogonal aspects of the world, and they represent two different points of view in mathematics. Thus the argument or dialogue between mathematicians in the past about the relative importance of geometry and algebra represents something very fundamental.

Of course, it does not pay to think of this as an argument in which one side loses and the other side wins. I like to think of it in the form of an analogy: "Should you just be an algebraist or a geometer?" is like saying "Would you rather be deaf or blind?" If you are blind, you do not see space, if you are deaf, you do not hear, and hearing takes place in time. On the whole, we prefer to have both faculties.

In physics, there is an analogous, roughly parallel, division between the concepts and the experiments. Physics has two parts to it: *theory*—concepts, ideas, words, laws—and *experimental apparatus*. I think that concepts are in some broad sense geometrical, since they are concerned with things taking place in the real world. An experiment, on the other hand, is more like an algebraic computation. You do something in time; you measure some numbers; you insert them into formulas, but the basic concepts behind the experiments are a part of the geometrical tradition.

One way to put the dichotomy in a more philosophical or literary framework is to say that algebra is to the geometer what you might call the "Faustian Offer." As you know, Faust in Goethe's story was offered whatever he wanted by the devil in return for selling his soul. Algebra is the offer made by the devil to the mathematician. The devil says: "I will give you this powerful machine, and it will answer any question you like. All you need to do is give me your soul: give up geometry and you will have this marvellous machine." [Nowadays you can think of it as a computer!] Of course we like to have things both ways: we would probably cheat on the devil, pretend we are selling our soul, and not give it away. Nevertheless the danger to our soul is there, because when you pass over into algebraic calculation, essentially you stop thinking; you stop thinking geometrically, you stop thinking about the meaning.

I am a bit hard on the algebraists here, but fundamentally the purpose of algebra always was to produce a formula that one could put into a machine, turn a handle and get the answer. You took something that had a meaning; you converted it into a formula; and you got out the answer. In that process you do not need to think any more about what the different stages in the algebra correspond to in the geometry. You lose the insights and this can be important at different stages. You must not give up the insight altogether! You might want to come back to it later on. That is what I mean by the Faustian Offer. I am sure it is provocative.

This choice between geometry and algebra has led to hybrids that confuse the two, and the division between algebra and geometry is not as straightforward and naive as I just said. For example, algebraists frequently will use diagrams. What is a diagram except a concession to geometrical intuition?

6. TECHNIQUES IN COMMON. Let me go back now to talk not so much about themes in terms of content, but perhaps in terms of techniques and common methods that have been used. I want to describe a number of common methods that have been applied in a whole range of fields. The first is

Homology Theory. Homology theory starts off traditionally as a branch of topology. It is concerned with the following situation. You have a complicated topological space and you want to extract from it some simple information that involves counting holes or something similar, some additive linear invariants you can associate to a complicated space. It is a construction, if you like, of linear invariants in a non-linear situation. Geometrically, you think of cycles that you can add and subtract and then you get what is called the homology group of a space. Homology is a fundamental algebraic tool that was invented in the first half of the century as a way of getting some information about topological spaces; some algebra extracted out of the geometry.

Homology also appears in other contexts. Another source of homology theory goes back to Hilbert and the study of polynomials. Polynomials are functions that are not linear, and you can multiply them to get higher degrees. It was Hilbert's great insight to consider "ideals," linear combinations of polynomials, with common zeros. He looked for generators of these ideals. Those generators might be redundant. He looked at the relations and then for relations between the relations. He got a hierarchy of such relations, which were called "Hilbert syzygies," and this theory of Hilbert was a very sophisticated way of trying to reduce a non-linear situation, the study of polynomials, to a linear situation. Essentially, Hilbert produced a complicated system of linear relations that encapsulates some of the information about non-linear objects, the polynomials.

This algebraic theory is in fact very parallel to the topological theory, and they have now got fused together into what is called "homological algebra." In algebraic geometry, one of the great triumphs of the 1950s was the development of the cohomology theory of sheaves and its extension to analytic geometry by the French school of Leray, Cartan, Serre, and Grothendieck, where you have a combination of the topological ideas of Riemann-Poincaré, the algebraic ideas of Hilbert, and some analysis thrown in for good measure.

It turns out that homology theory has wider applications still, in other branches of algebra. You can introduce homology groups, which are always linear objects associated to non-linear objects. You can take groups, for example finite groups, or Lie algebras: both have homology groups associated to them. In number theory there are very important applications of homology theory, through the Galois group. So homology theory has turned out to be one of the powerful tools to analyze a whole range of situations, a typical characteristic of 20th-century mathematics.

K-Theory. Another technique, which is in many ways very similar to homology theory, has had wide applications, and permeates many parts of mathematics, was of later origin. It did not emerge until the middle of the 20th century, although it is something that had its roots much further back as well. It is called "K-theory," and it is actually closely related to representation theory. Representation theory of, say, finite groups goes back to the last century, but its modern form, K-theory, is of more recent origin. K-theory can also be thought of in the following way: it is the attempt to take matrix theory, where matrices do not commute under multiplication, and try to construct Abelian or linear invariants of matrices. Traces and dimensions and determinants are Abelian invariants of matrix theory and K-theory is a systematic way of trying to deal with them; it is sometimes called "stable linear algebra." The idea is that if you have large matrices, then a matrix A and a matrix B that do not commute will commute if you put them in orthogonal positions in different blocks. Since in a big space you can move things around, then in some approximate way you might think this is going to be good enough to give you some information, and that is the basis of K-theory as a technique. It is analogous to homology theory, in that both try to extract linear information out of complicated non-linear situations.

In algebraic geometry, K-theory was first introduced with remarkable success by Grothendieck, in close relation to the story we just discussed a moment ago involving sheaf theory, and in connection with his work on the Riemann-Roch theorem.

In topology, Hirzebruch and I copied these ideas and applied them in a purely topological context. In a sense, while Grothendieck's work is related to Hilbert's work on syzygies, our work was more related to the Riemann-Poincaré work on homology, using continuous functions as opposed to polynomials. It also played a role in the index theory of linear elliptic partial differential equations.

In a different direction, the algebraic side of the story, with potential application to number theory, was then developed by Milnor, Quillen, and others, and has led to many interesting questions.

In functional analysis, the work of many people, including Kasparov, extended the continuous K-theory to the situation of non-commutative C^*-algebras. The continuous functions on a space form a commutative algebra under multiplication, but non-commutative analogues of those arise in other situations, and functional analysis turns out to be a very natural home for these kinds of questions.

So K-theory is another area where a whole range of different parts of mathematics lends itself to this rather simple formalism, although in each case there are quite difficult technical questions specific to that area, which connect up with other parts of the subject. It is not a uniform tool; it is more a uniform framework, with analogies and similarities between one part and the other.

Much of this work has also been extended by Alain Connes to "non-commutative differential geometry."

Interestingly enough, very recently, Witten in working on string theory (the latest ideas in fundamental physics) has identified very interesting ways in which K-theory appears to provide a natural home for what are called "conserved quantities." Whereas in the past it was thought that homology theory was the natural framework for them, it now seems that K-theory provides a better answer.

Lie Groups. Another unifying concept that is not just a technique is that of Lie groups. Now Lie groups, by which we mean fundamentally the orthogonal, unitary, and symplectic groups, together with some exceptional groups, have played a very important part in the history of 20th-century mathematics. Again, they date from the 19th century. Sophus Lie was a 19th-century Norwegian mathematician, and he, Felix Klein, and others pushed "the theory of continuous groups," as it was called. Originally, for Klein, this was a way of trying to unify the different kinds of geometry: Euclidean geometry and non-Euclidean geometry. Although this subject started in the 19th century, it really took off in the 20th century. The 20th century has been very heavily dominated by the theory of Lie groups as a sort of unifying framework in which to study many different questions.

I did mention the role in geometry of the ideas of Klein. For Klein, geometries were spaces that were homogeneous, where you could move things around without distortion, and so they were determined by an associated isometry group. The Euclidean group gave you Euclidean geometry; hyperbolic geometry came from another Lie group. So each homogeneous geometry corresponded to a different Lie group. But later on, following up on Riemann's work on geometry, people were more concerned with geometries that were not homogeneous, where the curvature varied from place to place and there were no global symmetries of space. Nevertheless, Lie groups still played an important role because they come in at the infinitesimal level, since in the tangent space we have Euclidean coordinates. Therefore, in the tangent space, infinitesimally, Lie group theory reappears, but because you have to compare different points in different places, you have to move things around in some way to handle the different Lie groups. That was the theory developed by Elie Cartan, the basis of modern differential geometry, and it was also the framework that was essential to Einstein's theory of relativity. Einstein's theory, of course, gave a big boost to the whole development of differential geometry.

Moving on into the 20th century, the global aspect, which I mentioned before, involved Lie groups and differential geometry at the global level. A major development, characterized by the work of Borel and Hirzebruch, gave information about what are called "characteristic classes." These are topological invariants combining the three key parts: the Lie groups, the differential geometry and the topology, and of course, the algebra associated with the group itself.

In a more analytical direction, we get what is now called non-commutative harmonic analysis. This is the generalization of Fourier theory, where the Fourier series or Fourier integrals correspond essentially to the commutative Lie groups of the circle and the straight line. When you replace these by more complicated Lie groups, then we get a very beautiful, elaborate theory that combines representation theory of Lie groups and analysis. This was essentially the lifework of Harish-Chandra.

In number theory the whole "Langlands program," as it is called, which is closely related also to Harish-Chandra's theory, takes place within the theory of Lie groups. For every Lie group, you have the associated number theory and the Langlands program, which has been carried out to some extent. It has influenced a large part of the work in algebraic number theory in the second half of this century. The study of modular forms fits into this part of the story, including Andrew Wiles' work on Fermat's Last Theorem.

One might think that Lie groups are particularly significant only in geometrical contexts, because of the need for continuous variation, but the analogues of Lie groups over finite fields give finite groups, and most finite groups arise in that way. Therefore the techniques of some parts of Lie theory apply even in a discrete situation for finite fields or for local fields. There is a lot of work that is pure algebra; for example, work with which George Lusztig's name is associated, where representation theory of such finite groups is studied and where many of the techniques that I have mentioned before have their counterparts.

7. FINITE GROUPS. This brings us to finite groups, and that reminds me: the classification of finite simple groups is something where I have to make an admission. Some years ago I was interviewed, when the finite simple group story was just about finished, and I was asked what I thought about it. I was rash enough to say I did not think it was so important. My reason was that the classification of finite simple groups told us that most simple groups were the ones we knew, and there was a list of a few exceptions. In some sense that closed the field, it did not open things up. When things get closed down instead of getting opened up, I do not get so excited, but of course a lot of my friends who work in this area were very, very cross. I had to wear a sort of bulletproof vest after that!

There is one saving grace. I did actually make the point that in the list of the so-called "sporadic groups," the biggest was given the name of the "Monster." I think the discovery of this Monster alone is the most exciting output of the classification. It turns out that the Monster is an extremely interesting animal and it is still being understood now. It has unexpected connections with large parts of other parts of mathematics, with elliptic modular functions, and even with theoretical physics and quantum field theory. This was an interesting by-product of the classification. Classifications by themselves, as I say, close the door; but the Monster opened up a door.

8. IMPACT OF PHYSICS. Let me move on now to a different theme, which is the impact of physics. Throughout history, physics has had a long association with mathematics, and large parts of mathematics, calculus, for example, were developed in order to solve problems in physics. In the middle of the 20th century this perhaps had become less evident, with most of pure mathematics progressing very well independently of physics, but in the last quarter of this century things have changed dramatically. Let me try to review briefly the interaction of physics with mathematics, and in particular with geometry.

In the 19th century. Hamilton developed classical mechanics, introducing what is now called the Hamiltonian formalism. Classical mechanics has led to what we call "symplectic geometry." It is a branch of geometry that could have been studied much earlier, but in fact has not been studied seriously until the last two decades. It turns out to be a very rich part of geometry. Geometry, in the sense I am using the word here, has three branches: Riemannian geometry, complex geometry, and symplectic geometry, corresponding to the three types of Lie groups. Symplectic geometry is the most recent of these and in some ways possibly the most interesting, and certainly one with extremely close relations to physics, because of its historical origins in connection with Hamiltonian mechanics and more recently with quantum mechanics. Now, Maxwell's equations, which I mentioned before, the fundamental linear equations of electromagnetism, were the motivation for Hodge's work on harmonic forms, and the application to algebraic geometry. This turned out to be an enormously fruitful theory, which has underpinned much of the work in geometry since the 1930s.

I have already mentioned general relativity and Einstein's work. Quantum mechanics, of course, provided an enormous input. Not only in the commutation relations, but more significantly in the emphasis on Hilbert space and spectral theory.

In a more concrete and obvious way, crystallography in its classical form was concerned with the symmetries of crystal structures. The finite symmetry groups that can take place around points were studied in the first instance because of their applications to crystallography. In this century, the deeper applications of group theory have turned out to have relations to physics. The elementary particles of which matter is supposed to be built appear to have hidden symmetries at the very smallest level, where there are some Lie groups lurking around that you cannot see, but the symmetries of these become manifest when you study the actual behavior of the particles. So you postulate a model in which symmetry is an essential ingredient and the different theories that are now prevalent have certain basic Lie groups such as $SU(2)$ and $SU(3)$ built into them as primordial symmetry groups. So these Lie groups appear as building blocks of matter.

Nor are compact Lie groups the only ones that appear. Certain non-compact Lie groups, such as the Lorentz group, appear in physics. It was physicists who first started the study of the representation theory of non-compact Lie groups. These are representations that have to take place in Hilbert space because, for compact groups, the irreducible representations are finite dimensional, but non-compact groups require infinite dimensions, and it was physicists who first realized this.

In the last quarter of the 20th century, the one we have just been finishing, there has been a tremendous incursion of new ideas from physics into mathematics. This is perhaps one of the most remarkable stories of the whole century. It requires perhaps a whole lecture on its own but, basically, quantum field theory and string theory have been applied in remarkable ways to get new results, ideas, and techniques in many parts of mathematics. By this I mean that the physicists have been able to predict that certain things will be true in mathematics based on their understanding of the physical theory. Of course, that is not a rigorous proof, but it is backed by a very powerful amount of intuition, special cases, and analogies. These results predicted by the physicists have time and again been checked by the mathematicians and found to be fundamentally correct, even though it is quite hard to produce proofs and many of them have not yet been fully proved.

So there has been a tremendous input over the last 25 years in this direction. The results are extremely detailed. It is not just that the physicists said, "this is the sort of thing that should be true." They said, "here is the precise formula and here are the first ten cases" (involving numbers with more than 12 digits). They give you exact answers to complicated problems, not the kind of thing you can guess; things you need to have machinery to calculate. Quantum field theory has provided a remarkable tool, which is very difficult to understand mathematically but has had an unexpected bonus in terms of applications. This has really been the exciting story of the last 25 years.

Here are some of the ingredients: Simon Donaldson's work on 4-dimensional manifolds; Vaughan Jones' work on knot invariants; mirror symmetry, quantum groups; and I mentioned the Monster just for good measure.

What is this subject all about? As I mentioned before, the 20th century saw a shift in the number of dimensions ending up with an infinite number. Physicists have gone beyond that. In quantum field theory they are really trying to make a very detailed study of infinite-dimensional space in depth. The infinite-dimensional spaces they deal with are typically function spaces of various kinds. They are very complicated, not only because they are infinite-dimensional, but they have complicated algebra and geometry and topology as well, and there are large Lie groups around, infinite-dimensional Lie groups. So, just as large parts of 20th-century mathematics were concerned with the development of geometry, topology, algebra, and analysis on finite-dimensional Lie groups and manifolds, this part of physics is concerned with the analogous treatments in infinite dimensions, and of course it is a vastly different story, but it has enormous payoffs.

Let me explain this in a bit more detail. Quantum field theories take place in space and time; and space is really meant to be three-dimensional but there are simplified models where you take one dimension. In one-dimensional space and one-dimensional time, typically the things that physicists meet are, mathematically speaking, groups such as the diffeomorphisms of the circle or the group of differentiable maps from the circle into a compact Lie group. These are two very fundamental examples of infinite-dimensional Lie groups that turn up in quantum field theories in these dimensions, and they are quite reasonable mathematical objects that have been studied by mathematicians for some time.

In such $1 + 1$ dimensional theories one can take space-time to be a Riemann surface, and this leads to new results. For example, the moduli space of Riemann surfaces of a given genus is a classical object going back to the last century. Quantum field theory has led to new results about the cohomology of these moduli spaces. Another, rather similar, moduli space is the moduli space of flat G-bundles over a Riemann surface of genus g. These spaces are very interesting, and quantum field theory gives precise results about them. In particular, there are beautiful formulas for the volumes, which involve values of zeta functions.

Another application is concerned with counting curves. If you look at plane algebraic curves of a given degree of a given type, and you want to know how many of them, for example, pass through so many points, you get into enumerative problems of algebraic geometry, problems that would have been classical in the last century. These are very hard. They have been solved by modern machinery called "quantum cohomology," which is all part of the story coming from quantum field theory, or you can look at more difficult questions about curves not in the plane, but curves lying on curved varieties. One gets another beautiful story with explicit results going by the name of mirror symmetry. All this comes from quantum field theory in $1 + 1$ dimensions.

If we move up one dimension, where we have 2-space and 1-time, this is where Vaughan Jones' theory of invariants of knots comes in. This has had an elegant explanation or interpretation in quantum-field-theory terms.

Also coming out of this is what are called "quantum groups." Now the nicest thing about quantum groups is their name. They are definitely not groups! If you were to ask me for the definition of a quantum group, I would need another half hour. They are complicated objects, but there is no question that they have a deep relationship with quantum theory. They emerged out of the physics, and they are being applied by hard-nosed algebraists who actually use them for definite computations.

If we move up one step further, to fully four-dimensional theory (three-plus-one dimension), that is where Donaldson's theory of four-dimensional manifolds fits in and where quantum field theory has had a major impact. In particular, it led Seiberg and Witten to produce their alternative theory, which is based on physical intuition and gives marvellous results mathematically as well. All of these are particular examples. There are many more.

Then there is string theory and this is already passé! M-theory is what we should talk about now, and that is a rich theory, again with a large number of mathematical aspects to it. Results coming out of it are still being digested and will keep mathematicians busy for a long time to come.

9. HISTORICAL SUMMARY. Let me just try to make a quick summary. Let me look at the history in a nutshell: what has happened to mathematics? I will rather glibly just put the 18th and 19th centuries together, as the era of what you might call classical mathematics, the era we associate with Euler and Gauss, where all the great classical mathematics was worked out and developed. You might have thought that would almost be the end of mathematics, but the 20th century has, on the contrary, been very productive indeed and this is what I have been talking about.

The 20th century can be divided roughly into two halves. I would think the first half has been dominated by what I call the "era of specialization," the era in which Hilbert's approach, of trying to formalize things and define them carefully and then follow through on what you can do in each field, was very influential. As I said, Bourbaki's name is associated with this trend, where people focused attention on what you could get within particular algebraic or other systems at a given time. The second half of the 20th century has been much more what I would call the "era of unification," where borders are crossed, techniques have been moved from one field into the other, and things have become hybridized to an enormous extent. I think this is an oversimplification, but I think it does briefly summarize some of the aspects that you can see in 20th-century mathematics.

What about the 21st century? I have said the 21st century might be the era of quantum mathematics or, if you like, of infinite-dimensional mathematics. What could this mean? Quantum mathematics could mean, if we get that far, understanding properly the analysis, geometry, topology, algebra of various non-linear function spaces, and by "understanding properly" I mean understanding it in such a way as to get quite rigorous proofs of all the beautiful things the physicists have been speculating about.

One should say that, if you go at infinite dimensions in a naive way and ask naive questions, you usually get the wrong answers, or the answers are dull. Physical application, insight, and motivation have enabled physicists to ask intelligent questions about infinite dimensions and to do very subtle things where sensible answers do come out, and therefore doing infinite-dimensional analysis in this way is by no means a simple task. You have to go about it in the right way. We have a lot of clues. The map is laid out: this is what should be done, but it is long way to go yet.

What else might happen in the 21st century? I would like to emphasize Connes' non-commutative differential geometry. Alain Connes has this rather magnificent unified theory. Again it combines everything. It combines analysis, algebra, geometry, topology, physics, and number theory, all of which contribute to parts of it. It is a framework that enables us to do what differential geometers normally do, including its relationship with topology, in the context of non-commutative analysis. There are good reasons for wanting to do this, applications (potential or otherwise) in number theory, geometry, discrete groups, and so on, and in physics. An interesting link with physics is just being worked out. How far this will go, what it will achieve, remains to be seen. It certainly is something that I expect will be significantly developed in the first decade at least of the next century, and it is possible it could have a link with the as-yet-undeveloped (rigorous) quantum field theory.

Moving in another direction, there is what is called "arithmetic geometry" or Arakelov geometry, which tries to unify as much as possible algebraic geometry and parts of number theory. It is a very successful theory. It has made a nice start but has a long way to go. Who knows?

Of course, all of these have strands in common. I expect physics to have its impact spread all the way through, even to number theory: Andrew Wiles disagrees and only time will tell.

These are the strands that I can see emerging over the next decade, but there is what I call a joker in the pack: going down to lower-dimensional geometry. Alongside all the infinite-dimensional fancy stuff, low-dimensional geometry is an embarrassment. In many ways the dimensions where we started, where our ancestors started, remain something of an enigma. Dimensions 2, 3, and 4 are what we call "low." For example, the work of Thurston in three-dimensional geometry aims at a classification of geometries one can put on three-dimensional manifolds. This is much deeper than the two-dimensional theory. The Thurston program is by no means completed yet, and completing that program certainly should be a major challenge.

The other remarkable story in three dimensions is the work of Vaughan Jones with ideas essentially coming from physics. This gives us more information about three dimensions, which is almost orthogonal to the information contained in the Thurston program. How to link those two sides of the story together remains an enormous challenge, but there are recent hints of a possible bridge. So this whole area, still in low dimensions, has its links to physics, but it remains very mysterious indeed.

Finally, I should like to mention that in physics what emerges very prominently are "dualities." These dualities, broadly speaking, arise when a quantum theory has two different realizations as a classical theory. A simple example is the duality between position and momentum in classical mechanics. This replaces a space by its dual space, and in linear theories that duality is just the Fourier transform. But in non-linear theories, how you replace a Fourier transform is one of the big challenges. Large parts of mathematics are concerned with how to generalize dualities in non-linear situations. Physicists seem to be able to do so in a remarkable way in their string theories and in *M*-theory. They produce example after example of marvellous dualities that in some broad sense are infinite-dimensional non-linear versions of Fourier transforms and they seem to work. But understanding those non-linear dualities does seem to be one of the big challenges of the next century as well.

I think I will stop there. There is plenty of work, and it is very nice for an old man like me to talk to a lot of young people like you; to be able to say to you: there is plenty of work for you in the next century!

Function

N. Luzin

What follows is a translation by Abe Shenitzer of an article by N. Luzin that appeared (in the 1930s) in the first edition of *The Great Soviet Encyclopedia*, Vol. 59, pp. 314–334. The article describes the evolution of the function concept.

In its most general form the term "function" denotes a connection between variable quantities. If a quantity x can take on arbitrary values and there is given a rule by means of which it is possible to associate with these values definite values of a quantity y, then we say that y is a function of x and denote this by symbolic notations such as $y = f(x)$, or $y = F(x)$, or $y = \varphi(x)$, and so on. We call the quantity x the independent variable, or argument, and y the dependent variable. However, this definition of the term function is somewhat vague and must be sharpened as follows: (1) concerning the variation of the independent variable x we must decide on an interval of variation $a < x < b$ and on whether x is to take on all values from a to b (the case of a continuous independent variable) or only some of them, say, only integral values; (2) we must make precise the nature of the rule that tells us how one is to associate a value y to a particular value x; (3) concerning the nature of the argument x it must be decided whether x is real or complex, and so on.

The function concept is one of the most fundamental concepts of modern mathematics. It did not arise suddenly. It arose more than two hundred years ago out of the famous debate on the vibrating string and underwent profound changes in the very course of that heated polemic. From that time on this concept has deepened and evolved continuously, and this twin process continues to this very day. That is why no single formal definition can include the full content of the function concept. This content can be understood only by a study of the main lines of the development that is extremely closely linked with the development of science in general and of mathematical physics in particular.

THE FUNDAMENTAL VIBRATIONS OF A MASS SYSTEM

Consider any mass system (say, a bridge) in a state of equilibrium. If it is slightly perturbed, then, in an effort to return to the state of equilibrium, the system begins to vibrate. A vibration is called *fundamental* if all points of the system pass through their respective positions of equilibrium at the same time. The study of the motion of a system with a single degree of freedom was essentially completed in the 17th century, and in the 18th century there began the study of the motions of systems with many degrees of freedom. The first steps in this direction were taken by the great Johann Bernoulli (1727). In order to study the motion of a vibrating string he performed a thought experiment in which he placed n equal and equally spaced weights on a stretched horizontal weightless string. He gives the periods of the fundamental vibrations when the number of weights is less than 8 and states the important principle that in a fundamental vibration the force acting on a material particle is always proportional to the distance of that particle from its equilibrium position. Using this principle he shows that the ratio $(y_{k+1} - 2y_k + y_{k-1})/y_k$ must be independent of k; here y_k is the distance of the k-th weight from the weightless thread when the latter is in the equilibrium position. It is assumed that at all times the amplitudes of the vibrating particles are infinitesimal. By means of this approach Bernoulli obtained a finite difference equation for y_k. The constants entering these finite difference equations are determined from an n-th degree algebraic equation. To each root of this equation there corresponds a definite fundamental vibration of the whole system. Bernoulli was unable to show that the roots of this equation are real and simple.

Somewhat later (1732–36), J. Bernoulli's son, Daniel, and his friend Euler tackled the analogous problem of the determination of the fundamental vibrations of a vertical weightless string attached at its upper end, fitted with n weights and free to swing in the air. Daniel Bernoulli was a superb experimenter. He first gave an experimental solution for $n = 2$ and 3 and then supplied a theoretical justification. Euler, who was just as superb a mathematician, treated the general case and showed that in a fundamental vibration the sides of a vibrating polygon intersect the vertical position of the string in fixed points. Then both of them began to investigate other systems, for example, a plate submerged in a liquid and swaying in it, the swinging of a heavy stick suspended at one end and, finally, a pendulum.

In all these problems Bernoulli and Euler investigated only fundamental vibrations. Whenever the force depended only on the location of the particle, the fundamental vibrations were harmonic, that is, the displacement of the k-th particle was given by the formula $y_k = f_k \cos at$, where f_k was specific for each particle and all particles had the same period $T = 2\pi/a$. D. Bernoulli explicitly formulated in the general case the existence of fundamental vibrations but was not able to show that the roots of the auxiliary equation are real and distinct. What is of greatest importance is the fundamental fact that, at the time, neither of them was able to express an arbitrary motion of the system in terms of fundamental vibrations alone. Much earlier (Rameau, 1726), music theorists pointed out that in addition to fundamental tones musical instruments also produce overtones. It is important to note that preoccupation with fundamental vibrations derived from the following error: beginning with the investigation of the great Taylor (1713), mathematicians clung to the erroneous view, shared by D. Bernoulli, that every composite vibration tends very rapidly to *status uniformis*, that is, to a fundamental vibration. To some extent this is true of physical situations in which friction, air resistance, and so on, cause dispersion of energy and thus give prominence to a fundamental component. The trouble was, however, that this conclusion was tacitly carried over to the mathematical apparatus, that is, to solutions of differential equations that are completely free of this side effect.

PASSING TO THE LIMIT FROM DISCRETE TO CONTINUOUS SYSTEMS.
D. Bernoulli and Euler passed without hesitation from finite systems of particles to continuous systems by thinking of the latter as composed of a great many, or of infinitely many, particles. The boldness of 18th-century mathematicians is well known. Except for Varignon, N. Bernoulli, and d'Alembert, none of them appreciated the difficulties involved in passing to the limit. They regarded it as obvious that a proposition that holds for every finite value of n continues to hold as n goes

to infinity. They had a blurred notion of the difference between "very large" and "infinitely large," and of the difference between results of limited accuracy and results whose accuracy can be indefinitely improved. They used finite differences instead of differentials and sums instead of integrals and ignored distinctions in either case. Usually, the transfer of conclusions from the finite to the infinite case was made either in the finished formulas or at the very beginning of an investigation. A very important example of the first approach is found in D. Bernoulli's paper dealing with the oscillation of a heavy homogeneous flexible string suspended at the top. He begins with a weightless string loaded by means of n weights, solves this problem, and, by assuming n in the answer to be infinitely large, obtains the solution of the problem of the vibrations of a heavy flexible string in the form $y = \cos(t/T)f(x)$, where x is the abscissa of a point of the string, y is the deviation from the equilibrium position, and $f(x) = 1 - x/a + (x/2!\,a)^2 - (x/3!\,a)^3 \ldots$. Here a is determined by the condition $f(l) = 0$, where l is the length of the string.

On the basis of an earlier result involving weights Bernoulli concludes that the equation $f(l) = 0$ has infinitely many roots $a_0, a_1, a_2, a_3, \ldots$, and that the root a_k corresponds to a fundamental vibration in which the heavy string has k fixed points in addition to the suspension point.

DIRECT DETERMINATION OF THE FUNDAMENTAL VIBRATIONS FROM THE DIFFERENTIAL EQUATION. A very important example of passing to the limit at the very beginning of an investigation is the replacing of a system of n ordinary differential equations

$$\frac{d^2 y_k}{dt^2} = f(y_{k+1}, y_k, y_{k-1}) \tag{1}$$

by the single partial differential equation

$$\frac{\partial^2 y}{\partial t^2} = F\left(y, \frac{\partial y}{\partial x}, \frac{\partial^2 y}{\partial x^2}\right). \tag{2}$$

This is done by putting $y_k = y_{k-1} + \Delta y_{k-1}$ and $y_{k+1} = y_{k-1} + 2\Delta y_{k-1} + \Delta^2 y_{k-1}$ and replacing finite differences by differentials. In this approach, in place of a system of algebraic equations linking the initial values of the y_k in a fundamental vibration—and thus in place of a single finite difference equation that combines all of these equations—comes a single differential equation for the initial shape. The latter equation is also obtained by putting $y = Y \cos at$ in equation (2) for the determination of the fundamental vibration and by requiring Y to depend on x alone (rather than on x and t). Also, one must take into consideration the special conditions that apply to the first and last points of the system and express them by means of particular equations (boundary conditions). The first to investigate a vibrating string in this way was Taylor (1713). He showed that the form of a vibrating string is that of a curve whose radii of curvature are to one another as the ordinates. In other words, he obtained the differential equation $y'' = -n^2 y$. After two integrations this equation yielded a quantity proportional to the sine of the argument, which, in turn, is proportional to the abscissa. Taylor did not write his solution explicitly because at that time the symbol "sin" for the sine function had not yet been introduced. That is why he could not formulate the question of the uniqueness of the integration constants satisfying all conditions. Here is where Taylor made his famous error to the effect that there is a unique fundamental vibration and that, for an arbitrary initial motion, every other motion of a vibrating string tends to the fundamental vibration found by him. J. Hermann and D. Bernoulli repeated Taylor's error. When he obtained Taylor's solution by means of his own approach, D. Bernoulli said that the form of a vibrating string is *socia trochoidis* (this was before the introduction of the term sine curve). Neither of these authors (1716 and 1728) suspected the possibility of the existence of other motions of the vibrating string. D. Bernoulli first anticipated the existence of many other fundamental vibrations when he began to treat the problem of the vibrations of a freely suspended heavy elastic string (1732 and 1739), which he regarded as an analog of a vibrating string. In this connection he experimented with a vibrating string and noted that it did not reject pieces of paper placed at its nodes. At that time (1734) Euler still mentioned only fundamental vibrations. It was only in 1744 that Euler, in the course of the investigation of the fundamental vibrations of an elastic plate fastened at one edge, showed that the auxiliary equation, with roots corresponding to the fundamental vibrations, has infinitely many solutions, which he tried to approximate.

THE DEBATE ABOUT THE VIBRATING STRING

THE PAPER OF D'ALEMBERT. While D. Bernoulli and Euler hinted in their papers at a multiplicity of fundamental vibrations of a vibrating string, it was d'Alembert who gave an almost exhaustive solution of this problem in his famous paper of 1747. He states directly that the aim of his paper is to prove that the problem of the shape of the vibrating string has infinitely many solutions other than the "companion of the cycloid." D'Alembert's method is as follows. He begins with the differential equation $\partial^2 y/\partial t^2 = \partial^2 y/\partial x^2$ and uses the identities $d(\partial y/\partial x) = (\partial^2 y/\partial x^2)\,dx + (\partial^2 y/\partial x\,\partial t)\,dt$ and $d(\partial y/\partial t) = (\partial^2 y/\partial x\,\partial t)\,dx + (\partial^2 y/\partial t^2)\,dt$ to obtain as consequences the relations

$$d\left(\frac{\partial y}{\partial t} + \frac{\partial y}{\partial x}\right) = \left(\frac{\partial^2 y}{\partial t^2} + \frac{\partial^2 y}{\partial x\,\partial t}\right)(dt + dx), \text{ and}$$

$$d\left(\frac{\partial y}{\partial t} - \frac{\partial y}{\partial x}\right) = \left(\frac{\partial^2 y}{\partial t^2} - \frac{\partial^2 y}{\partial x\,\partial t}\right)(dt - dx).$$

From this he directly concludes that $\partial y/\partial t + \partial y/\partial x$ depends only on $t + x$, and $\partial y/\partial t - \partial y/\partial x$ depends only on $t - x$, that is, $\partial y/\partial t + \partial y/\partial x = \Phi(t + x)$ and $\partial y/\partial t - \partial y/\partial x = \Delta(t - x)$. Hence $dy = (\partial y/\partial t)\,dt + (\partial y/\partial x)\,dx = \frac{1}{2}\Phi(t + x)d(t + x) + \frac{1}{2}\Delta(t - x)d(t - x)$.

Integrating the latter expression, d'Alembert obtains the final solution $y = \psi(t + x) + \delta(t - x)$, which he unhesitatingly calls the "general solution." In the case when the string, attached at the points $x = 0$ and $x = l$ of the $0X$-axis, passes through the equilibrium position (the $0X$-axis) at time $t = 0$, this solution becomes $y = \psi(x + t) - \psi(x - t)$, where ψ is an even periodic function with period $2l$. In the case when the form of the string at the initial moment $t = 0$ is given by

$y = \Sigma(x)$ and the velocity of its particles at that moment is given by the formula $\partial y/\partial t = \sigma(x)$, the solution takes the form $y = \psi(t + x) - \psi(t - x)$, where ψ is a periodic function with period $2l$ determined from the supplementary conditions $\psi(x) - \psi(-x) = \Sigma(x)$ and $\psi(+x) + \psi(-x) = \int \sigma(x)\,dx$. This, essentially, completes the paper.

EULER'S SOLUTION. In 1748, Euler, following d'Alembert, tackles the same problem. He notes that his solution differs inessentially from that of d'Alembert but stresses that it is *truly the general solution*. Euler assumes that the initial velocity (at $t = 0$) of the particles of the string is zero and that the initial form of the string (at $t = 0$) is $y = f(x)$. Under these conditions Euler's solution is $y = \frac{1}{2}f(x + t) + \frac{1}{2}f(x - t)$. Also, Euler is the first to note that the period of the vibration of the string is independent of the initial form as long as the latter cannot be subdivided into identical aliquot parts. At first sight it might seem that, apart from minor points, the solutions of Euler and d'Alembert are identical. But this is not at all the case. Both men use the same terminology but use the same words to denote different things. They agree on one thing, namely, that the term "equation" means equality of two analytic expressions (without entering on a discussion of what constitutes an analytic expression). Also, both agree that if two analytic expressions take on the same values at all points of an interval, they must be identical. But d'Alembert and Euler differ fundamentally in the meaning they assign to the word "function": *d'Alembert meant by it any analytic expression while Euler meant by it any curve drawn with a free hand.*

THE DEBATE BETWEEN D'ALEMBERT AND EULER. That Euler and d'Alembert subscribed to diametrically opposite views became clear during the lively debate that sharpened ideas and gave them exact formulations. D'Alembert was the first to look for contradictions in Euler's interpretation of the word "function." He writes: "One cannot imagine a more general expression for a quantity y than that of supposing it to be a function of x and t; in which case the problem of the vibrating string has a solution only if the different forms of that string are contained in the same equation." D'Alembert concludes that his own and Euler's solution make sense only if the given function $f(x)$ is *periodic*. Euler's objection takes the form of a question: "If the obtained solution is to be regarded as deficient in those special cases when the form of the string cannot be contained by a single equation, what is one to mean by a solution in such cases?" He insists that his "geometric construction is always correct, regardless of the initial form of the string," that "the different parts of the initial curve are not connected at all by an equation but are connected simply by their description," and that "knowledge of a geometric curve is entirely sufficient for the knowledge of the motion without recourse to computations." D'Alembert's reply was not long in coming. He insists on his interpretation of what is a solution and notes the frequently neglected fact that the very differential equation $\partial^2 y/\partial t^2 = \partial^2 y/\partial x^2$ demands that the ratio $\partial^2 y/\partial x^2$ have a definite (finite) value, that is, that the curve have a definite curvature at each point. In particular, this applies to the endpoints of the string where, in view of the equality $\partial^2 y/\partial x^2 = 0$, the radius of curvature must be infinitely large. Also, the presence of points such as corners, artificially linking curves of different nature, makes the force indeterminate there and consequently makes motion impossible: "here nature itself blocks computations." "We shall leave it to physics to worry" about the question of the motion of such a composite string. Euler declined to continue the debate but noted that it is possible to develop a theory of differential equations containing such "improper" or "mixed" functions. In response to d'Alembert's criticism he points out that his solution, employing such "improper" functions, confirms, for example, the propagation of shocks along a string—a fact observed by D. Bernoulli. D'Alembert insists on the validity of his viewpoint and repeats that the presence of a corner on the string makes a solution impossible.

THE IDEAS OF D. BERNOULLI. D. Bernoulli approached the problem in an altogether different manner. He had already acquired a measure of experience in the study of acoustical problems, and it dawned on him that a vibrating string has infinitely many fundamental vibrations. On the basis of his study of discrete systems he concluded that the most general motion of a string can be obtained by the composition of fundamental vibrations.

D. Bernoulli's ideas matured in 1753, and he concluded that the equation

$$y = \alpha \sin x \cos t + \beta \sin 2x \cos 2t + \gamma \sin 3x \cos 3t + \cdots$$

encompasses the solution of d'Alembert as well as that of Euler. Thus D. Bernoulli discovered an extremely important principle of mathematical physics and deserves to be honored not only for having formulated it but also for having clearly understood its far-reaching consequences. But while D. Bernoulli understood the importance and meaning of his principle of composition of vibrations, he was unable to justify it mathematically and thus provoked the strongest criticism of both d'Alembert and Euler. Euler pointed out that D. Bernoulli fails to note the totally unacceptable consequence implicit in his ideas, which is that an entirely arbitrary function of a variable x is representable by means of a series of sines of multiple arcs. Euler thought that such a function must be odd and periodic. We see that Euler again makes implicit use of the principle that if two analytic expressions have the same numerical values on some interval they must be identical everywhere. D. Bernoulli responded by pointing out that his formula contains infinitely many indeterminate coefficients that can be used to free the curve to pass through an arbitrarily large number of points of the given curve and thus to obtain an arbitrarily close approximation. Regarding the possibility of leaving out of this process one or another particular point, D. Bernoulli refers to d'Alembert's earlier criticism of Euler. To this Euler replied that it is extremely difficult, if not impossible, to choose the coefficients in the manner required by D. Bernoulli. As for d'Alembert, he stated that he fully agrees with Euler's criticism of D. Bernoulli and that he goes beyond it, for he is of the opinion that not every (analytic) periodic function can be represented by means of a sine series, that every function represented by a sine series must possess continuous curvature, and that the coincidence of two curves at infinitely many points does not necessarily make them identical. The nature of the controversy between d'Alembert and D. Bernoulli readily shows that, in modern terms, the former was an "arithmetizer" of mathematical analysis and the latter was a physicist who viewed things from a physicist's point of view.

THE ENTRANCE OF LAGRANGE. At the time when the most eminent mathematicians argued about the mathematical principles associated with the problem of the vibrating string, a young unknown, Lagrange, appeared on the stage. He immediately attracted attention by his "deft" computations (1759). Lagrange investigated the state of the problem of the vibrating string with utmost care and adopted a definite position in the controversy by completely siding with Euler and opposing d'Alembert as well as D. Bernoulli. In an effort to prove Euler's correctness Lagrange put first and foremost the *interpolation problem*. He takes one of Euler's "improper" functions, that is, a graphically given curve, composed in general of pieces of completely different curves, and subdivides the axis of abscissas into small equal segments. Then he erects at the division points perpendiculars, thereby determining a sequence of points on the graphical curve, and seeks an interpolation curve passing through these points. Lagrange works with *linear trigonometric interpolation* with a bounded number of terms. This made his interpolation curves "laws" even for d'Alembert, since they were given by simple analytic expressions. Having thus solved the interpolation problem, Lagrange looks for a solution of the problem of a vibrating string for the *interpolation curve*. By passing to the limit a number of times he obtains in the end Euler's formula for the form of a vibrating string. It is worth noting that Lagrange passed by a colossal discovery without noticing it. En route to the final derivation of Euler's formulas Lagrange obtains *Fourier's trigonometric series*. Had he merely interchanged limits, Lagrange would have discovered the formation law of Fourier coefficients, and this would have ended all debates. But Lagrange's efforts were aimed in a different direction, and, while virtually brushing against the discovery, he was so completely unaware of it that he aimed at D. Bernoulli the phrase: "It is a pity that so clever a theory is untenable." Ironically, it was precisely the ideas of D. Bernoulli, as eventually handled by Fourier, that essentially ended the controversy. In another paper (1760) Lagrange returns to the problem of the string, follows d'Alembert's method, obtains the latter's solution, and is certain that he "made no use of any continuity" ("continuity" in the sense of Euler, that is, in modern terms, "analytic continuability"). This was not quite the case, for, as is well known, Lagrange was fully convinced that every continuous (in the modern sense) function is infinitely differentiable and can be expanded in a Taylor series with the possible exception of isolated points. This being so, it is extremely difficult to decide where in Lagrange's arguments Euler-continuity does or does not enter. His critics objected only to certain special points but did not go into the fundamental side of his investigations and admitted that his computations were, by and large, "singularly deft." D'Alembert attacked, above all, Lagrange's frequent passing to the limit. His sharp mind began to fully appreciate the difficulties associated with this operation. D'Alembert also objected to Lagrange's use of divergent series. Lagrange's response was to the effect that "so far, no one has made a mistake by replacing the series $1 + x + x^2 + x^3 + \cdots$ by the formula $1/(1-x)$." In defending himself against d'Alembert's criticism (which the latter had also directed against Euler) to the effect that a vibrating string must have curvature at its points, Lagrange says that "nature cannot dwell on computations, for, in physical terms, there are no corners on a string; there is always a certain roundness to the stiffness of the string." In subsequent correspondence d'Alembert forces Lagrange to admit that his solution tacitly assumes the existence and finiteness of all derivatives. And since d'Alembert and Lagrange shared the prevalent contemporary conviction that the existence of derivatives of all orders implied that a function could be expanded in a Taylor series, Lagrange was forced to admit that he implicitly introduced Euler-continuity, that is, the representation of a function by means of equations—a point always insisted upon by d'Alembert.

Later Lagrange made one more attempt to bolster his considerations whose truth he was deeply convinced of. In this new exposition he passes through a definite number of points a curve that is a solution of the problem of the string and consists of m sine curves. It is important to note that these points are no longer on the given curve but near it. Lagrange calls the curve he introduced a *generatrix*. He notes that when m is very large, the generatrix deviates very little from the initial form of the string, so that the initial form may be regarded as a piece of the generatrix. He then poses the following question: does this not imply that the initial form of the string consists of sine curves? His answer is that when it comes to "geometric" identity such an assumption is unavoidable, but that in all other cases the initial curve is a kind of asymptote indefinitely approached by the generatrix without the two curves ever becoming one. And from the coefficients of his interpolation formula Lagrange deduces the conclusion that one can ignore the deviation of the generatrix only if the initial form has derivatives of all orders—a property that must be preserved throughout the time of the motion of the string. Motion of the string is possible under these conditions alone. Lagrange does not tell the reader that this assertion represents a complete refusal to defend Euler's viewpoint (which was his initial objective) and an acceptance of d'Alembert's position. The latter stubbornly insisted that the use of divergent series is inadmissible. It is worth pointing out that he quotes the function $\sqrt[3]{\sin x}$ as an example that an everywhere finite function need not have a Taylor series. It does not escape his penetrating vision that this very example goes against him. Indeed, on the one hand, we have an "equation," so that this form of a string admits a solution. On the other hand, we no longer have the finiteness of all derivatives. To help matters, d'Alembert says that infinitely large values of derivatives are admissible as long as there are no jumps. The debates lasted another 20 years without a final solution.

FOURIER'S DISCOVERY

At present the concept of function is not as finally crystalized and undeniably established as it seemed to be at one time at the end of the 19th century. It is no exaggeration to say that at present the function concept is still evolving and that the controversy about the vibrating string continues, except for the obvious fact that the scientific circumstances, the personalities involved, and the terminology are different. If we now look back at the 18th-century debate, what is especially striking is the remarkable perspicacity and intuitive power of the debating thinkers and the tremendous richness of deep analytic ideas connected with this controversy and largely generated by it. In this sense the debate was a motley tangle of profound and extremely difficult questions pertaining to the possibility of passing to the limit and the interchanging of limits; the conditions under which one may use diverging series; the convergence of the Taylor series of an infinitely differentiable function; the difference between a function and its analytic representation; the analytic continuability of a function; the concept of arbitrariness; infinite determinants; curves without curvature and curves consisting of just corner points; interpolation; discontinuities of functions and, in particular, of trigonometric series. The latter loomed so large during and after the debate that they have later been justly called "the axis of rotation of all of mathematical analysis." Even in the light of modern mathematical analysis it is not easy to get to know the particulars of the clash of all these ideas. What makes matters even more difficult is that we are not quite certain that we correctly understand the viewpoint of each of the debating thinkers. For example, as early as 1744 Euler communicated to Goldbach the formula $\sum_1^\infty (\sin nx)/n = (\pi - x)/2$ without at all concluding that *two analytic expressions that coincide on an interval need not coincide everywhere*. At that time such a conclusion would have been thought monstrous, and Euler, in possession of a fact confirming this conclusion, failed to take note of it for some, to us, inexplicable reason. In general, in the light of modern mathematical analysis, matters could, apparently, be described as follows. The key question of the debate pertained to the relation between an analytic definition of a function and a definition that is to some extent physical; does there exist a formula that gives the exact initial position of a string deflected from its position of equilibrium *in an arbitrary manner*? Neither the sophisticated analytic mind of d'Alembert, nor the creative efforts of Euler, D. Bernoulli, and Lagrange sufficed to solve this problem. The person fated to accomplish this task was Fourier. In 1807, to everybody's amazement, Fourier gave the rule for the coefficients a_n and b_n of the trigonometric series

$$a_0/2 + \sum_{n=1}^{\infty} (a_n \cos nx + b_n \sin nx)$$

representing an "arbitrarily given" function $f(x)$. The formulas

$$a_n = \frac{1}{\pi}\int_0^{2\pi} f(\alpha)\cos n\alpha \, d\alpha \quad \text{and} \quad b_n = \frac{1}{\pi}\int_0^{2\pi} f(\alpha)\sin n\alpha \, d\alpha,$$

now known as the Fourier formulas, categorically decided the controversy in favor of D. Bernoulli, the main objection to whose view was precisely the absence of a rule for the computation of the coefficients of a trigonometric series representing an "arbitrarily" given function $f(x)$. True, there remained the objection to Fourier's results fed by the fact that it was not known whether or not his series converged. An immediate argument in favor of Fourier was the extreme simplicity of his rule for the computation of the coefficients of the trigonometric series. A final argument in his favor was a succession of papers by Lejeune Dirichlet (1805–1859) in which he proved the convergence of the Fourier series of any function $f(x)$ with a finite number of minima and maxima [on an interval]. Fourier's discovery produced tremendous bewilderment and confusion among all mathematicians. It toppled all concepts. Up until that time everyone, including Euler and d'Alembert, thought that every analytic expression represents only curves whose successive parts depend on one another. Euler introduced his term "continuous function" to express this mutual dependence of the parts of a function (the modern meaning of this term is completely different). Under the influence of Euler's view of continuity Lagrange tried to show in his theory of analytic functions (1797) that every continuous function can be expanded in a Taylor series: already at that time one sensed that there was a connection between the different parts of a function that can be expanded in a Taylor series, for one was aware that knowledge of a small arc of the curve implied knowledge of the whole curve. Now Fourier showed that such claims are futile and impossible, for a physicist who draws a curve in an arbitrary manner is free to change its course at his whim; but once the curve has been drawn it can be represented by means of a single analytic expression. This suggested the paradoxical result that there is no organic connection between different parts of the same straight line or between different arcs of the same circle, since Fourier's discovery showed that one can subsume under a single analytic formula, a single equation, a continuous curve consisting of segments of different straight lines or arcs of different circles. True, some timid voices noted that the equation of a single straight line or a single circle looked "simpler" than a Fourier expansion. But it soon became clear that this criterion of "simplicity" was completely useless, since it compelled one to use only algebraic functions and banned the use of infinite expansions, compromised by Fourier's discovery, whose importance and usefulness grew from day to day.

THE FUNCTION CONCEPT AFTER FOURIER'S DISCOVERY

The modern understanding of function and its definition, which seems correct to us, could arise only after Fourier's discovery. His discovery showed clearly that most of the misunderstandings that arose in the debate about the vibrating string were the result of confusing two seemingly identical but actually vastly different concepts, namely that of function and that of its *analytic representation*. Indeed, prior to Fourier's discovery no distinction was drawn between the concepts of "function" and of "analytic representation," and it was this discovery that brought about their disconnection. After this, the efforts of mathematicians were channelled in two different directions. On the one hand, the desire to maintain the mutual dependence of the parts of a curve gave rise to the modern *theory of functions of a complex variable*. The prospect on this road was the complete separation of the concepts of a function and of its analytic representation. This was done by Weierstrass in his concept of an "analytic" ("holomorphic") function. On the other hand, Fourier's discovery and the study of the values of analytic expressions destroyed all connections between different parts of a curve. It seemed that the only property of the values of an analytic expression was their *determinacy*, and that they were otherwise completely arbitrary, each independent of the others. This was the sense of the definition of the function concept given by Dirichlet. This definition turned out to be of fundamental importance for the contemporary *theory of functions of a real variable*. For a time, the definitions of function, given by Dirichlet and Weierstrass, respectively, brought great clarity and a certain serenity into the mathematical milieu. It seemed that this clarity was final and that all that remained to do was to develop the consequences of the solid definitions achieved after so many difficulties and efforts. But quite recently it became clear that not all mathematicians are of one mind concerning the value and the very sense of these definitions. Ever more frequent hints, supported by incontestable facts, suggested that the Weierstrass definition of function is overly restrictive. On the other hand, mathematicians concluded with utmost consternation that they were not all of one mind concerning the sense of Dirichlet's definition of function. Some found it perfect, others overly broad, and still others devoid of all meaning. It thus became clear that, in our own time, the controversy about the vibrating string has been renewed in another light and with a different content. The grouping of names below suggests the general pattern of the evolution of the function concept.

FUNCTIONS OF A REAL VARIABLE. Fourier's discovery showed that it is possible to view as a single function the ordinate of a continuous curve composed of arcs of curves that have nothing in common and thus of completely different nature. This discovery utterly destroyed the notion of an organic (logical) connection [presumably] existing between different parts of a curve described by means of a single analytic expression, especially an expression as simple as a trigonometric series. This being so, it seemed that the only available option was to ignore analytic

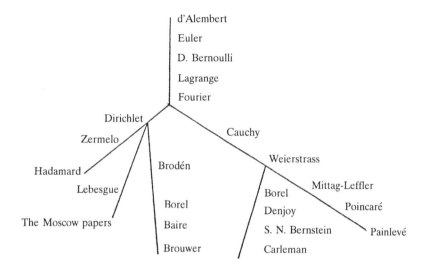

expressions and declare that all there is to the meaning of the function concept is that it is a collection of numerical values corresponding to different values of x that are, in general, completely independent of one another. This was the idea behind the famous Dirichlet definition of a function, still in use today, which states that

y is a function of a variable x, defined on an interval $a < x < b$, if to every value of the variable x in this interval there corresponds a definite value of the variable y. Also, it is irrelevant in what way this correspondence is established. This definition immediately clarified a great many hitherto at best vaguely understood phenomena of mathematical analysis. At first this definition seemed so perfect that it was virtually unanimously accepted. For a long time it was viewed as a genuine discovery. Its formulation was thought to be so exactly suitable that no thought was given to the very possibility of its modification. The established view was that from then on mathematical analysis was to concern itself with the discovery of properties of various special classes of functions obtained by restricting Dirichlet's general

definition of function. In this way there arose branches of analysis concerned with classes of functions such as *continuous* functions (in the sense of Cauchy); *monotonic* functions; functions *with a finite number of maxima and minima* [on an interval]; functions satisfying a *Lipschitz condition*; functions satisfying a *Dini condition*; *differentiable* functions, and so on. Only after these classes of functions had been singled out and investigated, voices arose that asked for clarification of the Dirichlet definition that was initially accepted without any reservations. The objections were directed against the clause: "*it is irrelevant in what way this correspondence is established.*" Later the arguments for and against this point linked up with the arguments for and against the so-called *axiom of choice*, first explicitly formulated by Zermelo. One of the first to clearly state his dissatisfaction with this "rider" to Dirichlet's definition of a function was Brodén (1897). Unfortunately, Brodén's argument was couched in rather general terms. As a result, not all contemporary mathematicians paid attention to his reservations. Brodén argued that the definition of a function must have a special property that would enable easy communication from mind to mind. To get an idea of what Brodén had in mind we subdivide the interval $[a, b]$ of definition of a function $y(x)$ into infinitely many subintervals $\delta_1, \delta_2, \delta_3, \ldots$. Suppose that our function coincides on δ_1 with the ordinate of some straight line L_1, on δ_2 with the ordinate of some cycloid L_2, on δ_3 with the ordinate of some lemniscate L_3, and so on. Brodén asks: when may $y(x)$ be said to be defined? His answer is that $y(x)$ is defined if and only if we are given a definite law of choice of the curves $L_1, L_2, L_3, \ldots$, that is, when these curves have something in common and thus are in some sense "homogeneous" [as a class]. Brodén claims that we cannot study a function made up of infinitely many absolutely "heterogeneous" curves, since such a function can neither be prescribed nor given. The only time one can prescribe or give absolutely different curves is when their number is *finite*, in which case they can be given as absolutely independent of one another. According to Brodén, then, one cannot study infinitely many curves that are absolutely independent of one another. Somewhat later—independently of Brodén—Borel, Baire, and Lebesgue (1905) supported the requirement of a definite law, always *tacitly implied*, whenever one deals with the function concept. Baire pointed out that one should, once and for all, banish the analogy of a bag with balls passed from hand to hand from all discussions involving the infinite. While it is true that a function is, essentially, the totality of numerical values corresponding to the different values of the variable x, this totality cannot be passed from hand to hand like the previously mentioned bag with balls; here the description of the *law of correspondence* that associates a $y(x)$ to an x is absolutely indispensable, and that law must be communicable to anyone who wants to investigate the function $y(x)$. Baire notes that "for our minds all reduces to the finite". To describe accurately the difference between his own views and those of Zermelo and Hadamard, Borel performs the following thought experiment. He notes first that the decimal expansion of $\pi = 3.1415926535\ldots$ must be viewed as *completely determined*, since every textbook of elementary geometry tells us how to compute as many of its decimal digits as we wish. This means that we may view each decimal digit, say the millionth one, as fully determined even if no one has as yet computed it. Then Borel makes a queue of a million people and makes each person name a decimal digit at random, thus obtaining a certain decimal expansion of a million digits. Borel regards this expansion as *fully determined*. Then he makes a queue of infinitely many, rather than a million, people and again makes each person name a decimal digit at random. Now Borel poses the question if one can continue to view the resulting infinite decimal as *fully determined*, that is, as fully determined as the infinite decimal expansion of π, say. Borel's reply is that mathematicians with the mental set of Zermelo and Hadamard will definitely regard this infinite decimal expansion as "fully determined", whereas he himself does not regard it as such. His reason is that the number obtained in this manner may not be ruled by any law, so that two mathematicians discussing this number will never be certain that they are talking *about the same number*; without a law of formation of its decimals they can never be certain of its identity. Lebesgue goes one step further and claims that a mathematician who does not have a law that realizes a function $y(x)$ he is considering cannot be certain that he is talking about the same function at different moments of his investigation; here we are no longer concerned about the common language of *two* mathematicians but about a mathematician agreeing *with himself*. In rebutting Borel's views Hadamard asserts that there is no difficulty in regarding a "lawless" decimal expansion as completely determined. Thus, for example, in the kinetic theory of gases one speaks of the velocities of the molecules in a given volume of a gas although no one will ever really know them. Hadamard points out that the requirement of a law that determines a function $y(x)$ under investigation strongly resembles the requirement of an *analytic expression* for that function, and that this is a throwback to the 18th century.

The mathematical papers of Baire and Lebesgue have shed a great deal of light on the question but have also made it extremely complex. Baire embarked on a systematic investigation of the representation of functions by means of analytic expressions. Since, by Weierstrass' theorem, every continuous function $f(x)$ is representable as the sum of a uniformly convergent series of polynomials $f(x) = \sum_{n=1}^{\infty} P_n(x)$, Baire calls all continuous functions *functions of class* 0. Baire defines the *functions of class* 1 as those discontinuous functions that are limits of continuous functions, that is, $f(x) = \lim_{n \to \infty} f_n(x)$. Baire calls functions not in the classes 0 and 1 that are limits of functions in class 1 *functions of class* 2, and so on. Baire's definition extends over all finite numbers and all countable transfinite numbers. Hence the famous *Baire classification of functions*:

$$K_0, K_1, K_2, \ldots, K_\omega, \ldots, K_\alpha, \ldots | \Omega.$$

Each function $f(x)$ in the Baire classification has a definite analytic representation by means of polynomials and finitely many or countably many symbols of passing to the limit. This is the kind of analytic expression considered by Baire. Lebesgue supplemented Baire's research in an essential way by showing that it is completely pointless to consider all other analytic operations, such as differentiation, expansion in series, integration, the use of transcendental functions such as $\sin x$, $\log x$, and so on; this because every function formed by using a finite or countable number of such operations is necessarily included in the Baire classification. Lebesgue also proved the fundamental fact that none of the Baire classes is empty and, finally, using a profound but extremely complex method, he found a specific function $f(x)$ outside the Baire classification. The impact of Lebesgue's discovery was just as stunning as that of Fourier's in his time. Lebesgue's result shows that a *logical* definition of a particular function is more extensive than a purely mathematical definition, since *a logical definition yielded a particular function f(x) that cannot be obtained from polynomials by passing to the limit a finite or countable number of times*. The Lebesgue function that is outside the Baire classification is extremely complex and its nature has not yet been fully investigated. But the Moscow papers showed that the most delicate point of Lebesgue's considerations gives rise to objections. When Lebesgue showed that *every* analytic expression consisting of mathematical symbols, finite or countable in number, can be transformed into a Baire expression composed of simple (countable) passages to the limit he had no actual complete catalogue of all possible analytic expressions. This meant that he was exposing his enterprise to a great danger, for there could always turn up an analytic expression not transformable into a Baire expression. In fact, the Moscow papers showed that already the analytic expression

$$f(x) = \overline{\lim_{y \to \infty}} \; \overline{\lim_{m \to \infty}} \; \overline{\lim_{n \to \infty}} P_{m,n}(x, y),$$

where $P_{m,n}(x, y)$ is a polynomial in x and y, where the passing to the limit on m and n is *simple* (countable), and where the passing to $\overline{\lim}$ on y is *continuous* (uncountable), is not reducible to a Baire expression for the right choice of the polynomial $P_{m,n}(x, y)$. At the same time it became clear that, as Borel anticipated, very frequently analytic expressions are entirely useless in the sense that, *apparently*, even functions in class 1 of the Baire classification confront us with problems that are unsolvable in principle. The indicated problems concerning the nature of analytic expressions are far from being solved. But it should be pointed out that there are marked and important nuances of opinion among mathematicians who object to Dirichlet's definition. Thus while Lebesgue is willing to accept any law (logical or mathematical) as long as it yields an individual function, Borel insists on the restriction that the law be *countable* (that is, that it involve the natural numbers but not a continuum). Brouwer seems to go still further, for he refuses to consider even the infinity of natural numbers.

FUNCTIONS OF A COMPLEX VARIABLE. A very different fate befell the definition of function that aimed at a formulation of the function concept such that "knowledge of a small arc of the curve under consideration implies knowledge of the whole curve." It is a fact that just as Dirichlet, working with real variables, gave a definition of function that was viewed as final, so too Weierstrass, working with complex variables, gave a definition of function so perfect that to this day the majority of mathematicians regard it as unique and, at any rate, as meeting all requirements of practical applications. Whereas criticism directed against Dirichlet's definition calls for it to be *narrowed*, criticism directed against Weierstrass' definition calls for it to be *broadened*. The papers of Weierstrass were preceded by those of Cauchy (1789–1857). Cauchy was the first to realize that the property of a curve to be determined by a small arc called for the use of a *complex variable*, and that while it might play an auxiliary role, it was an indispensable role. Cauchy's ideas and basic theorems were ordered and systematized by Weierstrass (1815–1897). His fundamental idea was that of *analytic continuation*. Cauchy's investigations showed that every series $P(x - a)$ of positive powers of the difference $x - a$ converges in the interior of a definite circle C with center a and diverges outside C. The sum of the series inside C is infinitely differentiable. Weierstrass regarded this sum of the series $P(x - a)$ as an "analytic function" defined in C and relied on a special process for extending its existence domain. This process is based on the following fundamental theorem: *If the circles of convergence of two given series $P(x - a)$ and $P(x - b)$ intersect and if this intersection contains a point at which the values of the two sums and the values of all their corresponding derivatives are equal, then the two sums have the same values throughout the intersection of the two circles.* In this case, Weierstrass regards each of the two series as a *direct continuation* of the other and calls each of them an "element" of the analytic function that is being defined. Weierstrass' definition of an analytic function is the following: *An analytic function $f(x)$ is the totality of elements obtained from a given one by means of successive direct continuations.* Volterra and Poincaré contributed a final clarification to this definition by showing that the complete definition of an analytic function throughout its domain of existence required no more than a *countable* number of direct continuations. An analytic function $f(z)$ is called *single-valued* if there is no point z at which two different elements $P(x - a)$ and $P(x - b)$ have different values. The set of points z in the interior of the circles of the elements of the single-valued function $f(z)$ under consideration is called its *natural domain of existence*. A point on the boundary of the natural domain of existence of a single-valued function is called a *singular point* of that function. A basic theorem is that the *circle of convergence of each element of an analytic function $f(z)$ contains a singular point*. Weierstrass' definition immediately shed a bright light on a great many hitherto dark areas of mathematical analysis. It explained a great many paradoxes and gave rise to a flood of papers (continuing to this day) devoted to the study of properties of analytic functions. It seemed that one had found so perfect a definition of function that all that remained to do was to study its implications. Above all, one seemed to have finally unriddled the property of a function that *"the values of a small part of a curve determine all of it"*: this property turned out to be just a consequence of the definition of function. In addition, many hitherto baffling properties of analytic expressions, primarily series and infinite products, became clear: it turned out that the sum of a uniformly convergent series of functions analytic in a domain D was an analytic function in D. The riddle of an analytic expression that converged to different functions in different domains was explained by noting that uniform convergence was disturbed between these domains. This explained why, for example, the series

$$\frac{1+z}{1-z} + \frac{2z}{z^2 - 1} + \frac{2z^2}{z^4 - 1} + \frac{2z^4}{z^8 - 1} + \cdots$$

converges to $+1$ inside the circle $|z| = 1$ and to -1 in its exterior. Thus the concepts of an analytic function and an analytic expression became unlinked. Borel (1895) was the first to point out definite shortcomings of Weierstrass' definition and made a number of attempts to construct a theory more general than the Weierstrass theory. The first two of these attempts were found wanting by Poincaré and Painlevé and only the third one (1917) must be judged satisfactory. Borel devoted a significant part of his scientific work to the search for a class of functions wider than the class of Weierstrass' analytic functions. In this area he stated a number of profound ideas that became the foundation of virtually all papers of his followers in this direction. Borel's key objection to Weierstrass' definition was the complete artificiality of the boundary of the "natural domain of existence of a single-valued analytic function." This boundary is truly natural if it consists of a finite or countable number of points. But if it is a closed curve, then "this boundary—writes Borel—is frequently entirely artificial in the sense that the analytic expression that yields a function with this boundary turns out to also converge uniformly outside the boundary, and so yields an external function. From Weierstrass' viewpoint these two functions, the internal and external, are *completely different*, for neither one is a continuation of the other. But this is, in fact, a single function cut in two by a singular curve, for it is possible to find a class of analytic expressions such that if one part satisfies an algebraic or a differential relation then so does the other." The analytic expressions Borel has in mind are series of rational fractions

$$\sum A_N / (Z - a_n),$$

where the series $\sum |A_n|$ converges and the singular points a_n ("the poles of the analytic expression") are everywhere dense on the closed curve under consideration or accumulate in its vicinity.

Poincaré and Wolf were critical of Borel's first attempt [to modify the Weierstrass definition]. Poincaré pointed out that it is always possible to divide the curve under consideration into two parts A and B and to define two analytic (from Weierstrass' viewpoint) functions $\varphi_1(z)$ and $\varphi_2(z)$ such that $\varphi_1(z)$ is analytic outside A, $\varphi_2(z)$ is analytic outside B, and yet $\varphi_1(z) + \varphi_2(z) = F_1(z)$ inside the curve and $\varphi_1(z) + \varphi_2(z) = F_2(z)$ outside the curve, where $F_1(z)$ and $F_2(z)$ are two *arbitrary* functions of which one is analytic inside the curve, the other is analytic outside the curve, and neither can be continued analytically anywhere across the curve. As for Wolf, he constructed a series $\sum A_n/(z - a_n)$ that converges to zero inside the curve, its poles a_n accumulate in a curve outside, and the series $\sum |A_n|$ converges. Following Poincaré's criticism Borel changed his theory and resorted to Mittag-Leffler star expansions. Mittag-Leffler star expansions are generalizations of Taylor series, for the n-th term of such an expansion is a linear combination of the first n coefficients $a_0, a_1, a_2, \ldots, a_{n-1}$ of a Taylor series. Borel expressed the conviction that the Weierstrass concept of an analytic function is too strongly tied to a particular class of analytic expressions, namely Taylor series, and that if one took instead of a Taylor series $K(x - a)$ a Mittag-Leffler star expansion, constructed for an interior point of the curve, then one could slide past the poles of the analytic expression, located everywhere densely on the singular curve, into outside space along its rays. We note that the domain of convergence of a Mittag-Leffler star expansion for an analytic function $f(z)$ is obtained as follows. One lights a source of light at the initial point a of a Mittag-Leffler star expansion $M(x - a)$, and one drives in the plane opaque pegs into all singular points of the analytic function that is being expanded. The domain of convergence of the star expansion $M(x - a)$ to $f(z)$ is all the lit places (the "star"). Borel's computations seemed to confirm his idea, for it turned out that the star expansion $M(x - a)$ for an interior point a turned out to converge on the infinite set of rays of the star to the magnitude of the external function on these rays. But Painlevé wrote a brilliant, detailed, and extremely subtle paper in which he pointed out to Borel that all this could be accidental, for there are Mittag-Leffler star expansions that converge to zero on a segment of a ray without the whole expansion representing zero. Then Borel made a third attempt—this time a successful one—by assuming that the series $\Sigma |A_n|$ *converges extremely strongly* (at least to the order of $e^{-e^{n^4}}$). He linked this assumption to the "monogenicity on the set" (that is, to the existence of $f'(z)$ on the set). Borel's new theory stood the test; for a certain class of (complex) functions (in the sense of Dirichlet) the star expansions *invariably* converge to $f(z)$. This means that knowledge of the magnitude of the function and of its derivatives completely determines the function in its entirety. This is certainly the case when the function is known on a segment. A somewhat delayed confirmation of Borel's third theory came from the Moscow papers (Privalov, Luzin). Specifically, it was shown that if a function is analytic near a rectifiable curve and vanishes almost everywhere on the curve when its points are approached along tangent paths, the function must necessarily be identically equal to zero. And since the external Borel function takes on the same values almost everywhere on the (rectifiable) singular curve as the internal function, it follows that there can be *only one such* external function. This uniqueness confirms Borel's ideas concerning the *organic* connection between the internal and external noncontinuable functions.

In their search for the most natural generalization of the notion of an analytic function, Denjoy, Bernstein, and Carleman followed an entirely different path. The most original feature of their investigations was their determination to work with *real* rather than complex *variables*.

Bernstein begins with his results on best approximation of analytic functions. His starting point is the following theorem. If $f(x)$ is holomorphic at all points of an interval $[a, b]$ then the best approximation $E_n f$ of $f(x)$ by means of an n-th degree polynomial must satisfy the inequality $E_n f < M \cdot \rho^n$, where $\rho < 1$. Bernstein calls a function (P) *quasi-analytic* if there exists an infinite sequence of natural numbers $n_1 < n_2 < \cdots < n_k < \cdots$ such that $E_{n_k} f < M \rho^{n_k}$. These functions turn out to be remarkable for, as Bernstein's fundamental theorem asserts, every (P) quasi-analytic function is completely determined on the whole interval $[a, b]$ by its values on any of its subintervals $[a', b']$. This proposition enabled Bernstein to define (P) *quasi-analytic continuation as the preservation of the inequality* $E_{n_k} f < M \cdot \rho^{n_k}$ *in a larger interval $[c, d]$ containing the given interval $[a, b]$*. Bernstein likened the observed fact that changing the basis $n_1 < n_2 < \cdots < n_k < \cdots$ of a (P) quasi-analytic continuation produced entirely different continuations of the given function $f(x)$ outside the interval $[a, b]$ to the *multivaluedness* of ordinary analytic functions.

Carleman gave a different definition of quasi-analyticity. Whereas Bernstein's (P) quasi-analytic function may not have a derivative, Carleman insists that the functions $f(x)$ he considers have derivatives of *all* orders. He denotes by C_A the class of all functions $f(x)$ that satisfy on a given interval $[a, b]$ the inequality $|f^{(n)}(x)| < k^n \cdot A_n$, where $A_1, A_2, \ldots, A_n, \ldots$ is a sequence of natural numbers and k is a positive constant independent of n.

The fundamental Carleman-Denjoy theorem is the following important proposition. *A necessary and sufficient condition for the family of functions C_A to be quasi-analytic* (that is, that its members have the property that their values on a subinterval $[a', b']$ of $[a, b]$ determine them on all of $[a, b]$) *is that every monotonic majorant of the series* $\Sigma 1/\sqrt[n]{A_n}$ *diverges*. Denjoy proved only the sufficiency of this condition. Carleman's definition has already been applied to the theory of moments. Its connection with Bernstein's definition is indeterminate in the sense that we have here neither the relation of sameness nor the relation of the general to the particular.

BIBLIOGRAPHY

N. N. Luzin, *The integral and the trigonometric series*, Moscow, 1915. (Russian)
_____, *Leçons sur les ensembles analytiques et leurs applications*, Paris, 1930.
H. Burkhardt, *Entwicklungen nach oszillierenden Funktionen*, Leipzig, 1901.
E. W. Hobson, *The theory of functions of a real variable*, 2 vols., Cambridge, 1921–26.
É. Borel, *Leçons sur la théorie des fonctions*, Paris, 1905.
_____, *Méthodes et problèmes de la théorie des fonctions*, Paris, 1922.
H. Lebesgue, Sur les fonctions représentables analytiquement, *J. Math. Pures Appl.*, Paris, 1905.
S. Bernstein, *Leçons sur les propriétés extrémales et la meilleure approximation des fonctions analytiques d'une variable réelle*, Paris, 1925.
T. Carleman, *Fonctions quasi-analytiques*, Paris, 1926.

Two Letters by N. N. Luzin to M. Ya. Vygodskiĭ*

with an introduction by S. S. Demidov
translated by A. Shenitzer

The two letters of Nikolaĭ Nikolaevich Luzin (1883–1950) to Mark Yakovlevich Vygodskiĭ (1883–1950) now being published are kept in Moscow's Central Municipal Archive (stock 2894, schedule 1, work 237). The letters are handwritten, in ink, the first on eight large cross-ruled sheets and the second on unruled sheets of standard size folded in two. There is also a copy of the first letter, probably typed by Vygodskiĭ. At the top of both versions of the first letter there is a handwritten note, which reads: [1930–1931] From acad. N. N. Luzin. The note is most likely Vygodskiĭ's. The first letter is Luzin's response after receiving from Vygodskiĭ a copy of his newly published course of mathematical analysis [1]. On the cover is the date of publication—1931. The book's introduction is dated 14 May 1931. This suggests that Luzin's first letter was written late in 1931 or early in 1932. This is corroborated by its beginning ("I have known for a long time about its publication and have heard about the impassioned debates it has provoked.... A lot of time has passed since the book's appearance, and most people have managed to speak their—very different—minds..."). The second letter was dated 20.XII.1933 by Luzin himself. The person who familiarized herself with these archive letters before V. A. Volkov and myself, was N. S. Ermolaeva.

Vygodskiĭ's book *Foundations of Infinitesimal Calculus* (there were three editions published in 1931, 1932, and 1933) was based on a course of lectures presented by him at the Moscow Chemico-Technological Institute. These lectures were a radically non-traditional account of the differential and integral calculus. Vygodskiĭ explained the essence of his approach as follows:

> I wrote this book because of my deep conviction that none of the existing textbooks puts the key ideas of infinitesimals before beginners with the necessary sharpness and clarity. The reason for this is that, while they differ in many details, all these textbooks share an approach which I regard as mistaken and harmful. Specifically, in all of them the fundamental concepts of analysis are presented in a formal logical manner. No matter how much individual authors try to simplify proofs, to avoid formal rigor, to introduce intuitive imagery and concrete problems, they invariably, and above all, attempt to explain the formal scheme of modern analysis. As a result, the fundamental concepts of analysis appear not in their evolution but in their congealed form.

---

*This paper appeared in the Russian journal *Historico-Mathematical Investigations*, 2 (37) 1997 128–152.

> This is the reason behind the depressing fact that the apparatus of analysis remains a dead apparatus in the hands of the students. While the learned arguments and proofs seem incontrovertible, they are nevertheless... unconvincing... [**1**, p. 4].

And further:

> after all, the notion of a differential did not come into being in the form in which it is presented in the textbook. What we find in the textbook is the result of the subsequent evolution of this concept. This evolution resulted from the need for logical "cleansing" due to the use of the concept of an infinitesimal change of a quantity. Here we encounter the process of "formalization" of a mathematical concept which is an historically unavoidable stage of [its (tr.)] evolution. But the only people who can properly understand and appreciate the need for this formalization are those who are familiar with the concept of a differential in its initial stage of development... [**1**, pp. 4–5]. The viewpoint underlying the present textbook is that the learner must be *introduced* to the study of analysis by getting acquainted with its fundamental notions at the stage at which they arise directly from practical needs. Their rigorization and cleansing must be a later issue, *initially* of secondary importance.
>
> In other words, I am attempting to replace the formal–logical scheme by the historical scheme, or rather, by the historico–logical scheme.
>
> Of course, this does not mean that I propose to take the reader through all the twists and turns of the historical development and reconstruct the chronological order of the evolution of the ideas of analysis. For this one must turn to a book on the history of analysis. The historical material in my book is not its subject matter but the basis for the exposition... [**1**, p. 5].

That is why, in this textbook, analysis is presented so that its fundamental ideas are shown in their evolution.

> Thus in the first stage of study I introduce the fundamental concepts of analysis in their crude form, in the form in which they are picked up when one studies the simplest facts of natural science and technology [this is taken from the introduction to the second edition (S.S. Demidov)]. In other words, I abandon the tradition of basing the presentation on the theory of limits. This theory will be given its proper place where it is truly indispensable not only as the formalizing apparatus but also as the basis for the development of new and more powerful methods.... I refrain from sharply dividing analysis into differential and integral calculus. Initially these two operations of analysis are regarded as independent, then the connection between them is established, and later both are studied together.
>
> The starting point is the problem of integration rather than that of differentiation. The reason for this is that integration offers a larger field for investigation of concrete questions. That is why, historically, the problem of integration arose before that of differentiation. [**1**, pp. 9–10].

Vygodskiĭ first introduces differentials in the way in which they were introduced at the birth of analysis, i.e., as actual infinitesimals! It is only later that he puts the house in order by explaining to the student how what was said earlier can be rigorized by using the concept of limit.

While Vygodskiĭ offers his approach primarily as a pedagogical device, he regards it as justified both psychologically (the learner will get a better understanding of the foundations of analysis) and historically (this is how the foundations evolved). The modern exposition is an indispensable contribution to the requirements of modern mathematics. But, in essence, the concepts of analysis (the differential and the integral) come easy to us intuitively and come to light in the course of history.

Vygodskiĭ's approach was not supported by the leading contemporary educational experts. In the book's dedication he mentions the sharp objections to his "method on the part of authoritative educational experts and scholars." The only eminent mathematician who supported him warmly was Luzin.

Luzin viewed the intuitive comprehension of mathematical truths as natural. The stage of axiomatization was the next (for Luzin routine) stage of activity. This is also how he approached the concepts of analysis. The first of the two published letters shows that when he tried, as a student, to learn analysis he encountered considerable psychological difficulties. He did not regard the twentieth-century "logical" method of presenting analysis, based on the notion of limit, as an ideal way of teaching its fundamental concepts. Bearing in mind his own difficulties, and trying to make the learning task easier for beginners, Luzin selected from the ocean of textbooks the one by the American W. E. Granville [2]. He translated and reworked it in an essential way, and this reworked translation became widely known in our country under the twin name of Granville and Luzin [3]. It later gave rise to Luzin's own textbook [4; 5] of which there were several editions.

The first of the two letters we are publishing strikes us as a remarkable document. On the one hand, Luzin sets down in it his own experiences. One can say with confidence that these experiences coincide with the experiences of many mathematicians who have thought about the foundations of the differential calculus during the more than three centuries of its history. From the vantage point of the mathematics of the end of the 20th century one can qualify these experiences as anticipations of the nonstandard analysis discovered in the 1960s. On the other hand, this letter lifts the screen hiding from us young Luzin's agonizing search (described by the mature Luzin) of the themes of his future investigations. It turns out that the reflections of Luzin the student on the nature of infinitesimals were for him the door to the kingdom of the theory of functions of a real variable. To quote (the mature) Luzin: "The major spiritual drama associated with the Weierstrass curve forced me to concentrate on the theory of functions and, more generally, on micromathematics. This is the term I used to denote the infinitesimal structure of functions or of sets." This letter, and Luzin's recently published, but unfortunately few, comments (such as his letter to O. Yu. Schmidt [6] or his review of the works of N. A. Vasil'ev [7]) on the nature of mathematics, of mathematical objects, and of mathematical creativity, shed light on his views on the foundations of mathematics. Given Luzin's role in the Moscow mathematical association between 1920 and 1940, we think that it is a problem of extreme importance to reveal the essence of his views and their evolution. Its solution would enable us to penetrate the universe of thought of contemporary Moscow mathematicians concerning problems of the foundations of their discipline, which determined to a large extent the flow of mathematical thought as we know it today. We should add that the implied secrecy surrounding this universe was to no small extent due to ideological dangers.

Historians have only begun to solve this problem. Here a pioneering step was taken by F. A. Medvedev in the paper "N. N. Luzin on non-Archimedean time" [8], published in 1993 (of the relatively recent papers dealing with Luzin's views on the foundations of mathematics we mention M. I. Panov's paper [9]). Medvedev's paper is based on Luzin's published papers, in particular on his Encyclopedia article "Differential Calculus" [10]. Medvedev explains that thinking about the formal possibility of constructing a constant infinitesimal suggested to Luzin the prospect of developing the idea of non-Archimedean time and shows how Luzin discussed it. We note that Luzin's Encyclopedia articles, whose study is just beginning, produced an unambiguous reaction of the contemporary Moscow mathematical association. The ending of the second of the two published letters, dated 20 December 1933, conveys an echo of these "encyclopedic battles". The content of the second letter is related to that of the first, and some of the issues in the first letter are developed in the second.

Luzin's second letter includes his agitated account of the reaction of the mathematicians of the turn of the 20th century to Weierstrass' approach to the foundations of mathematical analysis. The opposition to Weierstrass' ideology (on the part of Borel, Denjoy, and others), hidden from us by the curtain of time, reveals the sense of certain important episodes in the history of mathematical analysis (such as those dealing with the theory of divergent series and the theory of functions of a complex variable) and enables us to look at its newest history in a new way.

Two letters of N. N. Luzin to M. Ya. Vygodskiĭ
Published and annotated by V. A. Volkov and S. S. Demidov
Historico-Mathematical Investigation 2 (37) 1997 133–152

Translated by Abe Shenitzer (addenda in square brackets [...] are by the translator)

LETTER I

Dear Mark Yakovlevich,

Please accept my profound thanks for your wonderful and precious gift, the copy of your *Course of analysis*[1]. I have known for a long time about its publication and have heard about the impassioned debates it has provoked. It seems that none, or at best very few, consider it calmly. Familiarity with the work seems to turn some into its worshippers and others into its ardent opponents. This does not surprise me the least bit, for you had the courage to touch the most painful spot of analysis

---

[1] This refers to the first edition (1931) of M. Ya. Vygodskiĭ's *Foundations of Infinitesimal Calculus* [1].

in general and of modern analysis in particular; you tossed the proverbial stone in a hornets' nest.

After reading your book I concluded that it could not have been otherwise, and that, indeed, all kinds of people familiar with the foundations of analysis must react to it emotionally.

A lot of time has passed since the book's appearance, and most people have managed to speak their—very different—minds, and you, most likely, know their opinions about it. You must know that while most people in the pedagogical circles have reacted very favorably to the book, the reaction of most people in the theoretical circles has been restrained. I heard talk in Moscow about the restoration of the phlogiston theory in science and charges of decadence. In Leningrad, which is more uniformly trained, I heard talk to the effect that while Darwin drew the path of man's evolution from quadruped to biped, efforts are underway in mathematics to *reverse* this course... Briefly, almost all theoreticians are in favor of unconditionally basing analysis on the theory of limits. As a concession to "human frailty", they admit as a possibility a short stage of familiarity with analysis without the theory of limits. But this is intended for rank beginners, subject to the unalterable condition that "later all will be done properly..."

I think that what I have written is a fair description of the responses of the theoretical circles. They react to your presentation either purely negatively or admit it condescendingly as a pedagogical device. I have not encountered attempts to form other attitudes.

Having studied more than half of your book I also want to express my opinion, and I do this by writing you a letter. I prefer it this way. A letter gives one a chance to stop and think, and this is difficult to do in conversation. Also, this [mode of communication] suits better my temperament. Forgive me if this letter turns out to be a long one. In matters of this kind, length may ensure complete understanding of a person's position. I will probably be writing this letter for many days, with breaks, and in different states of mind.

Of course, I could spare you the trouble of reading my letter by joining one or another group of people who have expressed an opinion [about your book]. The trouble is that none of these opinions parallels my thoughts. Unlike my colleagues, I think that an attempt to reconsider the idea of an infinitesimal as a variable finite quantity is fully scientific, and that the proposal to replace variable infinitesimals by fixed ones, far from having purely pedagogical significance, has in its favor something immeasurably deeper, and that this idea is growing roots in modern analysis. In short, if a succinct statement were needed, I would put it as follows: "Modern science does not object to such an idea."

It is this that I wanted to write to you about at greater length. To begin with, I will just note that the idea of the actually infinitely small has certain deep roots in the mind. When the mind *begins* to acquire familiarity with analysis, in, as it were, its springtime, it always begins with the actually infinitely small, with what one might call "elements" of quantities. But gradually, step-by-step, as a result of accumulation of knowledge and of theories, of saturation with abstractions and of tiredness, the mind tends to forget its initial aspirations and is amused by their "childishness". Briefly, in its autumn period, the mind succumbs to the idea of the uniqueness of correct justification by means of limits.

Your book is sometimes faulted for its excess of passion, sarcasm, and acrimony. For me this is understandable; in my view, in pioneer attempts things cannot be different. The presumed acrimony is simply an echo of intense intellectual suffering and pain, and the more intense the suffering the better, for suffering is the source of creativity.

I do not want you to think that I am just trying to please you, and so will tell you something about my personal recollections. You will see that I understand a thing or two about these matters. This is shown by the fact that I have retained to this day such vivid recollections, recollections forever etched in my memory. People usually carefully hide such suffering and are reluctant to talk about it. And if they do, it is usually in the form of hints. I don't know why this it so. One must overcome such reluctance; it is necessary to do this.

I recall being a second-year student. The setup was the following. L. K. Lakhtin[2], who dictated the foundations of analysis, occupied a secondary position. I did not attend his dictation sessions. What for? One could buy his lithographed notes or borrow and read them. The tone was set by Boleslav Kornelievich Mlodzeevskiĭ[3], an ardent and imperious geometer of European class, and by the exact and rigorous D. F. Egorov[4] who, at the time, was just beginning to enter the life of the University. Lakhtin kept his distance from them. Mlodzeevskiĭ ruled, but respected Egorov's precise and rigorous mind.

I began to learn analysis by reading a variety of books in an avid and random manner. It was all a matter of chance: whether or not a certain book was in the library and what it looked like—solid or so so. Confusion and chaos were in my mind, fragments of threads that came together in random fashion. There was no guidance and no contact with the professors. I seemed to have fallen in the water and was thrashing about as best I could to avoid drowning. But it is conceivable that systematic guidance would have deprived me of the richness and multiplicity of colors which I sense in myself. Had such guidance existed, it could possibly have given a uniform coloration and boredom to the activity of the mind. But all this is of secondary importance.

The important thing is that *I knew neither Goursat[5] nor Jordan[6]*, and learned from ancient courses of analysis such as Lacroix[7] and the like. The newest one for me was the seven-volume *Traité d'Analyse* of Laurent[8], whose author took pride in

---

[2] Leonid Kuz'mich Lakhtin (1858–1927), professor at Moscow University [**11**, p. 540].

[3] Boleslav Kornelievich Mlodzeevskĭ (1858–1923), eminent geometer, professor at Moscow University [**11**, pp. 515–517].

[4] Dmitriĭ Fedorovich Egorov (1869–1931), eminent Russian mathematician, Luzin's teacher. Founder and head (together with Luzin) of the famous Moscow school of the theory of functions [**11**, pp. 517–519], [**12**].

[5] This refers to the famous *Cours d'Analyse mathématique* by E. Goursat, 4-me éd., Paris, Gauthier–Villars, 1923–24, V. 1–3, of which there were many editions in France and in Russia; see [**14**] and [**15**].

[6] This refers to the famous *Cours d'Analyse de l'Ecole Polytechnique* by C. Jordan, Paris, 1882, V. 1–3.

[7] This refers, most likely, to one of the editions of *Traité du calcul differéntiel et du calcul intégral* by S. F. Lacroix, Paris, 1797–1798, V. 1–2.

[8] See [**18**].

the fact that he had been a pupil of Cauchy. I encountered set theory and the theory of functions of a real variable at the time when I completed my university studies; more precisely, at the stage when I was preparing for a teaching position at the University[9]. My antiquated education was a matter of *pure chance*, for at the time when I was at the University the Goursat course was already in vogue abroad as well as in our country, in circles close to the professorial ones. I did not know this; I was completely self-taught.

My "antiquated" education—more correctly my self-education—is the basis for my originality and the freedom of my mathematical views. Be that as it may, during my university studies I was not exposed to the disciplining courses of Egorov, Goursat, and the logically inflexible Vallée-Poussin[10]; what was, was chaos of a possibly creative nature. The theory of limits entered my mind mechanically and crudely, not in a refined way but rather in a forced, policelike manner, in accordance with the formula: "shut up, I am telling you!" In general, I like old courses, which contain all that is relevant and, in addition, things that are not directly relevant but important for the absorption of the subject. As for modern courses, they remind me of French offices where they may shout at you if you forget to take off your hat and may send you away with a formal reply. When I worked on Granville[11], I involuntarily took a palette with colors and colored the theory of limits, for I well remembered the depressing effect that colorless theory, then being introduced, had on me.

I have a clear recollection of my ideas on infinitesimal analysis. I was a second-year student. When the professors announced that $\frac{dy}{dx}$ is the limit of a ratio, I thought: "What a bore! Strange and incomprehensible. No! They won't fool me: it's simply the ratio of infinitesimals, nothing else." It was with terrible pain that I "assimilated" the Weierstrass curve without a tangent[12]; I did not believe it, I tried to refute it, I read and reread Duhamel's[13] proof to the effect that every continuous function is differentiable. This pain—almost impossible to bear—associated with the Weierstrass curve is understandable: if $f(x)$ is the Weierstrass function then it is continuous, and as soon as dx exists dy must also exist. And yet there is no $\frac{dy}{dx}$! Why? In short, what happened was that the Weierstrass curve began to undermine the profoundly natural idea of the actually infinitely small and *the theory of limits entered my mind as the pathology of functions*.

The battle with the Weierstrass curve was for me a kind of nightmare. I dreamed of it. Now I understand that it was for me a monster that I wrestled with in an absurd manner. And since I was unable to win a victory I chose to circumvent it: I would show that the Weierstrass curve is not a marvel, and that it is possible to

---

[9] ca. 1905–1908; see [**11**, p. 566].

[10] The *Cours* [**19**] of Ch. La Vallée-Poussin was translated into Russian; see [**20**], [**21**].

[11] The textbook of W. E. Granville is discussed by S. S. Demidov in the paper preceding the two Luzin letters.

[12] This is a reference to Weierstrass' famous example of a continuous function without a finite derivative anywhere; see [**22**, p. 212].

[13] See [**22**, pp. 80, 208].

do the same by elementary constructions WHILE PRESERVING THE IDEA OF A FIXED INFINITESIMAL.

I plucked up my courage (I disliked contact with professors. I was shy and, except for Zhegalkin[14], was simply afraid of them), and after Mlodzeevskiĭ's geometry lecture I shyly approached him and asked permission to talk to him about curves without tangents.

I had *already* made the first mistake. I was a student. The professor seemed to be a higher being; I did not think that he could experience tiredness. I began to talk with him and failed to realize that while I, a young person, was rested, he was dead tired. This being so, when his voice rose to a shout, I blamed myself and not his tiredness. Such matters had to be discussed with a rested person.

I began by saying: "Boleslav Kornelievich, it seems to me that one can 'construct' a curve without a tangent in a very elementary way and I would like to know your opinion." He: "Hm..., go ahead—do you have a drawing?" I: "Yes, Boleslav Kornelievich, here it is:

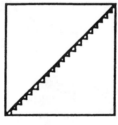

Figure 1

we divide the diagonal of a square into n equal parts and construct on each subdivision as base a right isosceles triangle. In this way we get a kind of delicate little saw. Now I put $n = \infty$. The saw becomes a continuous curve that is infinitesimally different from the diagonal. This means that its tangent must be the diagonal. But it is perfectly clear that its tangent is alternately parallel now to the X-axis and now to the Y-axis. I think that the same is true of the notorious Weierstrass curve." He: "Well, you know, if you actually ... such a circle of ideas ... what you say is nonsense. You must understand that *the actually infinite does not exist*. And ∞ is not a number. The sides of your triangles are not $1/\infty$. And there are no triangles... And all told there is nothing... There is only your failure to understand. Keep in mind that n is a finite number. And for every n there is, as you put it, its own delicate little saw. As n increases without bounds, things flash in your plane as on a movie screen, saw follows saw with ever increasing speed. That's all. There is no single saw. There is a series of saws. The tangent to which saw are you talking about now?" I: "The one obtained in the limit." He (clearly losing his patience): "Sounds like the story about the little white bull... Understand, ∞ is not a number. The series of saws comes ever closer to the diagonal. The limit is the diagonal itself. Of course, there is a tangent to the diagonal: it itself serves as its very own tangent." I said naively: "But I see that the

---

[14] Ivan Ivanovich Zhegalkin (1869–1947), professor at Moscow University [**11**, pp. 563–564].

tangents to the sides of the triangles are parallel to the *X*-axis and the *Y*-axis respectively, and are never parallel to the diagonal." He softened for a moment and said benevolently ... : "Ah ... Look ... remember: *the tangent of the limit is not the limit of the [moving] tangent*. The tangent itself is a limit. Well, you are permuting two passages to the limit, and this must not be done. See (here he picked up the chalk)

$$\lim_{n=\infty} \lim_{m=\infty} \frac{m}{m+n} = 1 \quad \text{and} \quad \lim_{m=\infty} \lim_{n=\infty} \frac{m}{m+n} = 0,$$

thus

$$\lim_{n=\infty} \lim_{m=\infty} \frac{m}{m+n} \neq \lim_{m=\infty} \lim_{n=\infty} \frac{m}{m+n}.$$

It's an old story: *the length of the limit is not equal to the limit of the [moving] length*. If you measure the length of your delicate little saw it turns out to be equal to 2. And the length of the diagonal is $\sqrt{2}$. One must not interchange limits. Give it some thought. You won't get it immediately. Goodbye." He went away and left me with the baffled impression that I had been misunderstood—or so it seemed to me at the time.

It must be borne in mind that I was a second-year student. A week later, by chance, a fourth-year student dragged me along to a lecture on real variables for fourth-year students by the same Boleslav Kornelievich. And I listened to his lecture, brilliant as usual, on the notion of cardinality and countable cardinality. For me this was almost a revelation. Afraid to move, I listened to his lecture on cardinality, on $\aleph_0$, and thought: "These are complete contradictions: in analysis they say that every number is finite and modestly pass over in silence points at infinity on straight lines. In geometry, on the contrary, they keep on talking about points at infinity and deduce marvelous things. A week ago Boleslav Kornelievich cut me short by exclaiming that 'the actually infinite does not exist.' And now he does it himself! No, there is something I don't understand. This is probably not my thing: I should become a physicist!" I rehearsed the words in my mind and after the lecture I approached him in a determined way and said: "Boleslav Kornelievich, I have a question about continuous curves without a tangent ... " He made a pained grimace and said dispiritedly: "You are back to the same thing ... You find the matter unclear ... what's the matter?" I: "You see, Boleslav Kornelievich, a week ago I constructed for you triangles and I may have done it clumsily. Now I would like to reconsider the question. Suppose that on the diagonal of a square there is a finite set (B. K. winced at the word 'set') of points. On the previous occasion

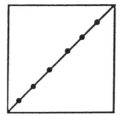 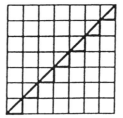

Figure 2

I constructed the required triangles on each segment one at a time. This was my mistake. Now I can do it *at once*. I take all straight lines passing through our points and parallel to the *X*- and *Y*-axes respectively. Then I cut off at once all 'superfluous' parts of the straight lines and obtain immediately a saw..."

He interrupted me and said: "I don't see what all this is for. It's clear that this is so. Well, what next?" I: "Well, Boleslav Kornelievich, instead of a *finite set* I take a *denumerably infinite* set, for example, all points on the diagonal that are a rational distance away from the beginning of the diagonal. Next I repeat all that I just told you: I take all straight lines passing through the points of this set and

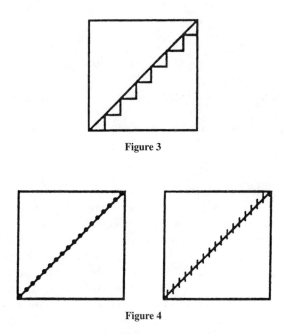

Figure 3

Figure 4

parallel to the *X*- and *Y*-axes respectively and, finally,
I eliminate all 'superfluous' parts of the straight lines. What will I obtain, *this time without any passage to the limit*? The same thing as before, i.e., a *saw*, but one with actually infinitely small teeth!

Figure 5

In other words, there is an *individual, fixed* 'curve' that is *infinitesimally different from the diagonal.* Last time, when I came with it, you, Boleslav Kornelievich, said that my construction is based on an error, since there is no actual infinity! And you forbade me to use ∞ as a number. But now, in your lecture, you yourself spoke of actual infinity, of denumerably infinite cardinality. So I changed the construction in accordance with your presentation by dispensing with ∞ as a number and starting with a denumerably infinite set, and I obtain the same thing: a saw with actually infinitely small teeth, i.e., a fixed, individual curve that differs infinitesimally from the diagonal." Boleslav Kornelievich suddenly fell silent and then, extremely gently and with a hint of a kind of respect, said: "I don't understand how you ended up in my lecture ... Listen: don't attend any more; for *you* it's bad, wait till you learn more. As for your curve, it is *logical, not real, not subject to intuition* ('good, good,' he said interrupting himself). It exists in logic but not geometrically. It is verbal but not real. One can talk about it but it is not genuine." I immediately realized that Boleslav Kornelievich got stuck (I did not take into account that the man was simply *tired* after a lecture outside his area and failed to understand my objections) and attacked him mercilessly: "Boleslav Kornelievich, what about the Weierstrass curve, is it genuine or logical? Is it verbal or does it exist in reality? Is it real like a sine curve or is it just conceivable? If it is the latter then, if we break off a trigonometric series at some term, we have a real curve. But if we take the whole series then we have just a verbal formation." But I was dealing with a formidable opponent who was tired, lost his head under pressure, but recovered instantly, simply by *gaining time* while talking about verbal curves and who, having gained heart, dealt me a terrible blow. "Listen —he said with a sudden burst of anger— what nonsense are you dishing out! Take any homofocal ellipse with foci at the ends of your diagonal.

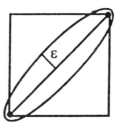

Figure 6

Please answer me straight without equivocation: 'Your saw is inside this ellipse, right?" I: "Of course, Boleslav Kornelievich!" He: "But the minor axis of the ellipse, denote it by ϵ, can be made arbitrarily small?" I: "Yes, B. K.!" He: "In other words, you agree that your curve is inside all ellipses with foci at the ends of the diagonal? I: "Of course, B. K." He: "But the *limit* of these homofocal ellipses as ϵ goes to zero is just the diagonal and not your saw. In other words, that saw of yours doesn't exist, and your reworked argument is not worth any more than the argument you showed me a week ago!" I fell silent and tried to cope with the wound I received. But now the roles were reversed, and he added mercilessly:

"You see, if some point M does *not* lie on the diagonal then,

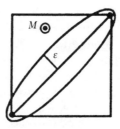

Figure 7

for ϵ small enough, it will be *outside* a homofocal ellipse. This means that the *limit* of the homofocal ellipses is just the diagonal, in the form of all of its points, and no other point M not on the diagonal. Clear?" But by then I had already recovered and said: "This is indeed so if ϵ is finite, but if ϵ is actually infinitely small..." But here I was interrupted by a storm of indignation: "Semper idem![15], he exclaimed. "I am talking to you for half an hour about *limits* and not about your actually infinitely small which don't exist in reality. I prove this in my course. Attend it—although for the time being I don't advise you to do so—and you will be convinced of this ... Is there anything else you want to tell me?" I was hurt to the quick and I began to talk, if I remember correctly, somewhat like this: "Boleslav Kornelievich, I think that it is possible to take as a starting point not only purely mathematical reasoning but also general scientific tendencies. For example, *in chemistry* there is no chemically pure water, for *in real water* there are always groups of molecules that do not enter into the composition of water. Nevertheless, in the science of chemistry there is H_2O and it tells us about *chemically pure water H_2O. What sort of thing is this? An abstraction? No, this is idealized water, an idealization! It seems that the process of idealization is inseparable from every stage of science. Granted there is no chemically pure water in reality, but there is H_2O in science*, and it is the job of science to talk about H_2O. Now take gypsum and the process of making a cast. For instance, you have a hollow, mathematically exact cone:

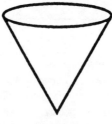

Figure 8

You pour in it a water solution of liquid gypsum in order to form a cast of the cone. What kind of process is this? The gypsum sets, and you take out of the empty

---

[15]*semper idem*—always the same.

cone a copy of it *which seems to you to be exact*.

Figure 9

But this is not the case. The process of hardening of the gypsum is a process of CRYSTALLIZATION, and what you take out of the cone is not a cone or its exact copy, but in reality only a crystal solid, merely similar to a cone, almost in the sense of the Cavalieri principle[16]. Now if we *idealize this process*, and think of the hardening of the ideal gypsum as a process of ideal crystallization then, after removing the cast, we find out that we have not a cone but rather a solid, actually infinitely small..."

Figure 10

But this conversation ended badly. I am ashamed to admit that B. K. suggested that next time I should bring him a jar of *such* gypsum and simply walked away... I never again discussed this topic with B. K. but two years later I encountered a situation that agitated me greatly.

Before I tell you about it, I will describe to you yet another characteristic feature of my mental state at the time. I recall being a third-year student, sitting in the first row, and attending Nikolaĭ Egorovich Zhukovskiĭ's[17] lecture on mechanics. More specifically, this was a lecture on *hydrodynamics*. The lecture hall was huge, the professor, a large and massive person, was extremely knowledgeable. As I recall today, he stood in front of a huge blackboard that took up the whole wall

---

[16] Cavalieri's principle can be stated as follows: If the areas of the sections of two solids by a plane parallel to a fixed plane are always equal, then so are the volumes of these solids.

[17] Nikolaĭ Egorovich Zhukovskiĭ (1847–1921), the great Russian mechanician and mathematician. Professor at Moscow University.

and, chalk in hand, talked to us about the forces acting on a fluid. He said: "we take a tiny element of the fluid..."

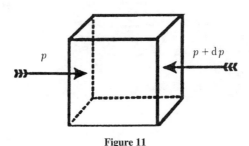

Figure 11

and, since the lecture hall was huge and chockful of students (for reasons of economy, the audience consisted of third- and fourth-year students), he drew a huge cube, *about a meter in size*. He continued: "Let p and $p + dp$ denote the pressures acting on the opposite walls of this tiny element..." And he began to carry out a complex geometric construction inside his cube.

It was as if someone had given me a shove. "Some tiny element", I thought. "One can easily put here a live goose. *And, in fact, the construct is* [*a*] *fixed* [*quantity*]. Here are the actually infinitely small that nobody wants to acknowledge for reasons of propriety, but which they use every now and again when they ignore propriety!" But then I concentrated on the lecture and forgot the scandalous thought.

Let me get back to Boleslav Kornelievich. This was about a year later, when I was a fourth-year student, i.e., about two years after my last conversation with Boleslav Kornelievich about saws. I recall that Byushgens[18] and I were invited to a meeting of the mathematical society. At that time students were not admitted to these meetings, and to attend a meeting one required a special invitation by the professors. There were about 12 members at the meeting. They sat at a long table covered with green cloth, drank tea, and ate rusks. (I was surprised by the prosaic combination of tea and science.) I don't remember who lectured, but the lecture dealt with *Pfaff equations*[19]. I sat in the second row behind Boleslav Kornelievich and D. F. Egorov, who sat at the table together with Mlodzeevskiĭ. At the time the lecturer was covering the blackboard with Pfaff equations involving *exact differentials*:

$$\omega'_i = \sum_{n,k} A_{ink}(dx_n \delta x_k - dx_k \delta x_n),$$

$$\omega^{(2m-6)} df_1\, df_2\, df = 0,$$

$$\omega'_i \equiv K_i(dx_{r+1} \delta x_{r+2} - dx_{r+2} \delta x_{r+1})(\mathrm{mod}\ \omega_1,\ldots,\omega_r)$$

Formula followed formula, marks of exact differentials d, δ flowed one after another like a swollen stream. The lecturer added, subtracted, multiplied, substi-

---

[18] Sergeĭ Sergeevich Byushgens (1882–1963), studied with Luzin at the university. He became an eminent geometer and a professor at Moscow University.

[19] We have been unable to ascertain whose lecture Luzin is referring to.

tuted one in another, expressed some linearly in terms of analogous marks d, δ of exact differentials. I looked attentively at the blackboard and tried at first to grasp the sense of what was done but (as I realize now), due to my ignorance of the theory of the Pfaff problem, I lost track of the reasoning and became aware that I was delighted with the externals of the stream of formulas. And it suddenly dawned on me: "Well, I don't know why it is so, but at least I know what d and δ are. If someone else came here with a good knowledge of algebra but ignorant of the differential calculus what would he think? He would think that the issue is algebraic transformations of certain unknown quantities dx, δf, and so on, and he would also express some of them in terms of others and solve for them. Indeed this is a genuine *algebra* of the quantities d and δ."

Just as I was thinking these thoughts I heard the animated voice of Boleslav Kornelievich talking to his neighbor Egorov: "I have always thought that the symbols for exact differentials are special symbols. Look at how he works with them! In his hands they are simply *constant* numbers: he adds, subtracts, multiplies, substitutes, and transforms them. One can completely forget their true origin and operate with them as if they were constant infinitely smalls. And you know, Dimitriĭ Fedorovich, it is not at all a hopeless attempt, in the spirit of Hilbert, to axiomatically..." At this moment the lecturer, disturbed by the animated voice, looked reproachfully at Boleslav Kornelievich who stopped his comments and said to him: "I am listening, I am listening!" He looked at D. F. and asked in a lower voice: "What is your opinion D. F.?" Egorov shook his head as if to say: "Indeed, Boleslav Kornelievich, and yet..."

A storm erupted in my mind: "So that's what it is! they teach us, kids, one thing, and they, the grownups, talk differently to one another. This means that, in fact, to judge by their conversations, things are not so absolutely determined." I looked at them with blazing eyes. I don't know what happened. Maybe my stool squeaked, or maybe this was one of those mysterious cases when people seem to sense a penetrating gaze in the back of them. Boleslav Kornelievich suddenly turned around, saw my blazing stare, leaned towards D. F. and said something to him in a low voice. The latter replied in an equally low voice and they fell silent.

I describe all this in such detail because I want to produce a psychological document of the state of my mind in the early stages of its development. You may find this of interest. It is possible that if I wandered patiently in the twilight of my recollections, I could bring to light more things of value (relating to mathematical education). But to tell the truth, I am *afraid* to enter this twilight. I am afraid to bring out of it and expose to daylight things intimately connected with the first stirrings of the mathematical mind or of mathematical consciousness, things that will immediately take a strong hold of me and deprive me of the possibility of following the chosen path in science. What I told you is so etched in my memory that it is always with me.

What happened to me later is the following. My education was based on the old treatises of analysis. From these I went *directly to set theory and the theory of functions*, bypassing the theory of limits. This occurred after I finished my university studies, when I declined Egorov's offer to remain at the University, studied medicine (for moral reasons) for 1/2 a year, and then studied for 1/2 a year at the former historical-philosophical faculty. I gave up on medicine because I could not look at corpses and would not have become a good medical doctor. It was only after the 1/2 year at the historical-philosophical faculty, where I listened to all the then famous philosophers and failed, for some reason, to acquire a passion for their arguments, that I returned to mathematics. I accepted Egorov's offer and worked on the completion of my mathematical education without asking for advice and following, as before, the guidance of chance.

The major spiritual drama associated with the Weierstrass curve forced me to concentrate on the theory of functions and, more generally, on micromathematics. This is the term I used to denote the infinitesimal structure of functions or of sets.

LETTER II

N. N. Luzin to M. Ya. Vygodskiĭ

20, XII, 33 Arbat, #25, apt.8. Moscow (2)

Dear Mark Yakovlevich,

I was very glad to get your letter. I am deeply grateful for your attention and for your most interesting and valuable gift—Zeuthen[1]. To tell the truth, when I heard about the publication of this book I very much wanted to acquire a copy but was unsuccessful—no matter where I went[2] the book was sold out. In the meantime, my need and interest in the book kept on growing, and I was about to embark on a long detour: to obtain the book in some provincial town. Judge for yourself how pleased and happy I was when the book was delivered directly to my house and came from you. I don't want you to think that these remarks are dictated by simple courtesy, and so will take the liberty of dwelling a bit on this matter.

You are not just the editor of the Zeuthen book[3] but also the author of extremely valuable supplements and of very profound notes. I saw these notes when I held in my hands somebody's copy of the Zeuthen book and was frustrated because I couldn't take it home with me. At the time, I appreciated the profundity of these supplements and notes and just wanted to think them over at leisure.

Also, your supplements to Zeuthen are very extensive and, due to my familiarity with Newton's *Principia*[4], I fully share your very clear[5] views. In particular, whenever you see in Zeuthen bias, onesidedness, or *preconceived* views and you argue against him, I am completely on your side. That's why getting a copy of Zeuthen from *you* was especially valuable.

I already wrote you a long letter on your and my views on infinitesimals[6]. If I will have the energy, I will once more, and perhaps more than once, return to this topic. All I can say now is just a few words.

---

[1] H. G. Zeuthen, *Geschichte der Mathematik im 16. und 17. Jahrhundert* (1903). Edited, commented on, supplied with an introduction, and translated into Russian by M. Vygodskiĭ.

[2] Unclear. Plausible variant.

[3] The text has "this book Zeuthen."

[4] Luzin took a great deal of interest in Newton's work and was familiar with A. N. Krylov's translation of the *Principia*.

[5] Word unclear. Vygodskiĭ's question mark in the margin indicates that he too could not make it out. Ours is a plausible variant.

[6] This is a reference to letter I.

To me, your views are *completely acceptable*. I attested this at one time publicly and am ready to do this whenever necessary. I am not in the least troubled by either the newness or boldness of these views. In this respect I am very different from many accomplished people whom I deeply respect and like, people who think highly of your activities, people who are practically, or actually, your friends, who are nevertheless put off by the boldness of your conceptions. On more than one occasion, when I expressed unconditional approval of your ideas and attempts, I was regarded with embarrassment. It was as if I was taking a risky or false step that was simply bad form. Such are the forces of *tradition*, such is the force of habit and the faith in the steadfastness of the notion of limit. I noticed this perfectly well but took no offense: one must always be aware of the power of published views. On the other hand, it would seem that *youth* should be versatile and *free*. This is amazing, but it is probably just a quality of human nature; and this is a great pity, because the sense of freedom in scientific conceptions is too valuable to sacrifice to an established trend.

Luckily, these are very different times and the earlier idols stand on unsteady legs. Imagine what would happen in physics if physicists held on to earlier views on atoms, i.e., if they imagined a small sphere, a little ball of matter covered with a shell on which one of their demons could happily stroll about. And yet the principle of transferring habits derived from the *finite* dimensions of our organisms to the world of intra-atomic distances is also a firmly established tradition, an *honorable* conception that has survived from the time of Democritus to the beginning of the 20th century. But it is of no use, and a modern quantum physicist says: "the transfer of our habits, of habits of perceiving oneself, into the world of intra-atomic structures is *impossible in principle*. Nouns such as 'attraction', 'pressure', 'repulsion', 'impact', 'rebound', and verbs such as 'moves', and 'undulates' must not be transferred *here*: this is a world different from the world of such crude, human, finite conceptions. *Here* one must use a completely different language, and if one wants to talk about the things that happen *there*, then it is simplest to use the language of mathematical symbols and to change them when necessary, so that what happens *there* corresponds to that language." It is a sobering thought that if we had adhered to tradition we would not have modern quanta!

It is much the same in mathematics. I look at the burning question of the foundations of infinitesimal analysis without sorrow, anger, or irritation. What Weierstrass—Cantor—did *was* very good. That's the way it *had* to be done. But whether this corresponds to what is in the depths of our consciousness is a very different question. I cannot but see a stark contradiction between the intuitively clear fundamental formulas of the integral calculus and the incomparably artificial and complex work of their "justification" and their "proofs". One must be quite stupid not to see this at once, and quite careless if, after having seen this, one can get used to this artificial, logical atmosphere, and can later on forget this stark contradiction.

I think that wherever people see the most complete harmony (in accordance with the claim that "all goes wonderfully well in this the best" etc.), there are deeply hidden mysterious facts not yet elucidated by consciousness and science.

Given my viewpoint, it is hardly necessary to say that I follow with deep satisfaction the genuine freedom of your thought and regard its movements as *genuine science*, far more valuable than the most recent, very proper, immaculately sewn together ("the way everybody does it") papers dealing with, say the theory of functions—to say nothing of other areas.

It is in these *free movements* of thought alone that I see a respectable pledge of the *dignity* of a scientist. If you want examples, here is the first that comes to mind: Weierstrass weighs down not only real variables, but his heavy, imperial hand also fettered for a long time the theory of functions of a complex variable. Who, when young, has not been thrilled by his notion of analytic continuation, by his conception of an analytic function, by his *purely formal* conception of singular points! Who was not delighted by his idea of a natural boundary beyond which a function *cannot be continued*. And it took time for mathematicians to realize that this noncontinuability involves a forced, artificial restriction. And Borel[7] was the first to choke in the narrow bounds of Weierstrass' theory of functions of a complex variable, and began to look for cracks in that formally beautiful theory. His attempts, his audacity, his impassioned language, and his efforts to break through the "natural" boundaries provoked amazement, horror, and mockery. They began by "reprimanding" him: as an old man, Borel himself described with amusement how Mittag-Leffler[8] asked him to sit down, took out of a trunk a manuscript of Weierstrass' lectures and began to "reprimand" him by pointing out that Weierstrass himself thinks that "what is not continuable is actually not continuable." "Magister dixit"[9], said Borel. Then two of Borel's attempts to "continue noncontinuable functions were demolished, the first by Poincaré[10] and the second by Painlevé[11]. The wish to break out of the charmed circle of the Weierstrass idea, an idea that was formally indestructible and essentially depressing, bothered Borel *throughout his life*. Is it not the case that we now have a theory of *quasianalytic functions*[12]? The secret of Weierstrass' uniqueness has not yet been completely puzzled out, but it is clear that (with the exception of textbooks) we do not need his continuations, and that, above all, the end of the tangled yarn has been found.

And what about Borel's divergent series[13]? What horror resulted when, after Abel, he *dared* say that one should not a priori reject divergent series as defective and throw them out of analysis. What cries of insult to the authority of Cauchy[14] and of disturbing the foundations! But Borel stubbornly pursued his aim and, when necessary, invoked the counter-authority of Euler. And now? Now we have the pronouncement of the academician A. N. Krylov to the effect that "except for

---

[7] E. Borel (1846–1927). Famous French mathematician. His contribution to the theory of functions of a complex variable is described in [**23**].

[8] M. G. Mittag-Leffler (1846–1927). Eminent Swedish mathematician. A devoted student of Weierstrass.

[9] *magister dixit*—thus said the teacher.

[10] H. Poincaré (1854–1912). French. One of the greatest mathematicians of the 19th and 20th centuries.

[11] P. Painlevé (1863–1933). Eminent French mathematician.

[12] Quasi-analytic functions. For Borel's investigations in the theory of analytic functions see [**23**, pp. 69–88].

[13] For divergent Borel series see [**23**, pp. 58–69].

[14] A. Cauchy (1789–1857). Great French mathematician.

progressions, convergent series are an empty abstraction of mathematical formalists. The series used by astronomers, physicists, mechanicians, and engineers are invariably *divergent*, and the mathematician's art is to obtain the mathematical essence of a phenomenon from 2–3 initial terms of such a series."

Every true scientist has in his lifetime *just one idea*, and his life is its embodiment. For Borel it was the battle with the stupid arithmetical restriction imposed by Weierstrass on functions of a complex variable. His whole life, his impassioned speech, his excursion into real variables—all these were the tools for the fight with Weierstrass' formalism. It is my guess that your impassioned language is the language of a fight against formalism, against that same evil genius Weierstrass (Denjoy[15] says publicly: "Weierstrass is the genius of evil in mathematics and not at all a mathematical genius"). I definitely do not interpret your statements as just a *pedagogical* fight: of course, you are right when you say that in pedagogy there is also a great deal of evil and ridiculous worship of formalism, and you fight against it. What is most important is your assertion between the lines: "The modern theory of infinitesimals is not the only possible theory." Pedagogical polemics are necessarily of very modest scope. But this idea of yours is on a universal scale and of extraordinary value! Much could be said along these lines.

In this connection there is a lot that we have in common. Like you, I think that the modern Weierstrass-Dedekind theory of irrational numbers is not the only possible theory. And this idea is probably also the cardinal idea of my life. Some time ago, the search for a new alternative made me analyze the works of Lebesgue[16] and brought me to the theory of analytic sets. But this search is extremely difficult.

It remains for me to reply to you concerning my purely personal problem involving the Encyclopedia. It seems that the attitude[17] of Sergeĭ Natanovich[18] has caused a major difficulty. *From a scientific point of view*, I personally insist on the modern independence of research on functions and sets in Euclidean spaces. But for me *personally* the question has lost its pointedness, and I have almost decided to leave the editorial group and to take no part in the Encyclopedia[19]. I must conserve my strength and *I am getting old*. Hence my departure from the Encyclopedia and the absence of my contributions will be a major *gain* for me!

You've just called and told me that you will come, so I end my letter.

With my most sincere regards, yours, *Nikolaĭ Luzin*

---

[15] A. Denjoy (1884—1974). Eminent French mathematician.

[16] H. Lebesgue (1875—1941). Eminent French mathematician.

[17] *attitude*—position (French).

[18] Sergei Natanovich Bernshteĭn (1880—1968).

[19] Luzin's articles in the *Great Soviet Encyclopedia* (Function, Differential Calculus, and others) are marked by a deep philosophical spirit that distinguishes them from the majority of articles in this work. In his philosophical quest Luzin followed in many respects roads close to those marked by the old Moscow philosophical-mathematical school. He had something in common with P. A. Florenskiĭ, who was inclined to look for the basis of the most important concepts of mathematics in the depths of intuition. His attitude was fundamentally different from that of the planners of the *Encyclopedia*, in particular, the mathematicians close to the Leningrad (Petersburg) school. This was true even of S. N. Bernshteĭn, who was sympathetic to Luzin.

REFERENCES

1. M. Ya. Vygodskiĭ *Foundations of Infinitesimal Calculus*, Moscow–Leningrad, 1931. (2nd ed.–1932; 3d ed.–1933)
2. W. E. Granville, *Elements of the Differential and Integral Calculus*. For technical schools and for self-study. 4th ed. Corrected by N. P. Tarasov. Edited and supplied with an introduction by Prof. N. N. Luzin. Moscow–Leningrad, 1926.
3. W. E. Granville and N. N. Luzin, *A Course of Differential and Integral Calculus*, parts 1–2, 6th ed., Moscow–Leningrad, 1938.
4. N. N. Luzin, *Differential Calculus*, 5th ed., Moscow, 1955.
5. _____, *Integral Calculus*, 5th ed., Moscow, 1955.
6. N. N. Luzin's letter to O.Yu. Shmidt. Published and supplied with an introduction and notes by S. S. Demidov, *Historico–Mathematical Investigations* **28** (1985) 278–287.
7. N. N. Luzin, Review of N. A. Vasil'ev's works on mathematical logic, in V. A. Bazhanov, *Nikolaĭ Aleksandrovich Vasil'ev*, Moscow, 1988, pp. 137–139.
8. F. A. Medvedev, N. N. Luzin on non-Archimedean time, *Historico–Mathematical Investigations*, **34** (1993) 103–128.
9. M. I. Panov, L. E. J. Brouwer and Soviet mathematics, in *The Laws of Evolution of Modern Mathematics*, ed. M. I. Panov, Moscow, 1989, pp. 250–278.
10. N. N. Luzin, Differential Calculus, in *The Great Soviet Encyclopedia*, 1st ed., V. 22, Moscow, 1934, pp. 622–642.
11. A. P. Yushkevich, *History of Mathematics in Russia up to 1917*, Moscow, 1968.
12. P. I. Kuznetsov, Dmitriĭ Ivanovich Egorov (on the occasion of the 100th anniversary of his birth), *Uspekhi Mat. Nauk*, 1971, V. 26, issue 5, pp. 169–206.
13. E. Goursat, *Cours d'Analyse mathématique*, 4-me éd., Paris, Gauthier–Villars, 1923–24, V. 1–3.
14. Russian translation of 13., Moscow, 1911–1923, V. 1–2.
15. Russian translation of 13., Moscow, 1923–1924, V. 1–3.
16. C. Jordan, *Cours d'Analyse de l'Ecole Polytechnique*, Paris, 1882, V. 1–3.
17. S. F. Lacroix, *Traité du calcul différentiel et du calcul intégral*, Paris, 1797–1798, V. 1–2.
18. M. P. H. Laurent, *Traité d'Analyse*, Paris, 1885–1891, V. 1–7.
19. Ch. La Vallée-Poussin, *Cours d'Analyse infinitésimale*, 5-me éd., Paris, 1925, V. 1–2.
20. Russian translation of 19. Translated and annotated by Ya. D. Tamarkin and G. M. Fikhtengolts under the ed. of V. A. Steklov, Petrograd, 1922, V. 1–2.
21. Russian translation of 19., Moscow–Leningrad, 1933, V. 1–2.
22. F. A. Medvedev, *Essays on the History of the Theory of Functions of a Real Variable*, Moscow, 1975.
23. E. M. Polishchuk, *Émile Borel*, Leningrad, 1980.

Riemann's Dissertation and Its Effect on the Evolution of Mathematics

Detlef Laugwitz
Translated from the German by Abe Shenitzer[†]

A SHORT ACCOUNT OF THE CONTENTS OF THE DISSERTATION. Riemann's doctoral dissertation of 1851 is titled *Grundlagen für eine allgemeine Theorie der Functionen einer veränderlichen complexen Grösse* (*Foundations for a general theory of functions of a variable complex quantity*) [**1**, 3–43]. It is of modest size. In discussing it we use modern terms.

Riemann defines holomorphic functions as complex single-valued functions on Riemann surfaces satisfying the Cauchy-Riemann differential equations. Riemann also worked with functions that were holomorphic except for finite poles in $\mathbb{C}$. Such meromorphic functions are viewed as conformal mappings between two Riemann surfaces. We must always think of the complex plane as extended by the addition of the point ∞ (as the Riemann complex number sphere or as a complex projective straight line).

Functions must be thought of not as given by expressions but as *determined* (to within arbitrary constants) *by the positions and nature of their singularities*. This leads to the question of the construction of functions with prescribed properties on a given Riemann surface. Here the topology of the surface is of decisive importance. The surface T is decomposed by means of n crosscuts into a system of m simply connected surface pieces. The number $n - m$, which is independent of the manner of decomposition, is called the order of connectivity of T [**1**, 10–11]; incidentally, in modern terms, this number is equal to the negative of the Euler characteristic of T.

In order to construct appropriate functions on T, Riemann uses a variational principle. (He called it later the Dirichlet principle because he came to know similar procedures in Dirichlet's lectures, and the historically unjustified name stuck.) First T is made into a simply connected surface T^* by means of crosscuts. Then, subject to suitable boundary conditions, the integral

$$\int \left[(u_x - v_y)^2 + (u_y + v_x)^2 \right] dx\, dy$$

is minimized on this surface. If there are singularities to be taken into consideration, then the integral is somewhat modified. With the possible exception of the boundary of T^*, the pair of functions u, v associated with the minimum is a

---

[†]*Translator's note.* Reprinted from "*Bernhard Riemann 1826–1866: Turning Points in the Conception of Mathematics*," by Detlef Laugwitz, Translated by Abe Shenitzer. Copyright 1999 Birkhäuser. This article is an excerpt (Section 1.2.2, pp. 108–110 and Section 1.2.5, pp. 124–130) from the author's book *Bernhard Riemann*, published by Birkhäuser Verlag in 1996. References such as Article 20 or §20 are to sections of Riemann's dissertation.

holomorphic function $f = u + iv$. It should be noted that the functional values on the two edges of a crosscut need not coincide; jumps ("periods") may occur.

The paper ends with an application of these methods to the Riemann Mapping Theorem. This theorem asserts that in certain cases the topological equivalence of two surfaces or regions implies their conformal equivalence, i.e., the existence of a conformal mapping between them. Here the theorem is first stated for regions in the complex plane that are homeomorphic to a circular disk.

We will examine the individual key words while considering further developments in the work of Riemann and others.

We explain briefly, in modern terms, the form of inference Riemann learned from Dirichlet. Let $I(\varphi, \psi)$ be the integral of $\varphi_x \psi_x + \varphi_y \psi_y$ over a region G and let $J(\varphi) = I(\varphi, \varphi)$. Let η be a function that vanishes on the boundary ∂G of G.

$$J(\varphi + t\eta) = J(\varphi) + 2tI(\varphi, \eta) + t^2 J(\eta)$$

implies that if $J(\varphi) \leq J(\varphi + t\eta)$ is to hold for all t, then we must have $I(\varphi, \eta) = 0$. Put $\Delta \varphi = \varphi_{xx} + \varphi_{yy}$. Our last result, the vanishing of η on ∂G, and the Gauss integral formula (Gauss' theorem) imply that

$$0 = \int_{\partial G} (\varphi_x \eta \, dy - \varphi_y \eta \, dx) = \int_G (\Delta \varphi) \eta \, dF + I(\varphi, \eta) = \int_G (\Delta \varphi) \eta \, dF.$$

Since this holds for every η, it follows that $\Delta \varphi = 0$. In other words, a function that minimizes $J(\varphi)$ is a solution of $\Delta \varphi = 0$. To be sure, the argument does not prove the *existence* of such a function, and this elicited justified criticism.

It is relatively easy to prove the uniqueness of the solution of the boundary-value problem. If ψ were another solution, then $\eta = \varphi - \psi$ would vanish on ∂G. Moreover,

$$J(\varphi) = J(\psi) + 2I(\psi, \eta) + J(\eta)$$

and

$$I(\psi, \eta) = \int_{\partial G} \eta(\psi_x \, dy - \psi_y \, dx) - \int_G (\Delta \psi) \eta \, dF = 0.$$

But then

$$J(\varphi) = J(\psi) + J(\eta) \geq J(\psi).$$

In view of the minimality of $J(\varphi)$, the inequality sign in the last expression must be replaced by an equality sign. But then $J(\eta) = 0$, i.e., $\eta_x = \eta_y = 0$. Since $\eta = 0$ on ∂G, it follows that $\eta = 0$, and therefore $\psi = \varphi$ throughout G.

THE EFFECT OF THE DISSERTATION. Today we are inclined to regard Riemann's dissertation as one of the most important achievements of 19th-century mathematics, but its immediate effect was rather slight. We saw that in the second part of Article 20 Riemann himself emphasized just one principle, namely the determination of a function by as few data as possible and the elimination of expressions as definitions of functions. Given its vague formulation, this principle must have struck his contemporaries as neither new nor interesting. Riemann was as restrained in his statement as he was in the specification of his sources.

The first person who had to read the paper carefully was the referee for the Göttingen faculty, that is, Gauss. His report read as follows: "The paper submitted by Herr Riemann is a concise testimony to its author's thorough and penetrating studies of the area to which the subject treated therein belongs; of a diligent and ambitious, truly mathematical spirit of investigation, and of praiseworthy and fertile independence. The report is prudent and concise, and in places even elegant; nevertheless, most readers might well wish for even greater transparency of arrangement in some of the parts. Taken in its entirety, it is a solid and valuable work which not only meets the requirements usually set for test papers for the attainment of the doctorate but exceeds them by far."

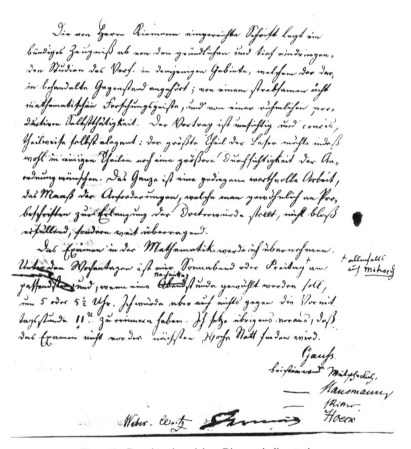

Figure 1. Gauss' testimonial on Riemann's dissertation

If one has a certain amount of experience with evaluations and forgets for a moment that here the *princeps mathematicorum* is writing about a person destined to become probably the most distinguished of his students, then one gets the following impression. The referee recognizes that the author has penetrated deep into a highly specialized field and has done this with great diligence, independently, and without the referee having to suggest the topic to him. There is no mention of the author's new ideas, of the solution of problems, or of new methods, but it is recognized that he may well be showing signs of independent research activity. The presentation is terse, elegant only in spots, and on the whole not clear enough. An objective reader must wonder what was the basis for the "Doktorvater's" (doctoral adviser's) very positive overall evaluation stated in the last sentence. Riemann wrote to his brother:

> When I visited Gauss he had not yet read my paper, but he told me that for years he had been preparing a paper (and is occupied with this right now) whose subject is the same, or partly the same, as the one I am treating

(Incidentally, this passage was quoted by Schering in his memorial address in 1866 [**2**, 835].) So far, no one has been able to find any indication that Gauss had discussed with Riemann the contents of his paper or had given him any hints or suggestions. Riemann would have reported such things. After all, he mentioned the rather disappointing conversation with Gauss which comes down, more or less, to this: right now I happen to be writing on a related topic, but your paper has not interested me enough that I should immediately and eagerly plunge into it.

Some (e.g., Remmert [**6**, Band 2, 158]) think that the old Gauss was "chary of praise" ("lobkarg"). But what argues against this is the fact that a few years earlier he had praised young Eisenstein to the skies. We will make no guesses about the great Gauss' admittedly baffling behavior toward Riemann.

We summarize the essential mathematical concerns that originated in Riemann's dissertation.

(1) The idea of a Riemann surface. Here, for the first time, the domain of definition of a function becomes one of the data that determine it. The complex plane is compactified by the addition of a single point ∞, the Riemann surfaces over it are precisely defined, the connectivity number is introduced and recognized as a topological invariant. (Complex) analysis is carried out not locally but on manifolds, which are compact in the case of algebraic functions. Local representability (by power series) is proved but is of secondary importance.

(2) In addition to poles, branch points are recognized as characteristic types of singularities, and the local series expansions in terms of (negative or fractional) powers are rigorously justified (Article 13/14, [**1**, 24-27]).

(3) The existence (together with the continuity) of $f'(z)$ is equivalent to the Cauchy–Riemann differential equations (together with the continuity of the occurring partial derivatives) and to the conformal character of f. It is also equivalent to the local expandibility, which implies the existence of all derivatives. (Holomorphic or analytic functions.)

(4) The transformation of surface integrals into line integrals is a tool for proving theorems (Articles 7-12, [**1**, 12-24]) of the "Cauchy type."

(5) The ("Dirichlet") principle of the existence of a function that minimizes a surface integral is used to solve boundary-value problems by means of holomorphic functions.

(6) The Riemann Mapping Theorem is a consequence of (5).

The response of contemporaries was amazingly slight; hardly any of the more than 500 titles in Purkert's list covering the period from 1851 to 1891 ([**2**, 869-895]) and relevant to Riemann's dissertation appeared before his death. This is all the more surprising if we keep in mind that two of Riemann's papers that presented the ideas of his dissertation in greater detail and applied them to the solution of problems appeared in 1857. Things were no different when it comes to textbooks. For example, Heinrich Weber's *Elliptische Functionen* of 1891 contains nothing relating to Riemann. Thus one can hardly speak of a significant impact of Riemann's ideas during his lifetime and in the first 25 years after his death. In the subsequent sections we will examine the question of the very special directions in which Riemann influenced research and the question of which elements of his essential ideas failed initially to attract attention.

Let us return to the year of the composition of the dissertation. Jacobi died on 18 February 1851. Dirichlet pushed Riemann in another direction, which led to his habilitation paper on trigonometric series. Representatives of the algorithmic direction could hardly be expected to approve of Riemann's dissertation. Eisenstein died on 11 October 1852 and Weierstrass had not yet appeared on the scene. The French mathematicians, whose contributions were not explicitly acknowledged in the dissertation, could at best be expected to recognize the concept of a Riemann surface as new. At the same time, they viewed it as too complicated and superfluous. Moreover, Cauchy's students soon got used to working with complex functions in the complex plane in much the same way as Cauchy, who had used complex formulations for his integral theorems and for his method of residues as early as 1831. They must have regarded the method of real partial differential equations as a backward step. At the time doubly periodic functions were in fashion, and they could be dealt with without the use of Riemann surfaces.

Of course, in time the six previously listed key issues associated with the dissertation exerted a powerful effect. What follows is a survey describing this effect.

The effect of (6) was later especially notable in applied mathematics. For a disk, the first boundary-value problem for the potential equation $u_{xx} + u_{yy} = 0$ is solved by the Poisson integral, which expresses the function u in terms of its boundary values. Since the differential equation is invariant under conformal mappings, we obtain a solution of this problem for any simply connected region bounded by a curve by mapping the disk conformally onto this region. But this is just an existence statement, and Riemann's theorem does not directly yield a formula representing the solution. Such representations were eventually obtained for regions of practical importance by H. A. Schwarz, E. B. Christoffel, and others.

The mapping theorem became effective in many respects independently of applications and of the other objectives and contents of the dissertation. It is an instance of Riemann's novel view of mathematics. For one thing, it illustrates the fruitfulness of the notion that functions are simply mappings. For another, it is a global proposition; all Gauss could prove was the conformal equivalence of small pieces of surfaces. Finally it was one of the deeper existence theorems to emerge after Cauchy's existence theorems about solutions of differential equations. For adherents of algorithms this was an unusual type of proposition; indeed, *they* took note of transformations only if they were associated with effective formulas. It is also noteworthy that the theorem shows that the theory of functions on a simply connected region with boundary is completely independent of the special choice of region. When investigating a special class of functions we can choose a convenient special region, say the upper halfplane.

Riemann's sketch of a proof in §21 is cryptic, and not just because of his use of the Dirichlet principle. Efforts to fully justify the idea of his proof failed. Given the importance of the theorem for applications, this failure stimulated attempts to develop new methods of proof. These remarks also apply to the uniformization theorem, which generalizes Riemann's mapping theorem. The geometric formulation promoted the acceptance of the notion of a Riemann surface. Riemann himself spoke [**1**, 40] of "geometric clothing" ("geometrische Einkleidung") used for "illustration and more convenient wording" (zur "Veranschaulichung und bequemeren Fassung"), formulations hardly ever encountered elsewhere in his writings. The use of complex methods for the computation of definite integrals opened up a new field for the applicability of complex function theory, and that is why complex analysis became a fixed component of the mathematical education of physicists and engineers. As for mathematics itself, the question of admissible boundaries of simply connected regions provided essential impulses for the evolution of point set theory.

For the effects of the dissertation in the first fifty years after Riemann, see [**5**]. For later developments see [**6**, Band 2, 157–163]. We recommend [**3**] and especially [**4**], a book saturated with Riemann's style of thinking. It is safe to say that, even had Riemann's dissertation consisted of just the mapping theorem, its influence would ultimately have been considerable.

The effect of (5) was unexpected. Riemann's justification of the existence of a minimal solution is inadequate. This was noted by Weierstrass, whose 1870 criticism was devastating and seemed to destroy the very basis of Riemann's justification of complex analysis. But this had also very positive consequences.

One consequence was that people tried, successfully, to prove the relevant results without using the Dirichlet principle. Actually they would have tried to find such proofs regardless of doubts about this principle. Such attempts reflect the wish to construct complex function theory in a "purely complex" way and to avoid the use of tools from real analysis, functions u and v of two real variables x and y. This too was achieved. Incidentally, this does not signify the rejection of Riemann's development of function theory. In view of its conceptual basis, it is closer to our way of thinking than is, say, the Weierstrass approach.

Another consequence of the criticism directed at Riemann's justification of the Dirichlet principle was even more important than the first one. Since there were no counterexamples and the principle itself was believable, people felt that it must be provable. Hilbert obtained a proof after 1900, and in doing so developed the so-called direct methods of the calculus of variations, which avoid the detour through the partial differential equations associated with the variational problem. One begins instead with a sequence of functions for which the values of the integral, or more generally of the functional, to be minimized approximate the infimum. One must show that the space of admissible functions has a compactness property which justifies the conclusion that a subsequence converges to a function for which the functional takes on its minimum. In this way a method was developed that not only saved the Dirichlet principle but has progressively become more important in the 20th century.

But let us go back briefly to the attempts to avoid the Dirichlet principle. Much was achieved by H. A. Schwarz and C. Neumann. As for the mapping theorem, the conclusive result was obtained independently by Poincaré and by Koebe in 1907. It asserts that every simply connected Riemann surface is holomorphically equivalent to one of following three surfaces: $\mathbb{C} \cup \{\infty\}$ (the number sphere or complex projective straight line), $\mathbb{C}$ (the number plane or complex straight line), or the open disk $|z| < 1$. The key that leads one to this group of problems in the literature is the uniformization theorem. This problem and its easy-to-formulate answer were almost obvious to Riemann, but half a century was needed to obtain it.

We do not know whether Riemann expected a stronger response. After all, he did say

> However, we now refrain from the realization of this theory...for we rule out, at present, consideration of an expression of a function

He set aside for a few years the task of investigating concrete functions and classes of functions, and tackled it in connection with lectures devoted to these matters. Of course, this did not happen during his first year as university instructor.

REFERENCES

1. (W.) *Bernhard Riemanns gesammelte mathematische Werke und wissenschaftlicher Nachlass*. Herausgegeben unter Mitwirkung von R. Dedekind von H. Weber. 2. Auflage: Teubner, Leipzig, 1892. Reprint: Dover, New York, 1953.
2. (N.) *Bernhard Riemann. Gesammelte mathematische Werke; wissenschaftlicher Nachlass und Nachträge. Coll. Papers*. Nach der Ausgabe von H. Weber und R. Dedekind neu herausgegeben von R. Narasimhan. Springer/Teubner, Berlin/Leipzig, 1990.
3. Ahlfors, L. V. "Development of the theory of conformal mapping and Riemann surfaces through a century." In *Contributions to the theory of Riemann surfaces. Centennial celebration of Riemann's dissertation*. Annals of Mathematics Studies, No. 30, 3–13, Princeton, 1953.
4. R. Courant, *Dirichlet's principle, conformal mapping and minimal surfaces*, with an appendix by M. Schiffer. Interscience, New York/London, 1950.
5. J. Gray, "On the history of the Riemann mapping theorem." Studies in the history of mathematics, I. *Supplemento ai Rendiconti del Circolo Matematico di Palermo*, ser. II, no. 34, 47–94, 1994.
6. R. Remmert, *Funktionentheorie* I, II. Springer, Berlin, 1991.

The Evolution of Integration

A. Shenitzer and J. Steprāns

THE GREEK PERIOD. The Greek problem underlying integration is the *quadrature problem*: Given a plane figure, construct a square of equal area.

It is easy to solve the quadrature problem for a polygon, a figure with rectilinear boundary. The first quadrature of a figure with curvilinear boundary was achieved by Hippocrates in the fifth century B.C. Hippocrates showed that the area of the lunule in Figure 1 (that is, the figure bounded by one-half of a circle of radius 1 and one-quarter of a circle of radius $\sqrt{2}$) is equal to the area of the unit square B.

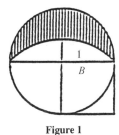

Figure 1

Figure 2

Hippocrates managed to square two other lunules.*

In the third century B.C. Archimedes effected the quadrature of a parabolic segment. He showed that its area is $\frac{4}{3}\Delta$, where Δ is the triangle of maximal area inscribed in the parabolic segment.

Archimedes effected a number of other quadratures (and cubatures). Some of his quadratures involved inventive constructions but most relied on the technique of wedging an area between ever closer upper and lower approximating sums. *Analogs of such sums are a key element of the definition of the Darboux integral (a variant of the Riemann integral introduced by Darboux in the 19th century) as well as of quadrature programs for computers.* We illustrate both of Archimedes' approaches next.

Consider Figure 3. Here the hypotenuse AB of the right triangle OAB is tangent to the spiral at A. It then turns out that the side AB is equal to the circumference of the circle with radius OA. (This is a special case of Archimedes' rectification of circular arcs by using tangents to spirals.) Since he knew that the area of a circle is half the product of its circumference by its radius, we can say that *Archimedes used (a tangent to) a spiral to rectify a circle and square its area.*

---

*Two more quadrable lunules were found by T. Clausen in the 19th century. In the 20th century, two Russian algebraists proved (independently) that these five lunules are the only quadrable ones. See "The Problem of Squarable Lunes," by M. M. Postnikov, p. 181.

Their brilliance notwithstanding, such constructions have been reduced to historical footnotes because they failed to yield general methods.

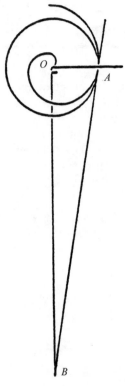

Figure 3

Figure 4 shows a turn of Archimedes' spiral $r = a\theta$ and the associated circle of radius $2\pi a$, and thus of area $K = 4\pi^3 a^2$. To compute the area S of the turn of the spiral in Figure 4 Archimedes approximates it from below and above by unions of circular sectors indicated in Figure 5.

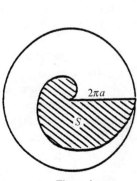

Figure 4

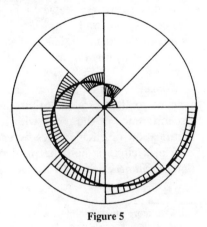

Figure 5

The areas of these approximating figures are, respectively,

$$S'_n = \frac{4\pi^3}{n^3}\left[1^2 + 2^2 + \cdots + (n-1)^2\right] = \frac{2\pi^3 a^2(n-1)(2n-1)}{3n^2}$$

and

$$S''_n = \frac{4\pi^3}{n^3}\left[1^2 + 2^2 + \cdots + n^2\right] = \frac{2\pi^3 a^2(n+1)(2n+1)}{3n^2}.$$

It is not difficult to see that

$$S'_n < \frac{4}{3}\pi^3 a^2 < S''_n$$

for all n. This double inequality can be rewritten as

$$S'_n < \frac{K}{3} < S''_n$$

for all n. Obviously,

$$S'_n < S < S''_n$$

for all n. To prove that $S = (K/3)$ Archimedes shows that $S''_n - S'_n = (4\pi^3 a^2/3n^2)$ and is thus small for large n. He can now show that the assumption $S \neq K/3$ leads to a contradiction and can conclude that $S = (K/3)$.

While Archimedes makes no explicit use of limits, he relies on the "method of exhaustion," and, in modern terms, the final part of the argument in a proof involving the method of exhaustion (in the above example it is disproving $S \neq K/3$) amounts to proving the uniqueness of the limit of a Cauchy sequence.

CONTINUATION IN THE 17TH CENTURY. Using nonrigorous infinitesimal techniques (rather than rigorous algebraic methods of the kind used by Archimedes) Cavalieri (1598–1647) managed to compute (what we now write as) $\int_0^1 x^k\,dx$ for $k = 1, 2, \ldots, 9$. His chief difficulty was the evaluation of $1^k + \cdots + n^k$. In about 1650 Fermat evaluated $\int_0^a x^{p/q}\,dx$ by means of a brilliant yet simple computation. Further progress was due to Torricelli, Wallis, and Pascal. In particular, Pascal interpreted Cavalieri's "sum of lines" (the equivalent of area) as a sum of infinitesimal rectangles.

If we combine Fermat's result with Cavalieri's understanding of the linearity of the definite integral (*our* terminology!) then we see that by the middle of the 17th century one could evaluate $\int_a^b P(x)\,dx$, $P(x)$ a "polynomial" with rational exponents.

In 1647 Gregory St. Vincent made a discovery that linked Napier's logarithm function and the area under the hyperbola $xy = 1$. This connection is now expressed as $\log_e(x) = \int_1^x (dt/t)$.

Newton and Leibniz invented the calculus and made it into a tool with countless applications but neither gave what we would call a rigorous definition of a definite integral (or saw the need for such a definition). Such concerns became dominant in the 19th century.

FROM CAUCHY TO LEBESGUE. The first rigorous definition of a definite integral was given by Cauchy in the 1820s. Cauchy dealt with continuous functions.

In view of the importance of Fourier series whose coefficients are given by integrals it was necessary to define the integral for more general functions. This was first done by Riemann. The limitations of the Riemann integral were remedied at the beginning of the 20th century by Lebesgue. An explanation follows.

With each theory of integration there is associated a theory of measure. Specifically, if f is a function on a set E and $f = f^+ - f^-$ (recall that $f^+(x) = \max\{f(x), 0\}$ and $f^-(x) = \max\{-f(x), 0\}$) then $\int_E f$ is defined as the difference $\int_E f^+ - \int_E f^-$ of the measures $\int_E f^+$ and $\int_E f^-$ of the ordinate sets of the nonnegative functions f^+ and f^- respectively.

The measure underlying the Riemann integral is Jordan measure and the measure underlying the Lebesgue integral is Lebesgue measure. How do they differ? In what way is one "better" than the other?

Consider the simple case of the ordinate set M of a bounded, nonnegative function f on an interval, $0 \le f(x) \le c$ for x in $[a, b]$. The Jordan measure of M is the common value, if any, of the outer and inner Jordan measures of M. The outer Jordan measure of M is the glb of the areas of the coverings of M consisting of *finite* unions of rectangles. The inner measure of M is the difference between the area $C(b - a)$ of the rectangle S with base $[a, b]$ and height C and the outer measure of the complement of M in S. Lebesgue replaced the word "finite" in the Jordan definition of the measure of a subset of S by "countable." This increased greatly the number of measurable subsets of S and led to a theory of integration far more comprehensive and mathematically flexible than Riemann's.

THE HK-INTEGRAL. Surprisingly, Henstock (in 1955) and Kurzweil (in 1957) came up with a new version of the Riemann integral—call it the HK-integral (see [7])—that is "as good as" the Lebesgue integral! Its definition and main characteristics follow (see [7]):

Definition:. A *tagged division* of $[a, b]$ is given by a finite ordered set $a = x_0 < x_1 < \cdots < x_n = b$ of points, together with a collection of *tags* z_i such that $x_{i-1} \le z_i \le x_i$ for $i = 1, \ldots, n$. We denote a tagged division by $D(x_i, z_i)$ and the corresponding Riemann sum by

$$S(D(x_i, z_i)) := \sum_{i=1}^{n} f(z_i)(x_i - x_{i-1}).$$

A *gauge* on $[a, b]$ is a function δ defined on $[a, b]$ such that $\delta(x) > 0$ for all $x \in [a, b]$. An important example of a gauge is a constant function. If δ is any gauge on $[a, b]$, we say that a tagged division $D(x_i, z_i)$ is δ-*fine* in case that $[x_{i-1}, x_i] \subseteq [z_i - \delta(z_i), z_i + \delta(z_i)]$; that is, in case $z_i - \delta(z_i) \le x_{i-1} \le z_i \le x_i \le z_i + \delta(z_i)$ for all $i = 1, 2, \ldots, n$. Finally, we say that the number A is an *HK-integral* of f if, for every $\varepsilon > 0$, there exists a gauge δ_ε such that if $D(x_i, z_i)$ is any tagged division of $[a, b]$ that is δ_ε-fine, then we have

$$|S(D(x_i, z_i)) - A| < \varepsilon.$$

It turns out that "the *HK*-integral of a function is uniquely defined when it exists and that a function is Riemann integrable if and only if the gauge δ_ε can be chosen to be constant." More importantly, "every Lebesgue integrable function is *HK*-integrable with the same value."

THERE IS NO PERFECT INTEGRAL. While in the eyes of some mathematicians the Lebesgue integral was the final answer to the difficulties associated with integration, there were others who were not willing to give up the search for the *perfect* integral, one which would make all functions integrable. Because Lebesgue's construction had shown that the key to a comprehensive theory of integration was the construction of an appropriate measure, the search now focussed on finding a total measure on $\mathbb{R}$, that is, one which assigns a measure to each subset of the real numbers.

Vitali [6] showed that a total measure on the reals cannot be countably additive *and* translation invariant. This being so, it is natural to ask which of these properties should be retained. This decision is, of course, somewhat arbitrary. While retaining translation invariance leads to some fascinating group theory and the Banach-Hausdorff-Tarski Paradox, we will consider what happens if countable additivity is retained instead.

In 1930 S. Ulam [1] showed that there is no such measure on ω_1, ω_2 or on any cardinal[1] which is the successor of some other cardinal. Ulam's proof was a spectacular advance in that it did not rely on any of the geometric assumptions, such as translation invariance, on which earlier proofs of the existence of non-measurable sets had relied.

By Ulam's theorem, the existence of a countably additive measure on $\mathbb{R}$ that measures all of its subsets implies that $2^{\aleph_0}$ is not the successor of any other cardinal, that is, it is a limit cardinal. By arguing a bit more carefully one can show that there must exist some limit cardinal $\lambda \leq 2^{\aleph_0}$ which is not the union of fewer than λ sets of size less than λ. The existence of such a cardinal has a profound influence on set theory.

In order to understand this influence, it is necessary to recall (a consequence of) Gödel's second incompleteness theorem which says that set theory can not prove its own consistency. One way to prove the consistency of a theory is to find a model of that theory, that is, a mathematical structure satisfying all of the axioms of that theory. We ask: What are the implications for set theory of the existence of a model of set theory? Recall the procedure for the construction of the hierarchy of sets. One begins with the empty set—call it V_0—and then defines V_k to be the power set of V_{k-1} for each integer $k \geq 1$. This is not the end, though, because one can then define V_ω to be the union of the sets V_k and then define $V_{\omega+1}$ to be the power set of V_ω. If one continues this as far as possible and takes the union one gets a model of set theory—or, at least, what would be a model of set theory if it were a set and not a proper class.

How soon, if ever, does this construction process lead to a model of set theory? It turns out that many of the axioms of set theory are satisfied at early stages of the construction. For example the axiom of infinity is satisfied as soon as a single infinite set is included and this is already true of $V_{\omega+1}$. The power set axiom is satisfied at any limit stage because any set which occurs, occurs at a stage before the limit and so all of its subsets are added at the very next stage. The power set itself is therefore added in no more than two stages and, in any case, before the

---

[1]Cantor introduced the notation ω to represent the next ordinal after the integers and it is still favored by set theorists today. The next cardinal after ω is denoted ω_1 and so on.

limit. For similar reasons, the pairing axiom is also satisfied at all limit stages. Well-foundedness and comprehension are also easy to deal with.

The problematic axiom is the axiom of replacement, which says, that the range of any function defined by a formula is a set. It has already been mentioned that $V_{\omega+\omega}$ will satisfy all of the axioms of set theory except for replacement. Replacement fails because the mapping which takes $2n$ to $\omega + n$ and $2n + 1$ to n is definable by a formula and its domain is ω which belongs to $V_{\omega+1} \subseteq V_{\omega+\omega}$. However, the range of this function is $\omega + \omega$ which does not belong to $V_{\omega+\omega}$. The same argument can be used to show that V_α is a model of set theory if and only if the following holds:

- if $\lambda < \alpha$ then $2^\lambda < \alpha$
- if $\lambda < \alpha$ then any function $F: \lambda \to \alpha$ (defined using only parameters from V_α) has range bounded in α.

Any cardinal satisfying these requirements is known as a *large* or *inaccessible* cardinal. Since the existence of a large cardinal implies that a model of set theory exists, it follows from Gödel's Theorem that it is impossible to prove the existence of inaccessible cardinals.

Ulam's argument shows that if there is a countably additive measure which measures every set of reals then there is a cardinal α which satisfies the second requirement of being an inaccessible cardinal. Such cardinals are known as weakly inaccessible. Another of Gödel's major contributions is the notion of the Constructible Universe, one of whose consequences is that any model of set theory contains a submodel which satisfies the generalized continuum hypothesis. This allows us to conclude that if there is a weakly inaccessible cardinal then, in the Constructible Universe, the weakly inaccessible cardinal is in fact an inaccessible cardinal; this is so because the cardinal arithmetic of this smaller model of set theory easily implies the first requirement for being a large cardinal.

In other words, if there is a countably additive measure which measures every set of reals then set theory is consistent. This and Gödel's theorem show that the existence of a *perfect* integral is not provable. On the other hand, it is conceivable that some day there may be a proof that it is *not* possible to have a perfect integral. The impact of this on set theory would be devastating. It would follow that many of the large cardinals which experts now consider quite innocuous, and which have played an important role in many important independence results, do not exist. While this would not show that set theory itself is inconsistent it would severely shake our faith in the assumption that it is.

* * *

We've told our story but would nevertheless like to tack on the following relevant "postscript":

In what sense does the integral solve the Greek quadrature problem and what is its conceptual significance? A telegraphic answer to these two questions follows.

The integral provides a direct "analytic" solution of the Greek quadrature problem for regions of the form.

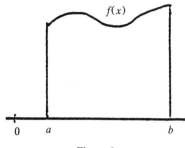

Figure 6

Indeed, the area of the region in the figure is

$$A = \int_a^b f(x)\, dx.$$

If we rewrite this as

$$A = \int_a^b f(x)\, dx = (b - a)\left(\frac{1}{b - a}\int_a^b f(x)\, dx\right),$$

then it is clear that our "integral region" has been replaced by a rectangle of equal area with base $b - a$ and height $(1/(b - a))\int_a^b f(x)\, dx$. The quantity $(1/(b - a))\int_a^b f(x)\, dx$ is the average of the functional values of f on $[a, b]$. *This averaging ability of the integral is the key to its importance in countless applications.*

REFERENCES

1. S. Ulam, Zur Masstheorie in der allgemeinen Mengenlehre, *Fund. Math.* 16 (1930), 140–150.
2. C. H. Edwards, Jr. *The Historical Development of the Calculus*, Springer-Verlag, 1979.
3. O. Toeplitz, *The Calculus—A Generic Approach*, University of Chicago Press, 1963.
4. A. Aaboe, *Episodes from the Early History of Mathematics*, the MAA, NML 13.
5. T. Jech, *Set Theory*, Academic Press, New York, 1978.
6. G. Vitali, *Sul problema della mesura dei gruppi di punti di una retta*, Bologna, 1905.
7. R. G. Bartle, review of R. Henstock's *The General Theory of Integration*, *BAMS*, v. 29, #1, July 1993, pp. 136–139.

On the Calculus of Variations and Its Major Influences on the Mathematics of the First Half of Our Century

Erwin Kreyszig

1. JOHANN BERNOULLI'S BRACHYSTOCHRONE. CARATHÉODORY'S METHOD. The calculus of variations evolved from the differential and integral calculus, "the calculus," for short. A key objective of the latter was the determination of extrema of *functions*, as shown by the title of the earliest relevant *published* paper *A new method for the determination of maxima and minima...* (Leibniz, 1684). However, the calculus had to be extended to the calculus of variations in order to take care of more general problems involving the determination of stationary values of *functionals*,[†] given in the simplest case by a definite integral involving an unknown function and boundary conditions. The earliest (not very simple) *solved* problem of this kind was Newton's determination of the shape of a gun shell of least air resistance (letter of Gregory of July 14, 1694).

The earliest problem that received general publicity [due to a rather bombastic advertisement in *Acta Eruditorum* by Johann Bernoulli (1667–1748)] in 1696 was the problem of determining the *brachystochrone*, the curve along which a particle will fall from one given point to another in the shortest time. This problem was solved by Newton, Leibniz, and Johann Bernoulli as well as by his brother Jacob (1654–1705), the solution being a *cycloid*. Thus 1696 can be called the birthyear of the calculus of variations. Johann Bernoulli not only posed that problem but also gave a solution capable of extensive generalization worked out in 1908 by Carathéodory. The resulting general method was later named after Carathéodory.

2. SIMPLEST GENERAL PROBLEMS. Although they arose from different geometric and physical applications, many of the early problems led to functionals that depended on real functions defined on an interval and satisfying boundary conditions, and all functionals were of the *same form* (as needed to create a general theory).

$$J[y] = \int_{x_0}^{x_1} L(x, y, y')\, dx, \quad y(x_0) = y_0, \quad y(x_1) = y_1, \quad x_0 < x_1. \quad (2.1)$$

The task was to determine a function $y(x)$ that satisfied the boundary conditions in (2.1) and rendered $J[y]$ stationary, possibly yielding a minimum or a maximum

---

*Abbreviated version of a paper with the same title.

[†]A functional is a function defined on a set of functions. A stationary value of a functional is its value at a "point" (= function) that satisfies the necessary conditions for an extremum (see Section 3 below).

of $J[y]$. At this early stage, the existence and uniqueness of solutions was "obvious" for physical reasons because a solution could be verified experimentally if desired. Also, with the concept of function not yet sharply defined, nobody made an attempt to characterize the set of functions in which such a $y(x)$ was to be found. This was accomplished well over one hundred years later in the works of Jacobi (1804–51) around 1835 and, especially, of Weierstrass (1815–97) around 1880.

3. EULER AND LAGRANGE. As the birthyear of the *theory* of the calculus of variations one usually considers 1744, the year in which Euler published his famous book *Methodus inveniendi lineas curvas maximi minimive proprietate gaudentes, sive solutio problematis isoperimetrici latissimo sensu accepti* (A method for discovering curved lines that enjoy a maximum or minimum property, or the solution of the isoperimetric problem taken in its widest sense). Thus Euler replaced "art of invention" (*ars inveniendi*), a very popular term in the works of Tschirnhaus and in other words of Leibniz's time, by "method of invention," a remarkable turn toward systematization. This book, a landmark in the development of the subject, contained the *Euler equation*

$$\frac{\partial L}{\partial y} - \frac{d}{dx}\left(\frac{\partial L}{\partial y'}\right) = 0, \tag{3.1}$$

(first published by Euler in 1736) as a necessary condition for $y(x)$ satisfying (2.1) to yield a minimum of $J[y]$. In more explicit form it is the equation

$$L_{y'y'}y'' + L_{y'y}y' + L_{y'x} - L_y = 0, \tag{3.2}$$

This equation suggests calling (2.1) a *regular problem* when $L_{y'y'}$ is never zero, and then assuming that $L_{y'y'} > 0$.

Euler's book also contains a fascinating collection of 66 problems. Carathéodory, the editor of the book as a volume of Euler's *Works*, said that it

> "is one of the most beautiful mathematical works ever written. We cannot emphasize enough the extent to which that *Lehrbuch* over and over again served later generations as a prototype in the endeavor of presenting special mathematical material in its [logical, intrinsic] connection."

Euler's inspiration came from geometry and even more from the *principle of least action*, according to which nature realizes all motions in the most economical manner; more precisely, among all possible ways of reaching a given goal, nature chooses the one which minimizes the *action integral* $\int mv\,ds$ over the path (m = mass, v = speed, s = arc length). The beginning of the principle is often dated back to Leibniz because of a (lost) letter he is supposed to have written on the principle in 1707, but the question is still an open one. The principle is usually named after de Maupertuis (1698–1759), president of the Berlin Academy under Frederick the Great. Actually, Euler most likely discovered it earlier, formulated it mathematically more rigorously, and applied it to a nontrivial problem (involving central forces). In contrast to this, Maupertuis published the principle (in 1744 and 1746) in a vague and almost theological form. He defended vigorously his (questionable) priority, but failed to realize that a rigorization of the principle would call for specification of conditions to be satisfied by the motions with which the actual motion was to be compared. Accordingly, his main merit seems to be that he was *searching for* a minimum principle.

The great significance of the calculus of variations in mathematical physics is due to the transparent and coordinate-free form that the laws of nature take in this calculus. That fact became apparent in the work of Euler, and even more impressively in that of Lagrange. In his path-breaking memoir *Essai d'une nouvelle méthode pour déterminer les maxima et les minima des formules intégrales indéfinies* (1760–61) Lagrange substantially overtook Euler (as Euler was well aware). His work was a milestone in the development of the field and in its application to geometry and analytical mechanics. In it he invented the "method of variations" together with the symbol δ. His new idea was to use "*comparison functions*,"

$$\bar{y} = y + \varepsilon\eta, \qquad \eta \in C^2([x_0, x_1]), \qquad \eta(x_0) = \eta(x_1) = 0, \qquad (3.3)$$

in (2.1) and to conclude from the vanishing of the first variation of (2.1),

$$\delta J = \varepsilon \frac{\partial J[\bar{y}]}{\partial \varepsilon}\bigg|_{\varepsilon=0} = \varepsilon \cdot \int_{x_0}^{x_1} (L_y \eta + L_{y'} \eta') \, dx = 0, \qquad (3.4)$$

(and an integration by parts) that Euler's equation (3.1) gave a necessary condition for $y(x)$ to render $J[y]$ stationary. (For a detailed and transparent explanation of this vital point see pp. 505–508 of G. F. Simmons, *Differential Equations*, second ed., McGraw-Hill, 1991.) In that paper Lagrange also started working on problems with variable endpoints, with application to brachystochrone and other problems. As another important step forward, he explicitly formulated his *multiplier rule* (without proof). This rule became a basic tool in his *Méchanique analitique*, in which he also included his theory of the calculus of variations and derived from the principle of least action his *equations of motion*, equivalent to *Newton's second law* and constituting analogues of Euler's equation. They are:

$$\frac{\partial V}{\partial x_i} + \frac{d}{dt}\left(\frac{\partial T}{\partial \dot{x}_i}\right) = 0, \qquad i = 1, 2, 3, \qquad (3.5)$$

where V and T are the potential and kinetic energy, respectively.

Whereas Euler's interest in the calculus of variations centered around applications, Lagrange's emphasis was on algorithmic aspects of analysis. Lagrange's entire work excels by its wealth of original discoveries as well as by the outstanding assimilation of the historical material.

> "By generalizing Euler's method he arrived at his remarkable formulas which in one line contain the solution of all problems of analytical mechanics.
>
> [In his Memoir of 1760–61] he created the whole calculus of variations with one stroke. This is one of the most beautiful articles that have ever been written. The ideas follow one another like lightning with the greatest rapidity..."

These enthusiastic lines are taken from lecture notes by C. G. J. Jacobi (1804–51).

4. MINIMAL SURFACES.
Euler's *Methodus inveniendi* of 1744 marked not only the beginning of the *theory* of the calculus of variations but also of one of its most fascinating geometric applications related to the creation of a remarkable class of

surfaces called *minimal surfaces*. These were originally obtained from the calculus of variations as (portions of) surfaces of least area among all surfaces bounded by a given space curve. Nowadays we define them as surfaces with vanishing mean curvature H,

$$H = \frac{1}{2}(\kappa_1 + \kappa_2) = \frac{1}{2}\left(\frac{GL - 2FM + EN}{EG - F^2}\right) = 0, \qquad (4.1)$$

using the discovery of Meusnier (1756–93) in 1776 that (4.1) is a necessary condition for least area. Here κ_1 and κ_2 are the principal curvatures, and E, F, G and L, M, N are the respective coefficients of the first and second fundamental forms of the surface.

What is most important to us here is Euler's discovery of the first non-trivial minimal surface, the *catenoid*. Euler obtained it by minimizing area, as the surface generated by rotating a *catenary* [a cosh curve, the curve of a hanging chain (*catena*) or cable], say,

$$r = A \cosh x, \qquad (4.2)$$

where r is the distance, in 3-dimensional space, from the x-axis.

Euler's discovery of the catenoid was a major accomplishment in his geometric work and marked the beginning of the study of minimal surfaces. It was followed by Lagrange's systematic theory developed in his Memoir of 1760–61. In this paper and in a subsequent one he extended his method to *double integrals* for functions of two variables.

$$J[z] = \int_\Omega \int L(x, y, z, p, q)\, dx\, dy \qquad (p = z_x, q = z_y) \qquad (4.3)$$

over a domain Ω in the xy-plane subject to given boundary conditions; the corresponding *Euler-Lagrange equation* [taking the place of (3.1)] is

$$L_z - \frac{\partial}{\partial x} L_p - \frac{\partial}{\partial y} L_q = 0. \qquad (4.4)$$

5. LEGENDRE, JACOBI, WEIERSTRASS. In the calculus, $y' = 0$ is only a necessary condition for a minimum of a function $y(x)$, and for a decision one must also consider y''. Similarly, in the calculus of variations Euler's equation is only a necessary condition for a minimum, and for a decision one must also consider the *second variation* of (2.1),

$$\delta^2 J = \frac{\varepsilon^2}{2} \frac{\partial^2 J[\bar{y}]}{\partial \varepsilon^2}\bigg|_{\varepsilon=0} = \frac{\varepsilon^2}{2} \int_{x_0}^{x_1} \left(L_{yy}\eta^2 + 2L_{yy'}\eta\eta' + L_{y'y'}\eta'^2 \right) dx, \qquad (5.1)$$

introduced by Legendre (1752–1833) in 1786. It was formally motivated by Taylor's theorem

$$J[y + \varepsilon\eta] = J[y] + \delta J + \delta^2 \bar{J}, \qquad (5.2)$$

where the bar means that the arguments are $y + \bar{\varepsilon}\eta, y' + \bar{\varepsilon}\eta'$ with $\bar{\varepsilon} \in (0, \varepsilon]$. Legendre obtained the condition $L_{y'y'} \geq 0$ along a minimizing curve and $L_{y'y'} \leq 0$ along a maximizing curve (very similar to the calculus!) but he did not justify his analysis completely.

In fact, it took another fifty years before Jacobi succeeded in rigorously demonstrating that $L_{y'y'} > 0$ and the so-called *Jacobi condition*, which asserts that

x_1 should be closer to x_0 than the so-called conjugate point* of x_0, suffice for a local minimum, that is a minimizing $\bar{y}$ among $y \in C^1[x_0, x_1]$ satisfying the boundary conditions in (2.1) and lying close to $\bar{y}$ in the C^1 sense, that is, satisfying

$$(a) \ |y - \bar{y}| < \rho, \qquad (b) \ |y' - \bar{y}'| < \rho \text{ for small positive } \rho. \qquad (5.3)$$

Jacobi's discovery of the conjugate point and of its significance closed a substantial gap. However, condition (5.3b) seemed to be too restrictive and in no way suggested by the nature of the problem. Weierstrass emphasized that one should extend the domain of (2.1) and consider *strong minima*, that is, one should drop (5.3b).

For this program of obtaining a sufficient condition for a strong minimum, Weierstrass set up entirely new machinery centering around two ingenious concepts. The first was a *field of extremals* of (2.1), which he defined as a domain Ω in the xy-plane such that through each of its points there passes precisely one extremal of a one-parameter family of extremals of (2.1) (solution curves of Euler's equation) depending continuously on the parameter, and the second was the so-called E-function, a turning point in the history of the calculus of variations.

To define the E-function, Weierstrass started from the *slope function* $p = p(x, y)$, the slope at (x, y) of the extremal of a field of extremals $y = h(x, \alpha)$; thus

$$p(x, y) = h'(x, \alpha)|_{\alpha = \alpha(x, y)}, \qquad (5.4)$$

where ' refers to differentiation with respect to x. Then he defined the E-*function* by

$$E(x, y, p, y') = L(x, y, y') - L(x, y, p) - (y' - p)L_{y'}(x, y, p), \qquad (5.5)$$

where $y = y(x)$ is any C^1-curve in the region covered by the field of extremals. Now he could prove that if, for an extremal $y = \bar{y}(x)$ of the field, the above sufficient conditions for a local minimum are satisfied, and if $E \geq 0$ at every point in the field and for every y', then $\bar{y}(x)$ gives a strong minimum of (2.1).

In addition to his path-breaking new method, Weierstrass also revolutionized the calculus of variations by stressing—practically for the first time—the importance of a precise definition of the domain $D(J)$ of the functional $J[y]$ and of *admissible functions*, the functions $y \in D(J)$ satisfying the side conditions.

6. IMPACT ON EARLY FUNCTIONAL ANALYSIS.
The proverbial *Weierstrassian rigor* (Felix Klein's term of 1885) had a profound influence on the theories of *functionals* and *function spaces* of our century. In fact, near the end of the last century, the central role of functionals in the calculus of variations may very well have directed attention to functionals in general. This may have been a kind of subliminal influence that affected primarily the younger generation represented by Volterra and later by Fréchet and F. Riesz.

Five notes of 1887 by Volterra on special classes of functionals, investigated as concepts *per se*, marked the *birth of functional analysis*. Influenced by Dini and

---

*The *conjugate point* of x_0 is the first value $x > x_0$ where a nonzero solution of

$$\frac{d}{dx}\left(L_{y'y'}\frac{dw}{dx}\right) - \left(L_{yy} - \frac{d}{dx}L_{yy'}\right)w = 0, \qquad w(x_0) = 0 \quad (x \geq x_0)$$

vanishes. Here $w(x) = \partial y/\partial \alpha|_{\alpha=0}$ and $\alpha = 0$ corresponds to $\bar{y}$ in the family of extremals $y = y(x, \alpha)$.

Betti, the latter a close friend of Riemann's, Volterra wanted to *generalize complex analysis*. His whole theory was based on the calculus of variations.

Hadamard was enthusiastic about Volterra's novel theory and contributed a basic result that attained final form in a celebrated paper by F. Riesz. (Riesz's theorem states that every continuous linear functional $U[f]$, $f \in C[a,b]$, can be expressed as a Stieltjes integral

$$U[f] = \int_a^b f(x) \, dw(x),$$

where $w(x)$ is determined by U and is of bounded variation on $[a,b]$.) Hadamard pointed to the variational-analytic roots of functional analysis—the *calculus* gave rise to the *calculus of variations*, which in turn produced the *functional calculus*, later called functional analysis.

Hadamard regarded the calculus of variations as

> "a first chapter of functional calculus, whose development will without doubt be one of the first tasks in the analysis of the future."

(This quotation is from Hadamard's book on the calculus of variations published in 1910.)

7. DIRICHLET'S PRINCIPLE. Besides its direct impact on developing functional analysis from 1887 to about 1903 the calculus of variations also had an indirect and, in the long run, an even greater impact on (classical and) functional analysis, an impact which dates back to about 1870 and reached functional analysis shortly after the turn of the century. This involved *partial differential equations*, for which the development proceeded from the search for general solution formulas for boundary and initial value problems (Green, Poisson, Kirchhoff, etc.) and on to existence (and uniqueness) proofs, notably for Laplace's equation for a function u of three variables

$$\Delta u = u_{xx} + u_{yy} + u_{zz} = 0 \tag{7.1}$$

(or of two variables x, y), which is basic in gravitation, electrostatics, stationary heat conduction, and fluid flow. For the corresponding *Dirichlet problem* in a general domain,

$$\Delta u = 0 \text{ in } \Omega; \quad u|_{\partial \Omega} = f, \quad u \in C^2(\Omega) \cap C^0(\overline{\Omega}), \quad \Omega \subset \mathbb{R}^2 \text{ or } \mathbb{R}^3, \tag{7.2}$$

a proof of existence of a solution u is not easy and attracted the efforts of many of the greatest mathematicians for quite some time. A (faulty) method of proof was provided by a principle taken from the *calculus of variations*. B. Riemann (1826–66) had first seen it in lectures of G. Lejeune Dirichlet (1805–59) in Berlin and named it after him:

Dirichlet's principle. There exists a function u that minimizes the functional (the so-called *Dirichlet integral*)

$$D[u] = \int_\Omega |\text{grad } u|^2 \, dV, \quad \Omega \subset \mathbb{R}^2 \text{ or } \mathbb{R}^3, \tag{7.3}$$

among all functions $u \in C^1(\Omega) \cap C^0(\overline{\Omega})$ which take on given values f on the boundary $\partial\Omega$ of Ω, and that function u satisfies (7.2).

Note that Laplace's equation is the Euler equation for (7.3).

We see that the Dirichlet integral is bounded below (by zero) and the claim of existence of a minimum is based on a conceptual error, namely the failure to distinguish *greatest lower bound* from *minimum*.

Dirichlet's principle was used earlier (in 1839) by C. F. Gauss (1777–1855) in his potential-theoretic investigations. Gauss claimed that if V is the potential of a mass distribution of density m on the surface S of a spatial region and U is a given function on S, then among all possible distributions there *obviously* exists one for which

$$\int (V - 2U) m \, dS$$

takes its minimum. If this is granted, then one can show, as Gauss did, that: (i) for this minimizing distribution, $V - U = const$ at all points of S that carry mass; (ii) if $U = 0$, there must be mass everywhere on S; (iii) one can obtain a distribution whose potential V on S equals U. Thus one could conclude the existence of the required *harmonic function* [a twice *continuously* differentiable solution of (7.1)] as well as its representation as the potential of a single layer of mass.

After Gauss, Dirichlet's principle was used in 1847 by W. Thomson (Lord Kelvin, 1824–1907) in order to "prove" the existence of a solution u of the differential equation

$$(\alpha^2 u_x)_x + (\alpha^2 u_y)_y + (\alpha^2 u_z)_z = 4\pi\zeta \tag{7.4}$$

that vanishes at infinity. Here α and ζ are given functions and ζ is zero outside a given bounded region. Because of this work, the principle is usually called in England *Thomson's principle*.

After attending Dirichlet's lectures in Berlin, Riemann used the principle in his famous thesis of 1851 as a key tool for obtaining fundamental results on *complex* analytic functions from *real* potential theory. Since the principle does not always hold, some of Riemann's proofs were not complete. All the results, however, turned out to be correct and were proved later by other methods. The same holds for Riemann's later use of Dirichlet's principle in his monumental paper on the theory of Abelian functions.

It was a strange situation. Dirichlet's principle had helped to produce exciting basic results but doubts about its validity began to appear, first in private remarks of Weierstrass—which did not impress Riemann, who placed no decisive value on the derivation of his existence theorems by Dirichlet's principle—and then, after both Dirichlet and Riemann had died, in Weierstrass's public address to the Berlin Academy:

"From Dirichlet's assumptions it can only be claimed that for (7.3) there exists a certain lower bound to which (7.3) can come arbitrarily close, without being forced to actually reach it. Dirichlet's argument appears invalid."

8. ANOTHER IMPACT ON FUNCTIONAL ANALYSIS.
The breakdown of Dirichlet's principle had an enormous positive effect on analysis because it led to the creation of three ingenious new methods for obtaining existence proofs for the

Dirichlet problem—by H. A. Schwarz (1843–1925), H. Poincaré (1854–1912), and C. Neumann (1832–1925)—as well as to the development of the direct methods of the calculus of variations initiated by D. Hilbert (1862–1943).

C. Neumann's *method of the arithmetic mean* (1870) was of great importance to analysis and to functional analysis because it sparked work on spectral theory, integral equations, and through it on Hilbert spaces. Neumann assumed the solution to be the potential of a *double layer*, a layer of dipoles normal to the boundary surface (or curve) $\partial\Omega$, of unknown density $\sigma(Q)$, $Q \in \partial\Omega$. Writing this potential in the form

$$u(P) = \frac{1}{2\pi} \int_{\partial\Omega} \sigma(Q) \frac{\partial}{\partial\nu}\left(\frac{1}{r}\right) dS(Q), \quad r = d(P;Q), \quad Q \in \partial\Omega, \quad (8.1)$$

(ν the outer normal of $\partial\Omega$) one obtains for the unknown density σ the integral equation

$$\sigma(Q) = \frac{1}{2\pi} \int_{\partial\Omega} \frac{\partial}{\partial\nu}\left(\frac{1}{r^*}\right) \sigma(Q^*) \, dS(Q^*) = \varphi(Q),$$

$\varphi(Q)$ the given values of u on $\partial\Omega$, $r^* = d(Q, Q^*)$. $\quad (8.2)$

Because of the term $\sigma(Q)$, the operator form of this "integral equation of the second kind" is

$$(I + K)\sigma = \varphi. \quad (8.3)$$

Its solution should obviously be

$$\sigma = (I + K)^{-1}\varphi = \varphi - K\varphi + K^2\varphi - K^3\varphi + \cdots. \quad (8.4)$$

Using this idea, Neumann was able to prove existence of a solution of the Dirichlet problem by integral equation methods. He solved (8.2) by successive approximation, defining $\sigma_0 = \varphi$ and

$$\sigma_n = (-K)\sigma_{n-1} = -\frac{1}{2\pi}\int_{\partial\Omega} \sigma_{n-1} \frac{\partial}{\partial\nu}\left(\frac{1}{r}\right) dS = (-K)^n \varphi.$$

This gave him the *Neumann series* (8.4) as the solution of the problem in a convex domain in space or in the plane.

We conclude this part of the section with a short list of events resulting from Neumann's work. In 1888, Weierstrass's former student du Bois-Reymond coined the term *integral equations* and expressed the desirability of a general theory of these equations, with which one could solve various problems such as that solved by Neumann. It did not take long for such theories—by Le Roux (1895), Volterra (1896), and the most famous one by Fredholm (1900, 1903)—to appear. Hilbert "caught fire at once" (as H. Weyl put it) and developed his spectral theory of integral equations with symmetric kernel published in six *Mitteilungen* between 1904 and 1910. Most important of these was the fourth *Mitteilung*, the earliest truly functional-analytic treatment of integral equations, in which he introduced continuous and compact (Hilbert said "completely continuous") forms (cast into operator language by F. Riesz in 1913).

After the breakdown of Dirichlet's principle there was no *general* principle for handling various problems of applied mathematics. It seems that in that situation Hilbert first put his hope in the calculus of variations, which had produced general principles in the past. In Problem 23 of his famous talk of 1900 on unsolved problems Hilbert had drawn attention to Weierstrass's work and to A. Kneser's book, the first presentation of the modern calculus of variations. Not intimidated by Weierstrass, he was able to re-establish the Dirichlet principle within proper limits as a valid method of proof. He did this in two papers of 1900 and 1901 (reprinted 1905). In the first of these papers he proposed the following more general formulation of Dirichlet's principle.

> "Every regular problem of the calculus of variations [Sec. 3] has a solution as soon as suitable restrictive assumptions with respect to the nature of the given boundary conditions are satisfied and, if necessary, the concept of a solution is suitably generalized."

This approach gave rise to the *direct methods* (methods without the use of the Euler-Lagrange equations), which became of basic importance in the existence theory of the calculus of variations. (A forerunner of these methods was Euler's almost forgotten "direct difference method.") Another solution of the Dirichlet problem by direct methods was given later (in 1907) by Lebesgue.

Apart from these splendid initial steps, Hilbert made no further attempts to uniformize analysis by methods of the calculus of variations. Instead, he turned to integral equations, perhaps as a more promising tool for the same purpose. But his idea of weakening the notion of solutions became a guiding principle in the calculus of variations of our century.

9. PLATEAU'S PROBLEM. By *Plateau's problem* one means the determination of a simply connected portion of a minimal surface S in R^3 bounded by a given curve in space. This problem is named after the Belgian physicist J. Plateau, who realized minimal surfaces experimentally by dipping wires (the boundary curves) into soap solutions, minimum area corresponding to minimum surface energy.

Plateau's problem has attracted great attention from around 1870 to the present. It is a problem genuinely belonging to the calculus of variations. The solution methods developed by a pleiad of researchers (Schwarz, Lebesgue, Korn, Bernstein, Haar, Garnier, Radó, Douglas, Courant, McShane) were interrelated with various branches of the mathematics of our century, which they fertilized immensely.

From a more general viewpoint we can regard the evolution of the theory of minimal surfaces and of Plateau's problem as particular cases of the development of the theory of partial differential equations. The first stage concerned special solutions of the minimal surface equation (special minimal surface), while the second stage dealt with general solutions (Weierstrass's general solution formulas). The third stage, the solution of boundary value problems (Plateau's problem), began in 1867 with Schwarz's work, slightly later than work on partial differential equations in general. For this work, Kelvin, Gauss, Riemann, and others had already switched from general solutions to the geometrically and physically more useful boundary and initial value problems.

10. GLOBAL CALCULUS OF VARIATIONS (MORSE THEORY). We have seen that early functional analysis owed much to the calculus of variations, and that *general topology* (*set-theoretic topology*) developed along with it, in a process of mutual give-and-take that extended over the first three decades of our century. This led to the creation of modern nonlinear analysis in connection with partial differential equations, and even more in connection with the calculus of variations, resulting in the so-called *calculus of variations in the large*, or *Morse theory*, for short.

In this calculus one is concerned with relations between properties of a "space" X (usually a topological space, often a Riemannian manifold) and a real-valued continuous function f defined on X.

The beginnings of this fascinating theory are due to Poincaré, who began his work on periodic solutions of the differential equations of celestial mechanics in his thesis of 1879 and knew of "Morse inequalities" for a surface as early as 1885, seven years before Morse was born.

The next stage of the development can perhaps best be seen from G. D. Birkhoff's book on Dynamical Systems (1927). In it Birkhoff emphasized the growing importance of topology in the calculus of variations. He also mentioned his Ph.D. student M. Morse, who worked out his calculations in the large in his book published by the AMS in 1934. "In the large" meant that Morse considered the whole manifold on which the variational problem was given and not just in a small neighborhood of an extremal ("calculus of variations in the small"). In the Preface he commented:

"Any problem which is nonlinear in character, which involves more than one coordinate system or more than one variable, or whose structure is initially defined in the large, is likely to require consideration of topology..."

Around 1930 topology was developing rapidly and eliciting general interest. This was evidenced, for example, by the wealth of new results contained in Alexandroff-Hopf's *Topologie I* of 1935 and by the papers presented at the Moscow Topology Congress of 1935. Thus it was just the right time for a marriage of the classical calculus of variations and topology, and Morse made ingenious use of the latter.

Define a *critical point* of a smooth function f on a smooth manifold M to be a point p at which
$$\text{grad } f = 0,$$
Morse classified the critical points in terms of the eigenvalues of the Hessian matrix
$$H(f, p) = \left[\frac{\partial^2 f}{\partial x_i \, \partial x_j} \right]_p \tag{10.1}$$
and obtained *topological* lower bounds (in terms of Betti numbers) for the number of critical points. These are the famous *Morse inequalities*. This work applied to *functions*. From *functions*, Morse proceeded to *functionals*, essentially n-dimensional analogs of our integral (2.1) and generalizations. Now for *functions* the topology at that time was sufficient. For *functionals*, that is, functions on a space of curves, Morse had to develop a topology in function spaces. Instead of critical *points* he had *critical curves*. He described this intuitively in his talk of 1932 at the Zurich International Congress. From it, one gains the impression that in Morse's theory the calculus of variations and topology had developed multiple relations. Incidentally, a similar theory was created simultaneously and independently by

L. A. Lyusternik and L. Schnirelman. This shows that at certain times certain things are *in the air*, in the sense that problems which extend known settings in a natural way become accessible as soon as basic theories (topology in the present case) have been sufficiently developed.

Let me conclude with the following remark. We started with the calculus, sketched briefly how calculus evolved into the calculus of variations, and outlined the most important ways in which the calculus of variations accompanied, influenced, or even initiated progress in various parts of analysis, geometry, functional analysis and, finally, topology. The whole process was heterogeneous, but I hope that we have seen traces of some intrinsic logic of the development here and there. The presentation can perhaps help to bring about a better understanding of certain features of present-day mathematics. The calculus of variations seems to have had a profound effect on the general development of mathematics. I hope that readers will see this as an invitation to further research pertaining to details as well as to larger issues that call for a more profound study than that presented in these pages.

REFERENCES

1. G. F. Simmons, *Differential Equations with Applications and Historical Notes* (second edition), McGraw-Hill, 1991. See Chapter 12.
2. N. I. Akhiezer, *Calculus of Variations*, Blaisdell, 1962.
3. I. G. Petrovsky, *Lectures on Partial Differential Equations*, Dover reprint, 1992.
4. G. M. Ewing, *Calculus of Variations with Applications*, Norton, 1969.

The Evolution of Literal Algebra

I. G. Bashmakova and G. S. Smirnova
Translated from the Russian by Abe Shenitzer

Part I. The Birth of Literal Algebra

1. MATHEMATICS IN THE FIRST CENTURIES AD. DIOPHANTUS. The Babylonians developed a kind of numerical algebra. Then came Greek geometric algebra.

The third—very important—stage of the development of algebra began in the first centuries AD and came to an end at the turn of the 17th century. Its beginning was marked by the introduction of *literal symbolism* by Diophantus of Alexandria and its end, by the creation of *literal calculus* in the works of Viète and Descartes. It was then that algebra acquired its own distinctive language, which we use today.

The first century BC was a period of Roman conquests and of Roman civil wars. Both took place in the territories of the Hellenistic states and the Roman provinces and were accompanied by physical and economic devastation. One after another, these states lost their independence. The last to fall was Egypt (30 BC). The horrors of war and the loss of faith in a secure tomorrow promoted the spread of religious and mystical teachings and undermined interest in the exact sciences, and in abstract problems in mathematics and astronomy. In Cicero's dialogue *On the state* one of the participants proposes a discussion of why two Suns were seen in the sky. But the topic is rejected, for "even if we acquired profound insight into this matter, we would not become better or happier."

In the second half of the first century BC mathematical investigations came to a virtual halt and there was an interruption in the transmission of the scientific tradition.

At the beginning of the new era, economic conditions in the Hellenistic countries, now turned Roman provinces, gradually improved, and there was a revival of literature, art, and science. In fact, the 2nd century came to be known as the Greek Renaissance. It was the age of writers such as Plutarch and Lucian and of scholars such as Claudius Ptolemy.

Alexandria continued its role as the cultural and scientific center of antiquity and, in this respect, Rome was never its rival. Nor did it ever develop an interest in the depths of Hellenistic science. As noted by Cicero in his *Tusculanae disputationes*, the Romans, unlike the Greeks, did not appreciate geometry; just as in the case of arithmetic, they stopped at narrow, practical knowledge of this subject.

Translator's note. Both parts of this chapter are taken from the Russian essay by I. G. Bashmakova and G. S. Smirnova devoted to the rise and evolution of algebra. The MAA published the English translation of this essay (by A. Shenitzer) under the title *The Beginnings and Evolution of Algebra* [Dolciani Mathematical Expositions 23, 2000].

They had little regard for all of mathematics. Even accounting, surveying, and astronomical observations were left to the Greeks, the Syrians, and other conquered nations. According to Vergil, the destiny of Romans was wise government of the world.

The revival of the Alexandrian school was accompanied by a fundamental change of orientation of its mathematical research. During the Hellenistic period geometry was the foundation of Greek mathematics; algebra had not, as yet, become an independent science but developed within the framework of geometry, and even the arithmetic of whole numbers was constructed geometrically. Now number became the foundation. This resulted in the arithmetization of all mathematics, the elimination of geometric justifications, and the emergence and independent evolution of algebra.

We encounter the return to numerical algebra already in the works of the outstanding mathematician, mechanician, and engineer Heron of Alexandria (1st century AD). In his books *Metrica*, *Geometrica*, and others, books that resemble in many respects our handbooks for engineers, one finds rules for the computation of areas and volumes, solutions of numerical quadratic equations, and interesting problems that reduce to indeterminate equations. In particular, they contain the famous "Heron formula" for the computation of the area of a triangle given its sides a, b, c:

$$S = \sqrt{p(p-a)(p-b)(p-c)},$$

where $p = (a + b + c)/2$. Here the expression under the square root sign is a product of four segments, and thus an expression totally inadmissible in geometric algebra. It is clear that Heron thought of segments as numbers, whose products are likewise numbers.

In his famous book, known under its Arabized name *Almagest*, Claudius Ptolemy, when computing tables of chords, identified ratios of magnitudes with numbers, and the operation of "composition" of ratios—defined in Euclid's *Elements*—with ordinary multiplication.

The new tendencies found their clearest expression in the works of Diophantus of Alexandria, who founded two disciplines: algebra and Diophantine analysis.

We know next to nothing about Diophantus himself. On the basis of certain indirect remarks, Paul Tannery, the eminent French historian of mathematics, concluded that Diophantus lived in the middle of the 3rd century AD. On the other hand, Renaissance scholars who discovered Diophantus' works, supposed that he lived at the time of Antoninus Pius, i.e., approximately in the middle of the 2nd century. The epigram in *Anthologia Palatina* provides the following information: "Here you see the tomb containing the remains of Diophantus, it is remarkable: artfully it tells the measures of his life. God granted him to be a boy for the sixth part of his life, and adding a twelfth part to this, He clothed his cheeks with down; He lit him the light of wedlock after a seventh part, and five years after his marriage He granted him a son. Alas! late-born wretched child; after attaining the measure of half his father's life, chill Fate took him. After consoling his grief by this science of numbers for four years he ended his life. By this device of numbers tell us the extent of his life." A simple computation shows that Diophantus died at the age of 84 years. This is all we know about him.

2. DIOPHANTUS' *Arithmetica*. ITS DOMAIN OF NUMBERS AND SYMBOLISM.

Only two (incomplete) works of Diophantus have come down to us. One is his *Arithmetica* (six books out of thirteen; four more books in Arabic, attributed to Diophantus, were found in 1973. They will be discussed in the sequel). The other is a collection of excerpts from his treatise *On polygonal numbers*. We discuss only the first of these works.

Arithmetica is not a theoretical work resembling Euclid's *Elements* or Apollonius' *Conic sections* but a collection of (189) problems, each of which is provided with one or more solutions and with relevant explanations. At the beginning of the first book there is a short algebraic introduction, which is basically the first account of the foundations of algebra. Here the author constructs the field of rational numbers, introduces literal symbolism, and gives rules for operating with polynomials and equations.

Already Heron regarded positive rational numbers as legitimate numbers (in classical ancient mathematics "number" denoted a collection of units, i.e., a natural number). While Diophantus defined a number as a collection of units, throughout *Arithmetica* he called every positive rational solution of one of his problems "number" ($\dot{\alpha}\rho\iota\theta\mu\acute{o}\varsigma$), i.e., he extended the notion of number to all of $\mathbf{Q}^+$. But this was not good enough for the purposes of algebra, and so Diophantus took the next decisive step of introducing negative numbers. It was only then that he obtained a system closed under the four operations of algebra, i.e., a field.

How did Diophantus introduce these new objects? Today we would say that he used the axiomatic method: he introduced a new object called "deficiency" ($\lambda\varepsilon\tilde{\iota}\psi\iota\varsigma$, from $\lambda\varepsilon\tilde{\iota}\pi\omega$—to lack) and stated rules for operating with it. He writes: "deficiency multiplied by deficiency yields availability (i.e., a positive number (*the authors*)); deficiency multiplied by availability yields deficiency; and the symbol for deficiency is ⋔, an inverted and shortened (letter) ψ" (Diophantus. *Arithmetica*. Definition IX). In other words, he formulated the rule of signs, which we can write as follows:

$$(-) \times (-) = (+),$$
$$(-) \times (+) = (-).$$

Diophantus did not formulate rules for addition and subtraction of the new numbers but he used them extensively in his books. Thus, while solving problem III_8 (i.e., Problem 8 in Book III), he needs to subtract $2x + 7$ from $x^2 + 4x + 1$. The result is $x^2 + 2x - 6$, i.e., here he carries out the operation $1 - 7 = -6$. In problem VI_{14}, $90 - 15x^2$ is subtracted from 54 and the result is $15x^2 - 36$. Thus here $15x^2$ is subtracted from zero; in other words, Diophantus is using the rule $-(-a) = a$.

We note that Diophantus used negative numbers only in intermediate computations and sought solutions only in the domain of positive rational numbers. A similar situation developed later in connection with the introduction of complex numbers. Initially they were regarded as just convenient symbols for obtaining results involving "genuine," i.e., real, numbers.

Diophantus also introduced literal signs for an unknown and its powers. He called an unknown a "number" ($\dot{\alpha}\rho\iota\theta\mu\acute{o}\varsigma$) and denoted it by the special symbol ς. It is possible that this symbol was introduced before him. We find it in the

Michigan papyrus (2nd century AD) as well as in a table appended to Heron's *Geometrica*. But Diophantus boldly breaks with geometric algebra by introducing special symbols *for the first six positive powers of the unknown, the first six negative powers, and for its zeroth power*. While the square and cube of the unknown could be interpreted geometrically, its 4th, 5th, and 6th powers could not be so represented. Nor could the negative powers of the unknown.

Diophantus denoted the positive powers of the unknown as follows:

$$x—\varsigma; \quad x^2—\Delta^v; \quad x^3—K^v; \quad x^4—\Delta^v\Delta; \quad x^5—\Delta K^v; \quad x^6—K^vK.$$

He defined negative powers as inverses of the corresponding positive powers and denoted them by adding to the exponents of the positive powers the symbol χ. For example, he denoted $x^{-2} = 1/x^2$ by $\Delta^{v\chi}$.

He denoted the zeroth power of the unknown by the symbol $\mathring{M}$, that is by the first two letters in Μόνας, or unity.

Then he set down a "multiplication table" for powers of the unknown that can be briefly written as follows:

$$x^m x^n = x^{m+n}, \quad -6 \leq m + n \leq 6.$$

He singled out two rules that correspond to the two basic axioms that we use for defining a group:

$$x^m \cdot 1 = x^m \quad \text{(definition VII)}; \tag{1}$$

$$x^m x^{-m} = 1 \quad \text{(definition VI)}. \tag{2}$$

In addition, Diophantus used the symbol $\overset{\prime}{\iota}\sigma$ for equality, and the symbol $\square$ for an indeterminate square. All this enabled him to write equations in literal form. Since he did not use a symbol for addition, he first set down all positive terms, then the minus sign (i.e., ⋔), then the negative terms. For example, the equation

$$x^3 - 2x^2 + 10x - 1 = 5$$

was written as

$$K^v \bar{\alpha} \varsigma \bar{\iota} \pitchfork \Delta^v \bar{\beta} \mathring{M} \bar{\alpha} \overset{\prime}{\iota}\sigma \mathring{M} \bar{\epsilon}.$$

Here $\bar{\alpha} = 1$, $\bar{\iota} = 10$, $\bar{\beta} = 2$, $\bar{\epsilon} = 5$ (we recall that the Greeks used the letters of the alphabet to denote numbers).

In the "Introduction" Diophantus formulated two basic rules of transformation of equations: 1) the rule for transfer of a term from one side of an equation to the other with changed sign and 2) reduction of like terms. Later, these two rules became well known under their Arabized names of *al − jabr* and *al − muqābala*.

Diophantus also used the rule of substitution in a masterly way but never formulated it.

We can say that in the introduction Diophantus defined the field **Q** of rational numbers, introduced symbols for an unknown and its powers, as well as symbols for equality and for negative numbers.

Before discussing the contents of *Arithmetica* we consider the possibilities and limitations of Diophantus' symbolism. Getting ahead of the story, we can say that, basically, Diophantus considered in his work indeterminate equations, i.e., equations with two or more unknowns. But he introduced symbols for just one unknown and its powers. How did he proceed when solving problems?

First he stated each problem in general form. For example: "To decompose a square into a sum of squares" (problem II$_8$). Now we would write this problem as
$$x^2 + y^2 = a^2.$$
How could Diophantus write this equation with just one symbol for an unknown and without symbols for parameters (in this case a)? He proceeded as follows: after the general formulation he assigned concrete values to the parameters—in the present case he put $a^2 = 16$. Then he denoted one unknown by his special symbol (we will use the letter t instead) and expressed the remaining unknowns as linear, quadratic, or more complex rational functions of that unknown and of the parameters. In case of the present example, one unknown is denoted by t and the other by $kt - a$ or, as Diophantus puts it, "a certain number of t's minus as many units as are contained in the side of 16," i.e., instead of a he takes 4 and instead of the parameter k—the number 2. But by saying "a certain number of t's" he indicates that the number 2 plays the role of an arbitrary parameter. Thus Diophantus' version of our equation is
$$t^2 + (2t - 4)^2 = 16,$$
so that
$$x = t = 16/5; \quad y = 2t - 4 = 12/5.$$

One might think that Diophantus was satisfied with finding a single solution. But this is not so. In the process of solving problem III$_{19}$ he finds it necessary to decompose a square into two squares. In this connection he writes: "We know that a square can be decomposed into a sum of squares in infinitely many ways."

The use of a concrete number to denote an arbitrary parameter has the virtue of simplicity. Sometimes it turned out that the parameter could not be selected arbitrarily, that it had to satisfy additional conditions. In such cases Diophantus determined these conditions. Thus problem VI$_8$ reduces to the system
$$x_1^3 + x_2 = y^3, \quad x_1 + x_2 = y.$$
Diophantus puts $x_2 = t$, $x_1 = \beta t$, where $\beta = 2$. Then from the second equation we obtain $y = (\beta + 1)t$, and from the first
$$t^2 = \frac{1}{(\beta + 1)^3 - \beta^3}.$$
Since $\beta = 2$, $t^2 = 1/19$, i.e., t is not rational. In order to obtain a rational solution Diophantus looks at the way t^2 is expressed in terms of the parameter β. The expression in question is a fraction whose numerator, 1, is a square. But then the denominator must also be a square:
$$(\beta + 1)^3 - \beta^3 = \square.$$
Diophantus took as the new unknown $\beta = \tau$ (he denoted it by the same symbol as the original unknown x_2) and obtained
$$3\tau^2 + 3\tau + 1 = \square.$$
Solving this equation by his method (which we will describe in detail in the next section) Diophantus obtained
$$\tau = \frac{3 + 2\lambda}{\lambda^2 - 3},$$

i.e., the parameter could only be chosen from the class of numbers $\{(3 + 2\lambda)/(\lambda^2 - 3)\}$. Diophantus takes $\lambda = 2$ and obtains $\beta = 7$. Then he goes back to solving the original problem.

Diophantus often deliberately chooses for parameters numbers that do not lead to solutions. He does this in order to show how to analyze problems.

Thus concrete numbers play two roles in *Arithmetica*. One role is that of ordinary numbers and the other is that of symbols for arbitrary parameters. Numbers were destined to play the latter role almost to the end of the 16th century.

Time to sum up. Diophantus was first to reduce determinate and indeterminate problems to equations. We may say that for a large class of problems of arithmetic and algebra he did the same thing that Descartes was later to do for problems of geometry, namely he reduced them to setting up and solving algebraic equations. Indeed, in order to solve problems—arithmetical in the case of Diophantus and geometric in the case of Descartes—both set up algebraic equations that they subsequently transformed and solved in accordance with the rules of algebra. Also, the transformations involved— such as elimination of unknowns, substitutions, and reduction of similar terms—had no direct arithmetic or geometric significance and were not subject to extensive interpretations. In both cases such interpretations were reserved for the final results. We are used to associating this important step with Descartes' creation of analytic geometry, but long before Descartes this step was taken by Diophantus in his *Arithmetica*. Before exposure to *Arithmetica*, none of the scholars of the period between the 13th and 16th centuries entertained the idea of applying algebra to the solution of number-theoretic problems.

3. THE CONTENTS OF *Arithmetica*. DIOPHANTUS' METHODS. We have already mentioned that *Arithmetica* is a collection of problems with solutions. This may create the impression that it is not a theoretical work. But a more careful reading makes it clear that the purpose of the painstaking choice and deliberate placement of problems was to illustrate the application of specific general methods. It is a characteristic of ancient mathematics that methods were not formulated apart from problems but were disclosed in the process of their solution. We recall that the famous "method of exhaustion"—the first variant of the theory of limits—was not set down in pure form either by its author Eudoxus of Cnidus or by Archimedes. It was mathematicians of the 16th and 17th centuries who isolated it by analyzing Euclid's *Elements* and Archimedes' quadratures and formulated it in general terms. The same applies to Diophantus' *Arithmetica*. As we show in the sequel, his methods were isolated in the 16th and 17th centuries by Italian and French mathematicians. Following them, we will try to isolate some of these methods and state them in general form.

In Book I Diophantus solved particular determinate linear and quadratic equations. The remaining books deal with the solution of indeterminate equations, i.e., equations of the form

$$F(x_1, \ldots, x_n) = 0, \quad n \geq 2,$$

$F(x_1, \ldots, x_n)$ a polynomial, or of systems of such equations:

$$\begin{cases} F_1(x_1, \ldots, x_n) = 0; \\ \cdots\cdots\cdots\cdots\cdots\cdots \\ F_k(x_1, \ldots, x_n) = 0, \quad k < n. \end{cases}$$

Diophantus looks for positive rational solutions $x_1^0, x_2^0, \ldots, x_n^0$, $x_i^0 \in \mathbf{Q}^+$, of such equations or of such systems.

It is clear that to solve his determinate equations Diophantus needed only symbols for x and x^2 and not for x^n, $-6 \leq n \leq 6$. In other words, he extended his domain of numbers and introduced most of his symbols to investigate and solve indeterminate equations, where he really needed higher powers of the unknown as well as its negative powers.

Thus the birth of literal algebra was connected not with determinate but with indeterminate equations.

Here we present just one of Diophantus' methods, namely his method for finding the rational solutions of a quadratic equation in two unknowns:

$$F_2(x, y) = 0, \tag{3}$$

where $F_2(x, y)$ is a quadratic polynomial with rational coefficients.

Basically, Diophantus proves the following theorem: if equation (3) has a rational solution (x_0, y_0) then it has infinitely many such solutions (x, y), and x and y are both rational functions (with rational coefficients) of a single parameter:

$$x = \varphi(k), \quad y = \psi(k). \tag{4}$$

When presenting his methods we use modern algebraic symbolism. This is by now a standard procedure in historical-mathematical literature.

Diophantus began by considering quadratic equations of the form

$$y^2 = ax^2 + bx + c, \quad a, b, c \in \mathbf{Q}, \tag{5}$$

and put $c = m^2$ (in other words, he assumed that the equation had two rational solutions $(0, m)$ and $(0, -m)$). To find solutions he made the substitution

$$y = kx \pm m \tag{6}$$

and obtained

$$x = \frac{b \mp 2km}{k^2 - a}, \quad y = \frac{b \mp 2km}{k^2 - a} \pm m.$$

By assigning to k all possible rational values (Diophantus took only values that yielded positive x and y) we obtain infinitely many solutions of equation (5).

We note that the substitutions (6) are the famous Euler substitutions that are applied to integrals of the form

$$\int \frac{dx}{\sqrt{ax^2 + bx + c}}.$$

We mentioned earlier problem II_8, which reduces to the equation

$$x^2 + y^2 = a^2, \tag{7}$$

and recall that Diophantus solved it by making the substitution
$$x = t; \quad y = kt - a, \tag{8}$$
and obtained (we are replacing his numerical values by appropriate letters)
$$x = t = a\frac{2k}{1 + k^2}; \quad y = a\frac{k^2 - 1}{k^2 + 1}.$$

To see the sense of this solution and to appreciate its generality we must look at its geometric interpretation. Equation (7) determines a circle of radius a centered at the origin, and the substitution (8) is the equation of a straight line with slope k passing through the point $A(0, -a)$ on that circle (Figure 1). It is clear that the straight line (8) intersects the circle (7) in another point B with rational coordinates. Conversely, if there is a point B_1 with rational coordinates (x_1, y_1) on the circle (7) then AB_1 is a straight line of the pencil (8) with rational slope k. Thus to every rational k there corresponds a rational point on the circle (7) and to every rational point on the circle (7) there corresponds a rational value of k. Hence Diophantus' method yields all rational solutions of equation (7).

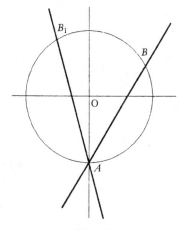

Figure 1

This argument shows that a conic with a rational point is birationally equivalent to a rational straight line.

Next Diophantus considered the more general case when equation (5) has a rational point but the coefficient c is not a square. He first considered this case in problem II$_9$, which reduces to the equation
$$x^2 + y^2 = a^2 + b^2 \tag{9}$$
(Diophantus put $a = 2$, $b = 3$). It is clear that equation (9) has the following four solutions: (a, b), $(-a, b)$, $(a, -b)$, and $(-a, -b)$. Diophantus makes the substitution
$$x = t + a, \quad y = kt - b \tag{10}$$
and obtains $t = 2(bk - a)/(1 + k^2)$. Applying a geometric interpretation analogous to the one just used we see that, essentially, he is leading a straight line with slope k through $(a, -b)$ on the circle (9).

Diophantus considered a more general case in lemma 2, proposition VI_{12} and in the lemma for proposition VI_{15}: assuming that equation (5) has a rational solution (x_0, y_0) he made the substitution $x = t + x_0$ and obtained

$$y^2 = at^2 + (2ax_0 + b)t + y_0^2,$$

i.e., he reduced the problem to the case $c = m^2$.

Finally, he considered equation (5) in the case when $a = \alpha^2$. He made the substitution (easily recognized as Euler's "second substitution" (*the authors*))

$$y = \alpha t \pm k \tag{11}$$

and obtained

$$t = \frac{c - k^2}{2\alpha k - b}.$$

This case calls for a separate discussion. To understand why the straight line (11) intersects the conic section (5) in just one point we introduce projective coordinates (U, V, W) by putting $x = U/W$, $y = V/W$, i.e., we consider our conic in the projective plane P^2. Then equation (5) takes the form

$$V^2 = \alpha^2 U^2 + bUW + cW^2. \tag{12}$$

The curve L so defined intersects the line at infinity $W = 0$ in two rational points: $(1, \alpha, 0)$ and $(1, -\alpha, 0)$. The straight line (11), whose equation in projective coordinates is

$$V = \alpha U + kW,$$

passes through the first of these points.

In summary, we can say that Diophantus carried out a complete investigation of a quadratic indeterminate equation in two unknowns. Later, his analysis served as a model for the investigation of the question of rational points on curves of genus 0.

Diophantus used more complex and more sophisticated methods to solve equations of the form

$$y^2 = ax^3 + bx^2 + cx + d,$$
$$y^3 = ax^3 + bx^2 + cx + d,$$
$$y^2 = ax^4 + bx^3 + cx^2 + dx + f,$$

and systems of the form

$$\begin{cases} ax^2 + bx + c = y^2, \\ a_1 x^2 + b_1 x + c_1 = z^2, \end{cases}$$

which he called "double equalities." Readers interested in getting a deeper understanding of Diophantus' methods should consult the book by Bashmakova and Slavutin: *A history of Diophantine analysis from Diophantus to Fermat* (Russian) which contains further references to the literature. The history of Diophantus' methods extends all the way to the papers of Poincaré that appeared at the beginning of the 20th century. It was on the basis of these methods that Poincaré constructed an arithmetic of algebraic curves—an area of mathematics that is being intensively developed at the present time.

We conclude our survey by considering Diophantus' problem III_{19}. This problem reduces to a system of 8 equations in 12 unknowns:
$$\begin{cases} (x_1 + x_2 + x_3 + x_4)^2 + x_i = y_i^2, \\ (x_1 + x_2 + x_3 + x_4)^2 - x_i = z_i^2; \quad i = 1,2,3,4. \end{cases}$$

Diophantus notes that "in every right triangle the square of the hypotenuse remains a square if we add to it, or subtract from it, twice the product of its legs." This means that he must find four right triangles with the same hypotenuse. Indeed, let the sides of the four triangles be a_i, b_i, c, $i = 1, 2, 3, 4$. Then it suffices to put $x_1 + x_2 + x_3 + x_4 = ct$, $x_i = 2a_i b_i t^2$, $i = 1, \ldots, 4$. Thus the problem reduces to finding a number c that can be written as a sum of two squares in four different ways. Diophantus solves this essentially number-theoretic problem as follows: he takes two right triangles with sides 3, 4, 5 and 5, 12, 13 respectively, and multiplies the sides of each of them by the hypotenuse of the other. As a result he obtains two right triangles with the same hypotenuse: 39, 42, 65 and 25, 60, 65. Now $5 = 1^2 + 2^2$ and $13 = 2^2 + 3^2$. Using the rule for composition of forms $u^2 + v^2$ known already to the Babylonians, namely

$$(u^2 + v^2)(\alpha^2 + \beta^2) = (\alpha u - \beta v)^2 + (\alpha v + \beta u)^2$$
$$= (\alpha u + \beta v)^2 + (\alpha v - \beta u)^2,$$

he obtains
$$65 = 5 \cdot 13 = (1^2 + 2^2)(2^2 + 3^2) = 4^2 + 7^2 = 8^2 + 1^2.$$

Using Euclid's formulas for the general solution of $x^2 + y^2 = z^2$ (i.e., $z = p^2 + q^2$; $x = p^2 - q^2$; $y = 2pq$) we obtain two more right triangles with hypotenuse 65: 33, 56, 65 and 63, 16, 65. This completes the solution of the problem.

In connection with this problem Fermat stated that a prime of the form $4n + 1$ could be written as a sum of squares in just one way. Then he gave a formula for the determination of the number of ways in which a given number can be written as a sum of squares. Thus problems involving indeterminate equations led to number-theoretic insights.

Did Diophantus know the theorems formulated by Fermat? It is possible that he did. Jacobi offered a reconstruction of Diophantus' conjectured proofs, but the answer to this question remains hypothetical.

One can hardly overestimate the significance of Diophantus' *Arithmetica* for the subsequent history of algebra. It is no exaggeration to say that its role was comparable to the role of Archimedes' treatises in the history of the differential and integral calculus. We will see that it was the starting point for all mathematicians up to Bombelli and Viète, and that its importance for number theory and for indeterminate equations can be traced up to the present.

4. ALGEBRA AFTER DIOPHANTUS. The period from the 4th to the 6th centuries AD was marked by the precipitous decline of ancient society and learning. But eminent commentators, such as Theon of Alexandria (second half of the 4th century) and his daughter Hypatia (murdered in 418 by a fanatical Christian mob), were still active. In the 5th century there was an exodus of scholars from Alexandria to Athens. Finally, in the 6th century, Eutocius and Simplicius, the last of the great commentators, were expelled from Athens and settled in Persia.

We can turn to the question of the Arabic translations of four books attributed to Diophantus. An analysis of these books, translated at the end of the 9th century from Greek to Arabic by Costa ibn Luca (i.e., the Greek Constantin, son of Luca) shows that they are a reworked version of Diophantus' *Arithmetica*. They contain problems, possibly due to Diophantus, as well as extensive additions and commentaries to them. According to Suidas' Byzantine dictionary, Hypatia wrote commentaries on *Arithmetica*. It is therefore very likely that the four books translated into Arabic are books edited and provided with commentaries by Hypatia. These books contain no new methods, but the material is presented in a complete and systematic manner. Their author went beyond Diophantus by introducing the 8th and 9th powers of the unknown.

The subsequent development of mathematics, including that of algebra, was connected with the Arabic East. Scholars from Syria, Egypt, Persia, and other regions conquered by the Arabs wrote scientific treatises in Arabic.

BIBLIOGRAPHY

1. I. G. Bashmakova, *Diophantus and Diophantine equations*. The Mathematical Association of America, 1997. Transl. by A. Shenitzer.
2. Th. L. Heath, *Diophantus of Alexandria*. New York, Dover, 1964.
3. Ø. Øre, *Number theory and its history*. New York, McGraw-Hill, 1948.
4. J. H. Silverman and J. Tate, *Rational points on elliptic curves*. New York, Springer, 1992.
5. B. L. van der Waerden, *Science awakening*. Transl. by A. Dresden. New York, Wiley, 1963.
6. A. Weil, *Number theory: An approach through history*. Basel, Birkhäuser, 1983.

Part II. The Literal Calculus of Viète and Descartes

5. THE CONTRIBUTION OF FRANÇOIS VIÈTE (1540–1603).
Viète tried to create a new science (he called it ars analytica, or analytic art) that would combine the rigor of the geometry of the ancients with the operativeness of algebra. This analytic art was to be powerful enough to leave no problem unsolved: nullum non problema solvere.

Viète set down the foundations of this new science in his *An introduction to the art of analysis* (In artem analyticem isagoge) of 1591.

In this treatise he created a literal calculus. In other words, he introduced the language of formulas into mathematics. Before him, literal notations were restricted to the unknown and its powers. Such notations were first introduced by Diophantus and were somewhat improved by mathematicians of the 15th and 16th centuries.

The first fundamentally new step after Diophantus was taken by Viète, who used literal notations for parameters as well as for the unknown. This enabled him to write equations and identities in general form. It is difficult to overestimate the importance of this step. Mathematical formulas are not just a compact language for recording theorems. After all, theorems can also be stated by means of words; for example, the formula

$$(a + b)^2 = a^2 + 2ab + b^2 \qquad (1)$$

can be expressed by means of the phrase "the square of the sum of two quantities is equal to the square of the first quantity, plus the square of the second quantity, plus twice their product." Shorthand also has the virtue of brevity. What counts is that we can carry out operations on formulas in a purely mechanical manner and obtain in this way new formulas and relations. To do this we must observe three rules: 1) the rule of substitution; 2) the rule for removing parentheses; and 3) the rule for reduction of similar terms. For example, from formula (1) one can obtain in a purely mechanical manner, without reasoning, formulas for $(a + b + c)^2$, for $(a + b)^3$, and so on. In other words, literal calculus replaces some reasoning by mechanical computations. In Leibniz' words, literal calculus "relieves the imagination".

We can hardly imagine mathematics without formulas, without a calculus. But it was such up until Viète's time. The importance of the step taken by Viète is so fundamental that we consider his reasoning in detail.

Viète adopted the basic principle of Greek geometry, according to which only homogeneous magnitudes can be added, subtracted, and can be in a ratio to one another. As he put it: "Homogena homogenei comparare." As a result of this principle, he divides magnitudes into "species": the 1st species consists of "lengths", i.e., of one-dimensional magnitudes. The product of two magnitudes of the 1st species belongs to the 2nd species, which consists of "plane magnitudes", or "squares", and so on.

In modern terms, the domain V of magnitudes considered by Viète can be described as follows:

$$V = \mathbf{R}_+^{(1)} \cup \mathbf{R}_+^{(2)} \cup \cdots \cup \mathbf{R}_+^{(k)} \cup \cdots,$$

where $\mathbf{R}_+^{(k)}$ is the domain of k-dimensional magnitudes, $k \in \mathbf{N}_+$. In each of the domains $\mathbf{R}_+^{(k)}$ we can carry out the operations of addition and of subtraction of a smaller magnitude from a larger one, and can form ratios of magnitudes. If $\alpha \in \mathbf{R}_+^{(k)}$ and $\beta \in \mathbf{R}_+^{(l)}$, then there is a magnitude $\gamma = \alpha\beta$ and $\gamma \in \mathbf{R}_+^{(k+l)}$. If $k > l$, then there exists a magnitude $\delta = \alpha : \beta$, and $\delta \in \mathbf{R}_+^{(k-l)}$.

After constructing this "ladder", Viète proposes to denote unknown magnitudes by vowels $A, E, I, O...$ and known ones by consonants $B, C, D, ...$ Furthermore, to the right of the letter denoting a magnitude he places a symbol denoting its species. Thus if $B \in \mathbf{R}_+^{(2)}$, then he writes B plan (i.e., planum−plane), and if an unknown $A \in \mathbf{R}_+^{(2)}$, then he writes A quad (square). Similarly, magnitudes in $\mathbf{R}_+^{(3)}$ get the indices solid or cub and those in $\mathbf{R}_+^{(4)}$ get the indices plano-planum or quadrato-quadratum, and so on.

For addition and subtraction Viète adopts the cossist symbols + and − and introduces the symbol $=$ for the absolute value of the difference of two numbers, thus $B = D$ is the same as $|B - D|$. For multiplication he uses the word "in", A in B, and for division the word "applicare".

Next he introduces the rules

$$B - (C \pm D) = B - C \mp D; \quad B \text{ in } (C \pm D) = B \text{ in } C \pm B \text{ in } D,$$

as well as operations on fractions, written by means of letters, e.g.,

$$\frac{B \text{pl}}{D} + Z = \frac{B \text{pl} + Z \text{ in } D}{D}.$$

Viète's next treatise was *Ad logisticam speciosam notae priores*, which appeared only in 1646 as part of his collected works. In it he set down some of the most important algebraic formulas, such as:

$$(A + B)^n = A^n \pm nA^{n-1}B + \cdots \pm B^n, \quad n = 2, 3, 4, 5;$$

$$A^n + B^n = (A + B)(A^{n-1} - A^{n-2}B + \cdots \pm B^{n-1}), \quad n = 3, 5;$$

$$A^n - B^n = (A - B)(A^{n-1} + A^{n-2}B + \cdots + B^{n-1}), \quad n = 2, 3, 4, 5.$$

Viète's literal calculus was perfected by René Descartes (1596–1650), who dispensed with the principle of homogeneity and gave the literal calculus its modern form.

6. THE CONTRIBUTION OF RENÉ DESCARTES (1596–1650).

The 16th century was marked by remarkable achievements in algebra and was followed by a period of relative calm in this area. Most of the energy of 17th-century mathematicians was absorbed by infinitesimal analysis, which was created at that time. Nevertheless, while inconspicuous at first sight, profound changes were taking place in algebra that can be characterized by one word—arithmetization.

The first steps in this direction were taken by the famous philosopher and mathematician René Descartes (1596–1650). In his *Geometry* (the fourth part of his 1637 *Discourse on method*), whose essential content was the reduction of geometry to algebra or, in other words, the creation of analytic geometry, he first of all transformed Viète's calculus of magnitudes (logistica speciosa). Descartes represented all magnitudes by segments, and constructed a calculus of segments that differed essentially from the one that was used in antiquity and that formed the basis of Viète's construction. Descartes' idea was that the operations on segments should be a faithful replica of (we would say "should be isomorphic to") the operations on rational numbers. Whereas the ancients and Viète regarded the product of two segment magnitudes as an area, i.e., as a magnitude of dimension 2, Descartes stipulated that it was to be a segment. To this end, he introduced a unit segment—which we will denote by e—and defined the product of segments a and b as the segment c that was the fourth proportional to the segments e, a, and b. Specifically (see Figure 1), he constructed an arbitrary angle ABC, and laid off the segments $AB = e$, $BD = b$ and $BC = a$. Then he joined A to C, drew $DF \| AC$, and obtained the segment $BF = c = ab$. This meant that the product belonged to the same domain of magnitudes (segments) as the factors. Division was defined

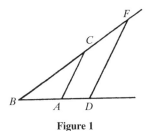

Figure 1

analogously: to divide $BF = c$ by $BD = b$ we lay off from the vertex B of the angle the segment $BA = e$, join F to D, and draw AC parallel to DF. The segment BC is the required quotient. In this way, Descartes made the domain of segments into a replica of the semifield $\mathbf{R}^+$. Later he also introduced negative segments (with directions opposite to those of the positive segments) but did not go into the details of operations with negative numbers. Finally, Descartes showed that the operation of extraction of roots (of positive magnitudes) does not take us outside the domain of segments. (We interpolate a comment. Long before Descartes, Bombelli introduced similar rules of operation with segments. Until recently it was thought that he did this in the fourth part of his manuscript published only in the 20th century. However, G. S. Smirnova showed recently that such operations on segments occur also in the parts of Bombelli's *Algebra* published in 1572, i.e., during his lifetime.) To extract the root of $c = BF$, Descartes extended this segment, laid off $FA = e$ on the extension, drew a semicircle with diameter BA, and erected at F the perpendicular to BA. If I is its intersection with the semicircle, then $FI = \sqrt{c}$.

Descartes' calculus was of tremendous significance for the subsequent development of algebra. It not only brought segments closer to numbers but also lent to algebra the simplicity and operativeness that we take advantage of to this day. Another convention introduced by Descartes and used to this day is denoting unknowns by the last letters of the alphabet: x, y, z, and knowns by the first letters: a, b, c. The only difference between Descartes' symbolism and modern symbolism is his equality sign: ∞.

Essentially, it was Descartes who established the isomorphism between the domain of segments and the semifield $\mathbf{R}^+$ of real numbers. However, he gave no general definition of number. This was done by Newton in his *Universal Arithmetic* in which the construction of algebra on the basis of arithmetic reached its completion. He wrote: "Computation is conducted either by means of numbers, as in ordinary arithmetic, or through general variables, as is the habit of analytical mathematicians." And further: "Yet arithmetic is so instrumental to algebra in all its operations that they seem jointly to constitute but a single, complete computing science, and for that reason I shall explain both together."

Newton immediately gives a general definition of number. We recall that in antiquity number denoted a collection of units (i.e., natural numbers), and that ratios of numbers (rational numbers) and ratios of like quantities (real numbers) were not regarded as numbers. Claudius Ptolemy (2nd century AD) and Arab mathematicians did identify ratios with numbers, but in 16th- and 17th-century Europe the Euclidean tradition was still very strong. Newton was the first to break with it openly. He wrote:

> By a 'number' we understand not so much a multitude of units as the abstract ratio of any quantity to another quantity which is considered to be unity. It is threefold: integral, fractional, and surd. An integer is measured by unity, a fraction by a submultiple part of unity, while a surd is incommensurable with unity.

With characteristic brevity, Newton goes on to define negative numbers:

> Quantities are either positive, that is, greater than zero, or negative, that is, less than zero. ... in geometry, if a line drawn with advancing motion in some direction be considered as positive, then its negative is one drawn retreating in the opposite direction.
>
> To denote a negative quantity... the sign $-$ is usually prefixed, to a positive one the sign $+$.

Then Newton formulates rules of operation with relative numbers. We quote his multiplication rule: "A product is positive if both factors are positive or both negative and it is negative otherwise."

He provides no "justifications" for these rules.

(*Translator's note*. The preceding quotations are taken from D. T. Whiteside's English translation of Newton's *Arithmetica universalis*.)

Thus Viète's elaborate domain of magnitudes was replaced in the 17th century by the field of real numbers and arithmetic formed the foundation of algebra.

REFERENCES

1. I. G. Bashmakova and G. S. Smirnova, *The rise and evolution of algebra*. In: *Essays on the history of mathematics*, Moscow State University, Moscow, 1997. (Russian)
2. H. L. L. Busard, François Viète, *Dictionary of Scientific Biography*, Vol. 14, pp. 18–25. New York, Scribners, 1970–1990.
3. Theodore M. Brown, René Descartes, *Dictionary of Scientific Biography*, Vol. 4, pp. 51–65. New York, Scribners, 1970–1990.
4. René Descartes, *Discourse on method and meditations*. Translated, with an introduction by L. J. Lafleur. New York, Macmillan, 1986.
5. _____, *Geometry*. Translated by D. E. Smith and M. L. Latham. New York, Dover, 1954.

The Evolution of Algebra 1800–1870*

I. G. Bashmakova and A. N. Rudakov

The first event of this period was the appearance, in 1801, of C. F. Gauss' *Disquisitiones Arithmeticae*. Of the seven parts of the book only one is devoted to an algebraic issue, namely the cyclotomic equation $x^n - 1 = 0$. But the author's brilliant algebraic thinking is apparent in all the other parts as well. *Disquisitiones*, an epoch-making work in algebraic number theory, was for a long time a handbook and source of ideas in algebra. In the course of his study of the cyclotomic equation Gauss shows that it is solvable for every n in the sense that the solutions are expressible in terms of radicals, gives a method for explicitly finding these expressions, and singles out the values of n for which the solutions are expressible in quadratic radicals and thus the values of n for which it is possible to construct a regular n-gon by means of ruler and compass. As always, his investigations are strikingly profound and detailed. They were continued by N. H. Abel, who proved the insolvability by radicals of the general quintic and singled out a class of equations, now named for him, that are solvable by radicals. The new notions of field (domain of rationality) and group (group of an equation) turned up in Abel's papers with greater definiteness. The next step in this direction that completed the theory was the papers of the young E. Galois, published in fragmentary form between 1830 and 1832, and, after his death, in more complete form by Liouville in 1846.

The papers of Abel, and especially of Galois, belong to the radically new trend of ideas now generally accepted in algebra. In his study of the ancient problem of solution of equations by radicals Galois shifted the center of gravity from the problem to the methods of its solution: he gave clear-cut definitions of the concepts of a field and of the group of an equation, established the correspondence between the subgroups of the group of an equation and the subfields of the splitting field of the polynomial on the left side of that equation, and, finally, singled out the normal subgroups of a group and studied its composition series. These were completely new and extremely fruitful methods of investigation and yet they were apprehended by mathematicians only in the 70s. The one exception was groups of substitutions. Such groups were considered by Galois and their investigation began already in the 40s.

---

*This article is a reprint of the major part of the introduction to an essay dealing with the evolution of algebra and algebraic number theory during the period of 1800–1870. The essay forms Chapter 2 of the book *Mathematics of the 19th Century* that deals with mathematical logic, algebra, number theory, and probability theory in the 19th century. Chapter 2 was written by I. G. Bashmakova and A. N. Rudakov with the assistance of A. N. Parshin and E. I. Slavutin. The book was published in 1992 by Birkhäuser Verlag and is a translation of a Russian book published by Nauka in 1978. A second, revised edition appeared in 2001. (The reprinted material is found on pp. 36–40 of the Birkhäuser Verlag book.) Reprinted with permission.

Another source of group theory was Gauss' theory of composition of classes of forms. In this theory one applied an operation analogous to addition (or multiplication) of numbers to objects very different from numbers. Gauss' study of forms of the same discriminant was in effect a study of the fundamental properties of cyclic and general abelian groups.

The two parts of Gauss' remarkable paper "The theory of biquadratic residues" appeared in 1828 and 1832, respectively. In it Gauss not only gave a geometric interpretation of the complex numbers (this was done before him) but also—and this is very important—transferred to complex numbers the notion of a whole number, a concept that seemed inseparable from the rational integers for more than 2000 years.

Gauss constructed an arithmetic of complex integers entirely analogous to the usual arithmetic and used the new numbers to formulate the law of biquadratic reciprocity. This opened for arithmetic boundless new horizons. Soon Eisenstein and Jacobi formulated and proved the law of cubic reciprocity and used for this purpose numbers of the form $K + m\rho$, $\rho^3 = 1$, $\rho \neq 1$, and in 1846 P. Lejeune-Dirichlet found all units (that is invertible elements) of the ring of integers of the field $\mathbb{Q}(\theta)$, where θ is a root of

$$x^n + a_1 x^{n-1} + \cdots + a_n = 0,$$

$a_i \in \mathbb{Z}$.[1] This paper, with its deep results in the theory of algebraic numbers, is also of interest from the point of view of group theory: in it Dirichlet constructed the first nontrivial example of an infinite abelian group and investigated its structure.

Further progress in algebraic number theory was linked to reciprocity laws and to Fermat's last theorem. Attempts to prove this theorem brought E. Kummer to the study of the arithmetic of fields $\mathbb{Q}(\zeta)$, $\zeta^p = 1$, $\zeta \neq 1$. In 1844–1847 Kummer discovered that if one defines a "prime" number to be an indecomposable integer in a field $\mathbb{Q}(\zeta)$, then the law of unique factorization into prime factors fails for the integers in $\mathbb{Q}(\zeta)$. To "save the day" and restore the possibility of constructing an arithmetic analogous to the usual (arithmetic) he introduced ideal factors. In so doing, he laid the foundations for the subtlest and most abstract theories of algebraic number theory. Kummer's methods were local. They were further developed by E. I. Zolotarev, K. Hensel, and others, and now form the core of commutative algebra.

Linear algebra continued to develop in the first half of the 19th century. In this connection, the first thing to be noted is that whereas no part of Gauss' *Disquisitiones* deals directly with linear algebra, its advance was bound to be furthered by the detailed study of integral quadratic forms in two variables contained in that work. A. Cauchy's "On an equation for the determination of the secular inequalities of planetary motions"(1826) dealt implicitly with the eigenvalues of matrices of arbitrary order. Somewhat later, in 1834, there appeared C. G. J. Jacobi's "On the transformation of two arbitrary homogeneous functions of the second order by means of linear substitutions into two others containing only squares of the variables; together with many theorems on the transformation of multiple integrals" in which he explicitly studied quadratic forms and their reduction to canonical

---

[1] $\mathbb{Z}$ is the ring of integers and $\mathbb{Q}$ is the field of rational numbers.

form. Jacobi also perfected the theory of determinants (1841). What was still lacking in this theory was geometric features and, above all, the all-important and fundamental notion of a linear space. The first, none-too-clear definition of a linear space was given by H. Grassmann in his *Die lineale Ausdehnungslehre* of 1844. This work, rich in new ideas but written in a muddled manner, first attracted attention when its author published a reworked and improved version in 1862. In particular, the work contains a construction of exterior products and the now famous Grassmann algebra. In 1843 there appeared A. Cayley's *Chapters in the analytical geometry of* (n) *dimensions*, a work less rich in ideas but better known to contemporary mathematicians. There is a close connection between the development of linear algebra and the theory of hypercomplex numbers (now known as the theory of algebras) which elicited considerable interest at the time. Years of fruitless attempts to generalize the complex numbers were crowned with success in 1843 by W. R. Hamilton's discovery of the quaternions. Hamilton studied the quaternions for over 20 years, for the rest of his life. His researches are summarized in two fundamental works: *Lectures on quaternions* (1853) and *Elements of the theory of quaternions* (1866). Their subsequent significance is due not so much to quaternions but to the new notions and methods of "vector calculus" introduced in this connection.

To resume our account of the further development of group theory we mention the series of A. Cauchy's papers, published between 1844–1846, in which he proves a great variety of theorems on groups of substitutions (subgroups of the symmetric group), including the famous theorem of Cauchy to the effect that a group whose order is divisible by a prime p contains an element of order p. A further major event in the history of group theory was the publication—in three parts (1854, 1854, 1859)—of Cayley's paper *On the theory of groups, as depending on the symbolic equation* $\theta^n = 1$. Following the spirit of the English school, Cayley views a group as an abstract set of symbols with a given law of composition and defines a number of fundamental notions of abstract group theory, chief among them being the notions of a group and of isomorphism. This was a notable step in the evolution of the new abstract mathematical thinking.

Of crucial importance for the further development of group theory was the appearance, in 1870, of C. Jordan's fundamental *Traité des substitutions et des équations algébriques*. This work contained the first systematic and complete exposition of Galois theory as well as a detailed presentation of results in group theory up to that time, including Jordan's own significant results in these areas. In it Jordan also introduced what is now known as the Jordan canonical form of matrices of linear transformations. The publication of Jordan's work was a major event in all of mathematics.

Mention must be made of the flourishing, in the middle of the 19th century, of an area of algebra intermediate between linear algebra and algebraic geometry known as the theory of invariants. On the one hand, its content consists in the generalization and development of topics in linear algebra such as reduction to canonical form of quadratic forms and matrices of linear transformations. On the other hand, it is the study, in concrete situations, of the answer to the following question: "Given a geometric object determined in some coordinate system by certain algebraic conditions, find a way of obtaining from the algebraic conditions geometric characteristics of the object that are invariant with respect to coordinate transformations." Between 1840–1870 many of the works of various mathematicians dealt with the determination of systems of invariants in different concrete situations. The best known are the works of Cayley, Eisenstein, Sylvester, Salmon and Clebsch. In this connection one must single out two papers by Hesse, published in 1844 and 1851, respectively, in which he introduced the notion of a hessian and applied it to geometry, and P. Gordan's famous 1868 paper in which he proved a general algebraic theorem on the existence of a finite system of base invariants. An important paper close to these investigations is Cayley's *A sixth memoir upon quantics* (1859). In it Cayley showed how to consider the metric properties of geometric figures from the single viewpoint of the theory of invariants. This paper was one of the sources of F. Klein's Erlangen Program that resulted in revolutionary changes in geometry.

At that time, an important achievement in linear algebra was Sylvester's 1852 proof of the law of inertia of quadratic forms, presented in the paper *Proof of the theorem that every homogeneous quadratic polynomial can be reduced by means of a real orthogonal substitution to the form of a sum of positive and negative squares*. It was proved, but not published, somewhat earlier by Jacobi. In 1858 there appeared Cayley's *Memoir on the theory of matrices*. In it Cayley introduced the algebra of square matrices and established the isomorphism between the algebra of quaternions and a certain algebra of second-order matrices (a subalgebra of the algebra of all square second-order complex matrices). This work was of great importance for the clarification of the relation between the theory of algebras and linear algebra.

In the sixties, the activities of K. Weierstrass had an important influence on the development of mathematics. He published virtually nothing but included the results of his investigation in his lectures at Berlin University. In his 1861 lectures Weierstrass introduced the notion of a direct sum of algebras and showed that every (finite dimensional) commutative algebra (over the field of real numbers) without nilpotent elements is the direct sum of copies of the fields of real and complex numbers. This was one of the earliest classification results in algebra.

One of the main problems of algebraic number theory in the sixties and seventies was the extension of Kummer's divisibility theory from cyclotomic fields to general algebraic number fields. This was accomplished in three different constructions due, respectively, to E. I. Zolotarev, R. Dedekind, and L. Kronecker. Of the three, it was Dedekind's work—the Xth Supplement to Dirichlet's lectures on number theory published in 1871 and the XIth Supplement to subsequent editions—that was accepted by all mathematicians as the solution of the problem. Dedekind's clear, algebraically transparent, account became the model of mathematical style for many decades to come. By this and other works Dedekind laid the foundations of the contemporary axiomatic presentation of mathematical theories.

In our survey of the evolution of algebra we have not touched on the theory of elliptic and abelian functions—one of the central lines of development of 19th-century mathematics, an area in which Gauss, Abel, Jacobi, Clebsch, Gordan, Weierstrass and many others invested great efforts. In the 19th century this area belonged primarily to analysis, more specifically to the theory of functions of a complex variable, and it was only gradually, especially at the end of the 19th century, that the role of algebraic ideas in it became very significant.

The algebraization of the area began with Dedekind's transfer of his theory, in a joint work with H. Weber (1882), to the field of algebraic functions. This established the deep parallelism between the theories of algebraic numbers and algebraic functions and was the decisive step for an abstract definition of the concepts of field, module, ring, and ideal. From the end of the last century ideas began to flow in the opposite direction, from the theory of algebraic functions to number theory. This resulted in the introduction of p-adic numbers and topology by means of p-adic metrics. But this is already part of the mathematics of the present century.

The evolution of the ideas, methods, and theories just described resulted in the creation of abstract "modern algebra" and, later, of algebraic geometry whose flourishing we witness today.

What Are Algebraic Integers and What Are They For?

John Stillwell

My title is stolen from Dedekind's *Was sind und was sollen die Zahlen?* (1888), the classic book on the meaning of the natural numbers. His book is actually about set theory rather than number theory or algebra but, as we shall see, Dedekind had a lot more up his sleeve. His reflections on the algebraic properties of the natural numbers would have filled a much larger book, which unfortunately he never had time to write. The following is a sketch of the book that might have been.

The natural numbers $0, 1, 2, 3, \ldots$ are the setting for the oldest and deepest mathematical problems. As early as 300 B.C., Euclid recognised that *divisibility* and *primes* are important concepts for natural numbers, and that even simple propositions about them involve subtle methods of proof. For example, to prove the seemingly obvious proposition that if a prime p divides the product of natural numbers a, b then p divides a or p divides b, Euclid had to introduce the concept of *greatest common divisor* (gcd), and use the euclidean algorithm to express the gcd in a suitable form. It follows almost immediately from this proposition about prime divisors that each natural number has a *unique prime factorisation*. That is, if n is a natural number, then n can be expressed as a product of primes in exactly one way, up to the order of factors. Unique prime factorisation was isolated as the "fundamental theorem of arithmetic" in the classic *Disquisitiones arithmeticae* of Gauss (1801).

Gauss organised much of earlier number theory into a cohesive structure by introducing unifying ideas, such as congruence, and proving unifying theorems, such as quadratic reciprocity. But Gauss also burst the bounds of the structure he had created with an array of new and baffling theorems. For example, he proved several theorems about the nth roots of 1, bearing on the geometric construction of regular n-gons. He showed in passing that these *irrational* numbers had application to the natural numbers; they were even connected with quadratic reciprocity (*Disquisitiones*, section 356). How come? Gauss's successors spent most of the 19th century searching for an answer to this question, for concepts to explain the effectiveness of irrational numbers in arithmetic. The most important of these was the concept of algebraic integer.

WHAT ALGEBRAIC INTEGERS ARE FOR. Actually, Gauss was not the first to use irrational numbers to answer questions about the natural numbers. Euler's *Algebra* (1770) contains several examples, the simplest being a proof of the following claim of Fermat: *among natural numbers, 27 is the only cube that exceeds a square by* 2. This claim is one of the famous marginal notes in Fermat's copy of Bachet's *Diophantus*, made around 1637. Fermat claimed to have a rigorous proof, but never revealed it. Euler's proof is *not* exactly rigorous, but it is distinguished by the wild idea of using the number $\sqrt{-2}$.

The problem amounts to showing that $x = 5, y = 3$ is the only solution of

$$y^3 = x^2 + 2.$$

Euler factorises the right hand side into $(x + \sqrt{-2})(x - \sqrt{-2})$, then confidently assumes that $x + \sqrt{-2}, x - \sqrt{-2}$ must both be cubes, since their product is a cube and he can show that their gcd is 1. This implies, presumably, that

$$x + \sqrt{-2} = (a + b\sqrt{-2})^3$$

for some integers a, b, and hence

$$x + \sqrt{-2} = a^3 - 6ab^2 + (3a^2b - 2b^3)\sqrt{-2}.$$

Equating imaginary parts we get

$$1 = 3a^2b - 2b^3 = b(3a^2 - 2b^2)$$

which implies $b = \pm 1$ and $3a^2 - 2b^2 = \pm 1$ since $+1$ and -1 are the only integer divisors of 1. It follows that $a = \pm 1$, which gives $x = 5$ as the only positive integer value of $a^3 - 6ab^2$. Q.E.D.!

Well OK, we don't know what these things $a + b\sqrt{-2}$ and their divisors really are, or how they work, but they seem amazingly good for answering questions about the natural numbers. The same goes for other irrational objects arising from the factorisation of polynomials. Euler gave many examples, the most spectacular being a proof of Fermat's last theorem for cubes:

$$x^3 + y^3 = z^3 \text{ has no solution in positive integers}$$

(Fermat himself had settled the case of fourth powers by his method of infinite descent.) He proved this using the factorisation

$$x^3 = z^3 - y^3 = (z - y)\left(z - \frac{-1 + \sqrt{-3}}{2}y\right)\left(z - \frac{-1 - \sqrt{-3}}{2}y\right)$$

and arguing, as above, that each factor in the right hand side must be a cube. This time the irrational objects involved are numbers of the form $a + b\omega$, where

$$\omega = \frac{-1 + \sqrt{-3}}{2}$$

is a complex cube root of 1. Gauss's work on the nth roots of 1 could be seen as a generalisation of this, involving objects of the form $a_0 + a_1 \zeta + \cdots a_{n-2} \zeta^{n-2}$, where ζ is a complex nth root of 1 and $a_0, a_1, \ldots, a_{n-2}$ are ordinary integers.

Thus by 1800 it was clear that irrational objects were good for number theory. What was not clear was the nature of these objects and the reasons they behaved like integers, if indeed they did. A lot more work had to be done before the right concept of "algebraic integer" was isolated.

ALGEBRAIC INTEGERS. Since we want to do arithmetic on algebraic integers, they must satisfy the requirement:

1. *The algebraic integers are closed under $+$, $-$ and $\times$.*

Also, since the aim in extending the concept of integer is to answer questions about ordinary integers, the relations between ordinary integers had better not change in the process of extension. In particular, an ordinary integer a should not become a divisor of an ordinary integer b if it is not one already. That is, the rational number b/a should not be an algebraic integer unless it is an ordinary integer. This gives us a second requirement:

2. *The rational algebraic integers are the ordinary integers.*

The numbers $a + b\sqrt{-2}$ satisfy these requirements, as do the numbers $a_0 + a_1\zeta + \cdots + a_{n-2}\zeta^{n-2}$. The only problematic operation is multiplication of numbers $a_0 + a_1\zeta + \cdots + a_{n-2}\zeta^{n-2}$, as this creates powers $\zeta^{n-1}, \zeta^n, \ldots$. The trick is to rewrite $\zeta^{n-1}, \zeta^n, \ldots$ as combinations of $1, \zeta, \ldots, \zeta^{n-2}$ with ordinary integer coefficients, using the equation

$$1 + \zeta + \cdots + \zeta^{n-2} + \zeta^{n-1} = 0$$

satisfied by ζ. Then a product of numbers $a_0 + a_1\zeta + \cdots + a_{n-2}\zeta^{n-2}$ is seen to be another number of the same form.

Now the numbers $\sqrt{-2}$ and ζ arise in the first place as the solutions of polynomial equations with ordinary integer coefficients, namely

$$x^2 + 2 = 0, \qquad x^{n-1} + \cdots + x + 1 = 0.$$

Is the same true for the combinations $a + b\sqrt{-2}$ and $a_0 + a_1\zeta + \cdots + a_{n-2}\zeta^{n-2}$? The answer is yes, and even more is true: the combinations satisfy *monic* polynomial equations with ordinary integer coefficients. This follows from a general property of monic polynomial equations pointed out by Eisenstein (1850).

If $f(x) = 0$ is a monic polynomial equation with ordinary integer coefficients and roots $\alpha_1, \alpha_2, \ldots, \alpha_n$, and if $g(\alpha_1, \ldots, \alpha_n)$ is any polynomial in the roots with ordinary integer coefficients, then $g(\alpha_1, \ldots, \alpha_n)$ is also the root of a monic polynomial equation $h(x) = 0$ with ordinary integer coefficients.

His proof applies Newton's theorem on symmetric polynomials to the product $h(x)$ of all terms $x - g(\alpha_{\sigma(1)}, \ldots, \alpha_{\sigma(n)})$, where σ is a permutation of $1, \ldots, n$, to conclude that the coefficients of $h(x)$ are ordinary integers. Obviously $h(x)$ is monic, and its roots include $g(\alpha_1, \ldots, \alpha_n)$. Since $\alpha_1 + \alpha_2$, $\alpha_1 - \alpha_2$ and $\alpha_1\alpha_2$ are particular cases of $g(\alpha_1, \ldots, \alpha_n)$, closure under $+, -$ and $\times$ will hold when algebraic integers are defined as follows.

Definition. An *algebraic integer* is the root of a monic polynomial equation with ordinary integer coefficients.

Just as easily, this definition meets the second requirement:

A rational algebraic integer is an ordinary integer.

Because if a/b is a rational number (in lowest terms) that satisfies the equation

$$x^n + c_1 x^{n-1} + \cdots + c_{n-1} x + c_n = 0$$

with ordinary integer coefficients, then substitution and rearrangement give

$$a^n/b = -c_1 a^{n-1} - \cdots - c_{n-1} ab^{n-2} - c_n b^{n-1},$$

which is possible only if $b = 1$, since the right hand side is an ordinary integer.

Eisenstein did not actually state the definition or the closure under $+, -$ and $\times$, but he probably didn't need to. The class of solutions of monic polynomial equations was under consideration by other mathematicians at the time, and its basic properties were implicit in known results about polynomials. In particular, the nature of rational algebraic integers was well known as a result about polynomials 50 years earlier (Gauss assumed it without proof in article 11 of the *Disquisitiones*). More important was Eisenstein's recognition that the algebraic integers shared properties with the ordinary integers. This pointed the way to a systematic explanation of the phenomena discovered by Euler and Gauss.

DIVISORS AND PRIMES. Closure of the algebraic integers under $+, -$ and $\times$ means that they inherit the ring properties of the complex numbers. Hence they are a ring, like the ordinary integers, and satisfy the same basic propositions such as the commutative, associative and distributive laws. Unfortunately, many useful propositions are not logical consequences of these laws and in fact are *false* in the larger ring. For example, there are no primes in the ring of all algebraic integers, because every algebraic integer α has the factorisation $\alpha = \sqrt{\alpha}\sqrt{\alpha}$, and $\sqrt{\alpha}$ is also an algebraic integer. This difficulty can be avoided by working in smaller rings tailored to particular problems, such as the ring of integers $a + b\sqrt{-2}$ where Euler looked for solutions of $y^3 = x^2 + 2$. "Sufficiently small" rings of algebraic integers do have primes, and also prime factorisation, but the big question is whether the factorisation is unique.

The first ring of algebraic integers to be examined in this light was
$$\mathbf{Z}[i] = \{a + bi: a, b \in \mathbf{Z}\},$$
the ring of *Gaussian integers*. Gauss (1832) showed that $\mathbf{Z}[i]$ has a divisibility theory like that of $\mathbf{Z}$, including unique prime factorisation, thanks to an analogue of the following *division property* of $\mathbf{Z}$. If $a, b \in \mathbf{Z}$ and $b \neq 0$ then there are $q, r \in \mathbf{Z}$ ("quotient" and "remainder") with $a = qb + r$ and $0 \leq |r| < |b|$. The analogous division property of $\mathbf{Z}[i]$ is that for any $\alpha, \beta \in \mathbf{Z}[i]$ with $\beta \neq 0$ there are $\mu, \rho \in \mathbf{Z}[i]$ with $\alpha = \mu\beta + \rho$ where $0 \leq |\rho| < |\beta|$ and the absolute value $|\ |$ is now distance in the complex plane.

The division property of $\mathbf{Z}[i]$ can be seen by viewing the Gaussian integer multiples of β as the corners of a lattice of squares in the plane. A typical square is the one with corners at $0, \beta, i\beta, (1 + i)\beta$. The "remainder" ρ is simply the difference between α and the nearest corner in the lattice, and $|\rho| < |\beta|$ because the distance from any point in a square to the nearest corner is less than the length of a side. With the division property established, one has a euclidean algorithm for gcd, and the rest of the route to unique prime factorisation is the same as in $\mathbf{Z}$.

A similar geometric argument establishes a division property for $\mathbf{Z}[\sqrt{-2}]$. Hence $\mathbf{Z}[\sqrt{-2}]$ also has a unique prime factorisation, and Euler's treatment of $y^3 = x^2 + 2$ is valid. Euler's proof that $x^3 + y^3 \neq z^3$ can likewise be justified by finding a euclidean algorithm for $\mathbf{Z}[\omega]$.

Alas, it is not always as simple as this. Unique prime factorisation *fails* in $\mathbf{Z}[\sqrt{-5}]$, and it also fails in $\mathbf{Z}[\zeta]$ for nth roots ζ of unity from $n = 23$ onwards (Kummer, 1844). The failure in $\mathbf{Z}[\sqrt{-5}]$ was encountered implicitly by Fermat in studying primes of the form $x^2 + 5y^2$ (see [2], p. 82). The failure in $\mathbf{Z}[\zeta]$ was a major stumbling block, though not the only one, in Kummer's attempt to prove Fermat's last theorem.

ALGEBRAIC NUMBERS AND FUNCTIONS. The story of how unique factorisation was lost, and then regained by Kummer and Dedekind's theory of ideals, has often been told (for example in [1], pp. 818–824). I shall not repeat it here. Instead I shall sketch how Dedekind built the theory of algebraic integers to make unique factorisation possible, and what he discovered in the process.

Dedekind's main idea was to embed rings of algebraic integers in algebraic number *fields*, where concepts of linear algebra come to the surface. "Sufficiently small" rings of algebraic integers lie in fields K of finite dimension over $\mathbf{Q}$, and any such field is of the form $K = \mathbf{Q}(\alpha)$ where the degree of α equals the dimension of K. The concept of field and the existence of the "primitive element" α were already implicit in Abel and Galois, but Dedekind made the crucial identification of degree with dimension by observing that $\{1, \alpha, \alpha^2, \ldots, \alpha^{n-1}\}$ is a basis for K, where n is the degree of α. This enabled him to give simple linear algebra proofs of theorems previously dependent on properties of symmetric functions, such as the closure of algebraic integers under $+, -$ and $\times$. It also enabled him to develop a "linear" approach to Galois theory, and to apply Galois theory to number theory.

These ideas were presented in the supplements Dedekind wrote for the 2nd, 3rd and 4th editions of Dirichlet's *Zahlentheorie* (1871, 1878, 1893). They were extended in Hilbert's *Zahlbericht* (1897) and presented there in almost modern form. Many 20th century algebraists, such as Emmy Noether, Artin and van der Waerden, learned their algebra from these works. In fact, Emmy Noether used to say "Es steht schon bei Dedekind" ("It's already in Dedekind").

A great benefit to flow from Dedekind's approach to algebraic numbers was a new approach to algebraic *functions*. Among algebraic functions of one variable, the polynomials play the role of the integers. This idea goes back at least as far as Stevin's *L'arithmétique* (1585), where the euclidean algorithm is used to find the gcd of polynomials. Dedekind took the idea that polynomials are "integers," and generalised to an analogy of algebraic integers. The ring $\mathbf{C}[z]$ of complex polynomials in z is extended to the field $\mathbf{C}(z)$ of rational functions, and each algebraic function lies in a finite-dimensional extension K of $\mathbf{C}(z)$. The functions analogous to algebraic integers are the *entire* functions—those with a finite value for each value of z. By carrying over concepts from number theory to function theory, Dedekind and Weber (1882) were able to give a completely algebraic definition of a *Riemann surface*. In doing so, they put many theorems about Riemann surfaces (that is, algebraic curves) on a sound basis for the first time, and laid the foundation of modern algebraic geometry.

This is very interesting, but is it number theory? Can this rarified form of geometry tell us anything about the ordinary integers? A full answer would cover most of 20th century mathematics, but the short answer of course is yes. All of these ideas, and much more, are needed for Andrew Wiles' proof of Fermat's last theorem.

REFERENCES

1. M. R. Kline, *Mathematical Thought from Ancient to Modern Times*, Oxford University Press, New York 1972.
2. A. Weil, *Number Theory: An Approach through History*, Birkhäuser, Boston 1984.

The Genesis of the Abstract Ring Concept

Israel Kleiner

Algebra textbooks usually give the definition of a ring first and follow it with examples. Of course, the examples came first, and the abstract definition later—much later. So we begin with examples.

Among the most elementary examples of rings are the integers, polynomials, and matrices. "Simple" extensions of these examples are at the roots of ring theory. Specifically, we have in mind the following three examples:

(a) The integers Z can be thought of as the appropriate subdomain of the field Q of rationals in which to do number theory. (The rationals themselves are unsuitable for that purpose: every rational is divisible by every other (nonzero) rational.) Take a simple extension field $Q(\alpha)$ of the rationals, where α is an algebraic number, that is, a root of a polynomial with integer coefficients. $Q(\alpha)$ is called an algebraic number field; it consists of polynomials in α with rational coefficients (e.g., $Q(\sqrt{3}) = \{a + b\sqrt{3} : a, b \in Q\}$). The appropriate subdomain of $Q(\alpha)$ in which to do number theory—the "integers" of $Q(\alpha)$—consists of those elements that are roots of *monic* polynomials with integer coefficients (the integers of $Q(\sqrt{3})$ are $\{a + b\sqrt{3} : a, b \in Z\}$). This is our first example.

(b) The polynomial rings $\mathbb{R}[x]$ and $\mathbb{R}[x, y]$ in one and in two variables, respectively, share important properties but also differ in significant ways. In particular, while the roots of a polynomial in one variable constitute a discrete set of real numbers, the roots of a polynomial in two variables constitute a curve in the plane—a so-called algebraic curve. Our second example, then, is the ring of polynomials in two (or more) variables.

(c) Square $m \times m$ matrices (for example, over the reals) can be viewed as m^2-tuples of real numbers with coordinate-wise addition and appropriate multiplication obeying the axioms of a ring. Our third example consists, more generally, of n-tuples $\mathbb{R}^n$ of real numbers with coordinate-wise addition and appropriate multiplication, so that the resulting system is a (not necessarily commutative) ring. Such systems are extensions of the complex numbers—in the 19th and early 20th centuries they were called hypercomplex number systems.

In what contexts did these examples arise? What was their importance? The answers will lead us to the genesis of the abstract ring concept.

The abstract ring concept emerged in the context of a theory—in fact, in the context of two theories: commutative ring theory and noncommutative ring theory. The abstract theories of these two categories came from distinct sources and developed in different directions. Commutative ring theory originated in algebraic number theory and algebraic geometry. Central to the development of these subjects were, respectively, the rings of integers in algebraic number fields and the rings of polynomials in two or more variables. Noncommutative ring theory began with attempts to extend the complex numbers to various hypercomplex number systems. We consider first the evolution of the "simpler" theory of noncommutative rings.

A. NONCOMMUTATIVE RING THEORY. In a strict sense, noncommutative ring theory originated from a single example—the quaternions—invented (discovered?) by Hamilton in 1843. These are "numbers" of the form $a + bi + cj + dk$ (a, b, c, d real numbers) that are added componentwise and in which multiplication is subject to the relations $i^2 = j^2 = k^2 = ijk = -1$. This was the first example of a noncommutative number system, obeying all the (algebraic) laws of the real and complex numbers except for commutativity of multiplication. Such a system is now called a skew field or a division algebra. Hamilton's motivation was to define an algebra of vectors in 3-space so that multiplication would represent composition of rotations (just as multiplication of complex numbers represents composition of rotations in the plane). Having failed in this task, he turned to quadruples of reals and created the algebra of quaternions. The quaternions did, in fact, yield the required computing tool for rotations in 3-space.

Examples. Hamilton's invention of the quaternions was conceptually groundbreaking, but like all revolutions, it was initially received with less than universal approbation. Most mathematicians, however, soon came around to Hamilton's point of view. The quaternions acted as a catalyst for the exploration of diverse "number systems", with properties that departed in various ways from those of the real and complex numbers. Among the examples of such hypercomplex number systems are octonions, exterior algebras, group algebras, matrices, and biquaternions. See [6].

Structure. The first example of a noncommutative algebra was given by Hamilton in 1843. During the next forty years mathematicians introduced other examples, began to bring some order into them, and singled out certain types for special attention. For example, Frobenius and C. S. Peirce showed in 1880 that the reals, the complex numbers, and the quaternions are the only finite-dimensional (associative) division algebras over the reals. The stage was set for the founding of a general theory of finite-dimensional, noncommutative, associative algebras (important examples of rings).

In the 1890s, Cartan, Frobenius, and Molien proved (independently) the following fundamental structure theorem for finite-dimensional semi-simple algebras over the real or complex numbers: Any such algebra is a finite unique direct sum of simple algebras. These, in turn, are isomorphic to matrix algebras with entries in division algebras. An algebra is "semi-simple" if it has no nontrivial nilpotent ideals and it is "simple" if it has no nontrivial ideals.

In 1907, Wedderburn extended that result to algebras over arbitrary fields. This was no mere generalization as it necessitated the introduction of such fundamental algebraic concepts as ideal, quotient algebra, nilpotent algebra, radical, semi-simple and simple algebra, direct sum, and tensor product. Wedderburn's theorem is one of the basic results in ring theory, and it served as a model for many ring-theoretic structure theorems. It is also central in group-representation theory. See [6].

B. COMMUTATIVE RING THEORY. *Commutative* ring theory originated in algebraic number theory and algebraic geometry and has in turn been applied mainly to these two subjects. Invariant theory, with roots in both number theory and geometry, also had a role in these developments.

Algebraic number theory. Several of the central topics of number theory, mainly Fermat's Last Theorem, binary quadratic forms, and reciprocity laws, were instrumental in the emergence of algebraic number theory. Although the key problems in these areas were expressed in terms of integers, it gradually became apparent that the solutions called for embedding the integers in domains of what came to be known as algebraic integers. The following examples give an idea of what is involved.

(i) To show that $x^3 + y^3 = z^3$ has no nonzero integer solutions, factor the left side as $(x + y)(x + yw)(x + yw^2) = z^3$, $w^3 = 1$, $w \neq 1$. This is now an equation in the domain $D_3 = \{a + bw: a, b \in Z\}$. Assuming the existence of a solution in D_3 one can arrive at a contradiction, showing in particular that $x^3 + y^3 = z^3$ has no solutions. A similar approach turned out to be fruitful in the general case of Fermat's Last Theorem, $x^p + y^p = z^p$. See [5].

(ii) A conceptual way to determine which integers are sums of two squares is to factor the right side of $n = x^2 + y^2$ and consider the equation $n = (x + yi)(x - yi)$ in the domain $G = \{a + bi: a, b \in Z\}$ of Gaussian integers. (Gauss introduced G in order to *state* the biquadratic reciprocity law.) This problem is but an instance of the problem of representing integers by binary quadratic forms $ax^2 + bxy + cy^2$ ($a, b, c \in Z$). The general approach is to factor $ax^2 + bxy + cy^2$ and to consider the resulting equation in a domain of "complex integers". See [1].

(iii) The diophantine equation $x^2 + 2 = y^3$ is a special case of the famous Bachet equation $x^2 + k = y^3$. It was in the margins of Bachet's Latin translation of Diophantus' *Arithmetica* that Fermat made his famous remark about the equation $x^n + y^n = z^n$. While the general Bachet equation is still a topic of intensive investigation, the equation $x^2 + 2 = y^3$ can readily be solved using "complex integers". It is easy to see that $x = \pm 5$, $y = 3$ are solutions. To find *all* solutions, we write $x^2 + 2 = y^3$ as $(x + \sqrt{2}i)(x - \sqrt{2}i) = y^3$. This is now an equation in the domain $D = \{a + b\sqrt{2}i: a, b \in Z\}$. We can show that D is a unique factorization domain (UFD) and that $x + \sqrt{2}i$ and $x - \sqrt{2}i$ are relatively prime in D. Since their product is a cube, each factor must be a cube (in D). In particular, $x + \sqrt{2}i = (a + b\sqrt{2}i)^3$, $a, b \in Z$. Cubing and equating coefficients, we can easily show that $x = \pm 5$, $y = 3$ are the *only* solutions of $x^2 + 2 = y^3$—no easy feat to accomplish without the use of complex integers. See [1].

What is common to these examples? Additive problems in Z have been transformed to multiplicative problems in the domains D_3, G, and D respectively (these domains are important examples of rings). In the latter settings the problems can be dealt with effectively provided that the domains in question are unique factorization domains (UFDs). Now, D_3, G, and D *are* UFDs, but the domains arising from the respective general problems (Fermat's Last Theorem, binary quadratic forms, the Bachet equation) are (as a rule) not. For example, $\{a_0 + a_1 w + a_2 w^2 + \cdots + a_{22} w^{22} : a_i \in Z$, w a primitive 23rd root of unity$\}$, arising from the equation $x^{23} + y^{23} = z^{23}$, is not a UFD; neither is the domain $\{a + b\sqrt{5}\,i : a, b \in Z\}$ resulting from factoring the left side of the Bachet equation $x^2 + 5 = y^3$. The problem then becomes one of restoring, *in some sense*, unique factorization in such domains. Kummer dealt with it by means of ideal numbers, Dedekind by means of ideals, and Kronecker by means of divisors. We consider briefly Dedekind's contribution.

Dedekind's rings and ideals. The main result of Dedekind's groundbreaking 1871 work, which appeared as Supplement X of Dirichlet's *Vorlesungen über Zahlentheorie*, was that every nonzero ideal in the domain of integers of an algebraic number field is a unique product of prime ideals. Before one could state this theorem one had, of course, to define the concepts in its statement, namely "the domain of integers of an algebraic number field", "ideal", and "prime ideal". It took Dedekind about twenty years to formulate them.

Given an algebraic number field $Q(\alpha)$, all its elements are roots of polynomials with integer coefficients. Dedekind defined the *domain of integers* of $Q(\alpha)$ to be the subset of elements that are roots of *monic* polynomials with integer coefficients. This notion is an extension of the domain of integers (of Q), whose elements are the roots of monic *linear* polynomials. He showed that these elements "behave" like integers—they are closed under addition, subtraction, and multiplication. For example, the integers of $Q(\sqrt{5}) = \{a + b\sqrt{5} : a, b \in Q\}$ are $\{\frac{a + b\sqrt{5}}{2} : a, b \in Z, \ a \equiv b \pmod{2}\}$ rather than $\{a + b\sqrt{5} = a, b \in Z\}$, as seems perhaps more natural. See [1].

Having defined the domain R of algebraic integers of $Q(\alpha)$ in which he would formulate and prove his result on unique decomposition of ideals, Dedekind considered, more generally, sets of integers of $Q(\alpha)$ closed under addition, subtraction, and multiplication. He called these "orders". The domain R of integers of $Q(\alpha)$ is the largest order. Here, then, was an algebraic first for Dedekind—an essentially axiomatic definition of a (commutative) ring, albeit in a concrete setting. The *term* ring, also in the setting of domains of algebraic integers, was coined by Hilbert in 1897.

The second fundamental concept of Dedekind's theory, that of ideal, derived its motivation (and name) from Kummer's ideal numbers (see [2]). Dedekind defined it essentially as we do today. Having then defined the notion of prime ideal, he proved his fundamental theorem that every nonzero ideal in the ring of integers of an algebraic number field is a unique product of prime ideals. See [2].

How did Dedekind's theory relate to the number-theoretic problems (e.g., Fermat's Last Theorem, reciprocity laws, binary quadratic forms) from which it drew inspiration? It did shed important light on these problems and resolved special cases (see [1], [5]). But as often happens, the ideas Dedekind put forth acquired great significance independent of the original problems that stimulated their introduction. (Galois theory also far superseded in importance the problem of solution of equations that gave it birth.)

Algebraic geometry. Algebraic geometry is the study of algebraic curves and their generalizations to n dimensions, algebraic varieties. An algebraic curve is the set of roots of an algebraic function; that is, a function $y = f(x)$ defined implicitly by the polynomial equation $P(x, y) = 0$.

Several approaches were used in the study of algebraic curves, notably the analytic, the geometric-algebraic, and the algebraic-arithmetic. In the analytic approach, to which Riemann (in the 1850s) was the major contributor, the main objects of study were algebraic functions $f(w, z) = 0$ (of a complex variable) and their integrals, the so-called abelian integrals, which are closely related to the important notion of the genus of an algebraic curve. It was in this connection that Riemann introduced the fundamental notion of a Riemann surface, on which algebraic functions become single valued. Riemann's methods were brilliant but nonrigorous, and relied heavily on the physically obvious Dirichlet Principle, which was mathematically incorrect in its unrestricted form.

(i) Algebraic function fields. Dedekind and Weber, in their groundbreaking 1882 paper "Theory of algebraic functions of a single variable", proposed to "provide a basis for the theory of algebraic functions, the major achievement of Riemann's researches, in the simplest and at the same time rigorous and most general manner". The fundamental idea of their algebraic-arithmetic approach was to carry over to algebraic function fields the ideas that Dedekind had earlier introduced for algebraic number fields.

Just as an algebraic number field is a finite extension $Q(\alpha)$ of the field Q of rationals, so an algebraic function field is a finite extension $K = \mathbb{C}(z)(w)$ of the field $\mathbb{C}(z)$ of rational functions (in the indeterminate z). That is, w is a root of a polynomial $a_0 + a_1\alpha + a_2\alpha^2 + \cdots + a_n\alpha^n$, where $a_i \in \mathbb{C}(z)$ (we can take $a_i \in \mathbb{C}[z]$). Thus $w = f(z)$ is an algebraic function defined implicitly by the polynomial equation $P(z, w) = a_0 + a_1w + a_2w^2 + \cdots + a_nw^n = 0$. In fact, all elements of $K = \mathbb{C}(z)(w)$ are algebraic functions.

Now let A be the "ring of integers" of K over $\mathbb{C}(z)$; that is, A consists of the elements of K that are roots of *monic* polynomials over $\mathbb{C}[z]$. As for algebraic numbers, here too every nonzero ideal of A is a unique product of prime ideals. Incidentally, in the case of the field of meromorphic functions on a Riemann surface, the role of the integers is played by the entire functions.

Dedekind and Weber were now ready to give a rigorous, algebraic definition of the Riemann surface S of the algebraic function field K: it is (in our terminology) the set of nontrivial discrete valuations on K. The finite points of S correspond to ideals of A; to deal with points at infinity of S Dedekind and Weber introduced the notions of "place" and "divisor". Many of Riemann's ideas about algebraic functions were here developed algebraically and rigorously. In particular, a rigorous proof was given of the important Riemann-Roch theorem. See [3].

Beyond Dedekind and Weber's technical achievements in putting major parts of Riemann's algebraic function theory on solid ground, their conceptual breakthrough lay in pointing to the strong analogy between algebraic number fields and algebraic function fields, hence between algebraic number theory and algebraic geometry. This analogy proved extremely fruitful for both theories. For example, the use of power series in algebraic geometry inspired Hensel in 1897 to introduce p-adic numbers ("power series" in the prime p). The resulting idea of p-adic completion proved important in both algebraic number theory and algebraic geometry. Another noteworthy aspect of Dedekind and Weber's work was its generality and applicability to arbitrary fields, in particular Q and Z_p, which were important in number-theoretic contexts. Thus, ideas from algebraic geometry could be applied to number theory.

(ii) Polynomial rings and their ideals. Polynomial ideals in algebraic geometry had their implicit beginnings in M. Noether's work in the 1870s. Important advances were made by Kronecker in the 1880s and especially by Lasker and Macauley in 1905 and 1913, respectively.

The need for polynomial ideals in the study of algebraic varieties is manifest. An algebraic variety V is defined as the set of points in $\mathbb{R}^n$ (or $\mathbb{C}^n$) satisfying a system of polynomial equations $f_i(x_1,\ldots,x_n) = 0$, $i = 1, 2, 3, \ldots$. The Hilbert Basis Theorem implies that finitely many equations will do. But different systems of polynomial equations may give rise to the same set of roots. For example, the circle V in $\mathbb{R}^3$ of radius 2 lying in the plane parallel to the (x, y) plane and two units above it may be described as $V = \{(x, y, z): x^2 + y^2 - 4 = 0, z - 2 = 0\}$, as $V = \{(x, y, z): x^2 + y^2 + z^2 - 8 = 0, z - 2 = 0\}$, or as $V = \{(x, y, z): x^2 + y^2 - 4 = 0, x^2 + y^2 - 2z = 0\}$. Is there a canonical set of polynomials that describes the variety (circle) V?

It is easy to see that if $f_1,\ldots,f_m$ are polynomials that vanish on the points of V, then so do all polynomials of the set $I = \{g_1 f_1 + \cdots + g_m f_m: g_i \in \mathbb{R}[x, y, z]\}$. But I is an ideal of the polynomial ring $\mathbb{R}[x, y, z]$. In fact, the set of *all* polynomials of $\mathbb{R}[x, y, z]$ that vanish on the points of V is also an ideal—and it is evidently the "canonical" set of polynomials to describe V.

Note that the preceding remarks point to a correspondence between ideals of $\mathbb{R}[x_1,\ldots,x_n]$ (or of $\mathbb{C}[x_1,\ldots,x_n]$) and varieties in $\mathbb{R}^n$ (or $\mathbb{C}^n$): If V is a variety, let $I(V) = \{f(x_1,\ldots,x_n) \in \mathbb{R}[x_1,\ldots,x_n]: f(a_1,\ldots,a_n) = 0 \text{ for all } (a_1,\ldots,a_n) \in V\}$, and if J is an ideal of $\mathbb{R}[x_1,\ldots,x_n]$, let $V(J) = \{(b_1,\ldots,b_n) \in \mathbb{R}^n: g(b_1,\ldots,b_n) = 0 \text{ for all } g \in J\}$. The Hilbert Nullstellensatz, in one of its incarnations, says that $V(J) \neq 0$ if the variety is in $\mathbb{C}^n$, or K^n for any algebraically closed field K. This correspondence is central in algebraic geometry. It is, in fact, a one-to-one correspondence between varieties (over an algebraically closed field K) and their largest defining ideals (the so-called radical ideals). Under this correspondence, prime ideals correspond to irreducible varieties (those that cannot be non-trivially decomposed into finite unions of other varieties). See [3].

Lasker and Macauley exploited this correspondence in the early 20th century by undertaking a thorough study of ideals in polynomial rings in order to shed light on algebraic varieties. Lasker's major result was the "primary decomposition" of ideals: Every ideal in a polynomial ring $F[x_1,\ldots,x_n]$ is a finite intersection of primary ideals. Primary ideals, first defined by Lasker, are generalizations of prime ideals; the former are to the latter what prime powers are to primes in the ring of integers. Translated into the language of algebraic geometry, the result says that every variety is a finite union of irreducible varieties. Macauley proved the uniqueness of the primary decomposition, which implied that every variety can be expressed uniquely as a union of irreducible varieties—a type of fundamental theorem of arithmetic for varieties. By the way, it is no easy matter to determine *geometrically* when a curve is irreducible; it is the algebra that comes to the geometer's aid here.

C. THE ABSTRACT DEFINITION OF A RING. In the first decade of the 20th century there were well-established, flourishing, concrete theories of both commutative and noncommutative rings and their ideals. Their roots were in algebraic number theory, algebraic geometry, and the theory of hypercomplex number systems. Moreover, abstract (axiomatic) definitions of groups, fields, and vector spaces had then been in existence for about two decades. The time was ripe for the abstract ring concept to emerge.

The first abstract definition of a ring was given by Fraenkel (of set-theory fame) in a 1914 paper entitled "On zero divisors and the decomposition of rings" [**4**]. He defines a ring as "a system" with two (abstract) operations, to which he gives the names addition and multiplication. Under one of the operations (addition) the system forms a group (he gives its axioms). The second operation (multiplication) is associative and distributes over the first. Two axioms give the closure of the system under the operations, and there is the requirement of an identity in the definition of the ring. Commutativity under addition does *not* appear as an axiom but is proved! So are other elementary properties of a ring such as $a \times 0 = 0$, $a(-b) = (-a)b = -(ab)$, and $(-a)(-b) = ab$.

Fraenkel's work exerted little influence since it was not grounded in the major concrete theories that had earlier been established. Its main significance was that rings now began to be studied as independent, abstract objects, not just as rings of polynomials, as rings of algebraic integers, or as rings (algebras) of hypercomplex numbers.

D. EMMY NOETHER AND EMIL ARTIN. Yet rings of polynomials, rings of algebraic integers, and rings of hypercomplex numbers remained central in ring theory. In the hands of the master algebraists Noether and Artin their study was transformed in the 1920s into powerful, abstract theories. Noether's two seminal papers of 1921 and 1927 extended the decomposition theories of polynomial rings on the one hand and of the rings of integers of algebraic number fields and algebraic function fields on the other, to abstract commutative rings with the ascending chain condition—now called *noetherian rings*.

More specifically, Noether showed in her 1921 paper, "Ideal theory in rings", that the results of Hilbert, Lasker, and Macauley on primary decomposition in polynomial rings hold for *any* ring with the ascending chain condition. Thus results which seemed inextricably connected with the properties of polynomial rings were shown to follow from a single axiom! In her 1927 paper, "Abstract development of ideal theory in algebraic number fields and function fields", she characterized abstract commutative rings in which every nonzero ideal is a unique product of prime ideals. These are now called *dedekind domains*.

Artin, inspired by Noether's work on commutative rings with the ascending chain condition, generalized Wedderburn's structure theorem in his 1927 paper,

"On the theory of hypercomplex numbers", to noncommutative semi-simple rings with the descending chain condition. In particular, he showed that such rings (now called *artinian rings*) can be decomposed into direct sums of simple rings that, in turn, are matrix rings over division rings.

While Fraenkel gave the first abstract definitions of a ring, Noether and Artin made the abstract ring concept central in algebra by framing in an abstract setting the theorems that were its major inspirations. In this context they introduced, and gave prominence to, such fundamental algebraic notions as ideal (including one-sided ideal), module, and chain conditions (both ascending and descending). Ring theory now took its rightful place alongside the by then well-established theories of groups and fields as one of the pillars of abstract algebra.

REFERENCES

1. Adams, W. W., and Goldstein, L. J., *Introduction to Number Theory*, Prentice-Hall, 1976.
2. Edwards, H. M., "The genesis of ideal theory", *Arch. Hist. Ex. Sci.* 23 (1980), 321–378.
3. Eisenbud, D., *Commutative Algebra, with a View Toward Algebraic Geometry*, Springer-Verlag, 1995.
4. Fraenkel, A., "Über die Teiler der Null und die Zerlegung von Ringen", *Jour. für die Reine und Angew. Math.* 145 (1914), 139–176.
5. Ireland, K., and Rosen, M., *A Classical Introduction to Modern Number Theory*, 2nd ed., Springer-Verlag, 1982.
6. van der Waerden, B. L., *A History of Algebra*, Springer-Verlag, 1985.

Field Theory: From Equations to Axiomatization

Israel Kleiner

Part I

1. INTRODUCTION. The evolution of field theory spans a period of about 100 years, beginning in the early decades of the 19th century. This period also saw the development of the other major algebraic theories, namely group theory, ring theory, and linear algebra. The evolution of field theory was closely intertwined with that of the other three theories, as we shall see.

Abstract field theory emerged from three concrete theories—what came to be known as Galois theory, algebraic number theory, and algebraic geometry. These were founded, and began to flourish, in the 19th century. Of some influence in the rise of the abstract field concept were also the theory of congruences and (British) symbolical algebra. The 19th century's increased concern for rigor, generalization, and abstraction undoubtedly also had an impact on our story.

In this paper we discuss the sources of field theory as well as some of the main events in its evolution, culminating in Steinitz's abstract treatment of fields.

2. GALOIS THEORY. For three millennia (until the early 19th century) algebra meant solving polynomial equations, mainly of degrees up to 4. Field-theoretic ideas are implicit even here. For example, in solving the linear equation $ax + b = 0$, the four algebraic operations come into play and hence implicitly so does the notion of a field. In the case of the quadratic equation $ax^2 + bx + c = 0$, its solutions, $x = (-b \pm \sqrt{b^2 - 4ac})/2a$, require the adjunction of square roots to the field of coefficients of the equation. The concept of adjunction of an element to a field is fundamental in field theory.

Field-theoretic notions appear much more prominently, even if at first still implicitly, in the modern theory of solvability of polynomial equations. The groundwork was laid by Lagrange in 1770, but the field-theoretic elements of the subject were introduced by Abel and Galois in the early decades of the 19th century. Ruffini's 1799 proof of the insolvability of the quintic had a major gap because he lacked sufficient understanding of field-theoretic ideas [16].

Such ideas were starting points in Galois's 1831 "Mémoire sur les conditions de résolubilité des équations par radicaux" [16, p. 305]:

> One can agree to regard all rational functions of a certain number of determined quantities a priori. For example, one can choose a particular root of a whole number and regard as rational every rational function of this radical. When we agree to regard certain quantities as known in this manner, we shall say that we adjoin them to the equation to be resolved. We shall say

that these quantities are adjoined to the equation. With these conventions, we shall call rational any quantity which can be expressed as a rational function of the coefficients of the equation and of a certain number of adjoined quantities arbitrarily agreed upon.... One can see, moreover, that the properties and the difficulties of an equation can be altogether different, depending on what quantities are adjoined to it.

It is clear that Galois has a good insight into the fields that we would denote today by $F(u_1, u_2, \ldots, u_n)$, obtained by adjoining the quantities $u_1, u_2, \ldots, u_n$ to the (field of) coefficients of an equation. In the specific example mentioned, he has in mind a quadratic field, $Q(\sqrt{d}\,)$.

Galois was the first to use the term "adjoin" in a technical sense. The notion of adjoining the roots of an equation to the field of coefficients is central in his work [9], [16].

One of the fundamental theorems of the subject proved by Galois is the Primitive Element Theorem. This says (in our terminology) that if E is the splitting field of a polynomial $f(x)$ over a field F, then $E = F(V)$ for some rational function V of the roots of $f(x)$. Galois used this result to determine the Galois group of the equation $f(x) = 0$ [1], [16]. The Primitive Element Theorem was essential in all subsequent work in Galois theory until Artin bypassed it in the 1930s by reformulating Galois theory, for he felt that the theorem was not intrinsic to the subject [9].

3. ALGEBRAIC NUMBER THEORY. The central field-theoretic notion here, due independently to Dedekind and Kronecker, is that of an algebraic number field $Q(a)$, where a is an algebraic number. How did it arise? Mainly from three major number-theoretic topics: Fermat's Last Theorem (FLT), reciprocity laws, and representation of integers by binary quadratic forms. Although all three topics have to do with the domain of (ordinary) integers, in order to deal with them effectively it was found necessary to embed them in domains of what came to be known as algebraic integers. The following examples illustrate the ideas involved.

(a) To prove FLT for (say) $n = 3$, that is, to show that $x^3 + y^3 = z^3$ has no nonzero integer solutions, one factors the left side to obtain the equation $(x + y)(x + yw)(x + yw^2) = z^3$, where w is a primitive cube root of unity, $w = (-1 + \sqrt{3}\,i)/2$. This is now an equation in the domain $D = \{a + bw : a, b \in Z\}$ of algebraic integers. This approach to FLT (for $n = 3$) was essentially used by Euler and later by Lamé and others [5].

(b) Gauss's quadratic reciprocity law appeared in his *Disquisitiones Arithmeticae* of 1801. It says that $x^2 \equiv p \pmod{q}$ is solvable if and only if $x^2 \equiv q \pmod{p}$ is solvable, unless $p \equiv q \equiv 3 \pmod 4$, in which case $x^2 \equiv p \pmod{q}$ is solvable if and only if $x^2 \equiv q \pmod{p}$ is not. Here p and q are odd primes [8].

Gauss and others tried to extend this result to "higher" reciprocity laws. For example, for cubic reciprocity one asks about the relationship between the solvability of $x^3 \equiv p \pmod{q}$ and $x^3 \equiv q \pmod{p}$. These higher reciprocity-type problems are much more difficult to deal with than quadratic reciprocity. Gauss remarked that [8, p. 108]:

> The previously accepted laws of arithmetic are not sufficient for the foundations of a general theory [of higher reciprocity]... Such a theory demands that the domain of arithmetic be endlessly enlarged.

His comments were no idle speculation. In fact, he himself began to implement the above "programme" by formulating and proving a law of *biquadratic reciprocity*. To do that he extended the domain of arithmetic by introducing what came to be known as the *gaussian integers* $G = \{a + bi : a, b \in Z\}$. He could not even formulate such a law without introducing G [8].

(c) The problem of representing integers by binary quadratic forms, namely determining when $n = ax^2 + bxy + cy^2$ ($a, b, c \in Z$), goes back to Fermat. In particular, Fermat asked and answered the question: which integers n are sums of two squares, $n = x^2 + y^2$? In the *Disquisitiones* Gauss studied the *general* problem very thoroughly, developing a comprehensive and beautiful, but very difficult, theory. To gain a deeper understanding of Gauss's theory of binary quadratic forms, Dedekind found that he, too, needed to extend the domain Z of integers. For example, even in the simple case of representing integers as sums of two squares, it is the equation $(x + yi)(x - yi) = n$ rather than $x^2 + y^2 = n$ that yields conceptual insight [1], [10].

Dedekind's ideas. The fundamental question in extending the domain of ordinary arithmetic to "higher" domains is whether such domains behave like the integers, namely whether they are unique factorization domains (UFDs). It is this property that facilitates the solution of problems (a)–(c). While the domains D and G introduced above are UFDs, most domains that arise in connection with the three number-theoretic problems we have described are not. For example, when we factor the left side of $x^n + y^n = z^n$ for $n \geq 23$, the resulting domains are never UFDs. To rescue unique factorization in such domains Dedekind introduced (in Supplement X (1871) to Dirichlet's *Vorlesungen über Zahlentheorie*) ideals and prime ideals, and showed that every ideal in these domains is a unique product of prime ideals [10].

But what *are* the domains with restored unique factorization? To answer that—one of the fundamental questions of his theory—Dedekind needed to introduce fields, in particular *algebraic number fields* $Q(a)$, where a is a root of a polynomial with integer coefficients. These were the natural habitats of his domains, just as the rationals are the natural habitat of the integers. The domains in question were then defined as "the integers of $Q(a)$," namely those elements of $Q(a)$ that are roots of *monic* polynomials with integer coefficients. Dedekind showed that they form a commutative ring with identity and without zero divisors whose field of quotients is $Q(a)$ [3], [10], [13].

Given Dedekind's predisposition for abstraction—a rather rare phenomenon in the 1870s, he placed his theory in a broader context by giving axiomatic definitions of rings, fields, and ideals. Here is his definition of a field [1, p. 117]:

> By a field we will mean every infinite system of real or complex numbers so closed in itself and perfect that addition, subtraction, multiplication, and division of any two of these numbers again yields a number of the system.

To Dedekind, then, fields were subsets of the complex numbers, which is, of course, all he needed for his theory of algebraic numbers. Still, an axiomatic definition in number theory/algebra, even in this restricted sense, is remarkable for that time. Also remarkable are Dedekind's use of infinite sets ("systems"), which predates Cantor's, and his "descriptive" rather than "constructive" definition of a mathematical object as a set of all elements of a certain kind satisfying a number of properties.

The field concept was a unifying mathematical notion for Dedekind. Before his definition of a field he says [4, p. 131]:

> In the following paragraphs I have attempted to introduce the reader into a higher domain, in which algebra and number theory interconnect in the most intimate manner.... I became convinced that studying the algebraic relationship of numbers is most conveniently based on a concept that is directly connected with the simplest arithmetic properties. I had originally used the term "rational domain," which I later changed to "field."

Hilbert remarked that Gauss, Dirichlet, and Jacobi had also expressed their amazement at the close connection between number theory and algebra, on the grounds that these subjects have common roots in (as Dedekind would put it) the theory of fields [4].

Dedekind produced several editions of his groundbreaking theory of ideal decomposition in algebraic number fields. In his mature 1894 version (4th edition of Dirichlet's *Zahlentheorie*) he included important concepts and results on fields —nowadays standard—such as [9, pp. 130–132]:

(i) If S is any subset of the complex numbers containing the rationals, the intersection of all fields containing S is a field; it is called "rational with respect to S."
(ii) He defines field isomorphism, calling it "permutation of the field," as a mapping of a field E onto a field F that preserves all four operations of the field. He observes that if F is nonzero, the mapping is one-one. He also notes that the mapping is the identity on Q.
(iii) If E is a subfield of K, he defines the *degree* of K over E as the dimension of K considered as a vector space over E. He shows that if the degree is finite then every element of K is algebraic over E.

Kronecker's ideas. Kronecker's work was broader but much more difficult than Dedekind's. He developed his ideas over several decades, beginning in the 1850s, trying to frame a general theory that would subsume algebraic number theory and algebraic geometry as special cases. In his great 1882 work *Grundzüge einer arithmetischen Theorie der algebraischen Grössen* he developed algebraic number theory using an approach entirely different from Dedekind's. One of his central concepts was also that of a field—he called it "domain of rationality," defined as follows [9, p. 127]:

> The domain of rationality $(R', R'', R''', \ldots)$ contains every one of those quantities which are rational functions of the quantities $R', R'', R''', \ldots$ with integer coefficients.

Note how different Kronecker's "definition" of a field is from Dedekind's! It is a constructive description rather than the kind of definition that would be acceptable to us today. But it was dictated by Kronecker's views on the nature of mathematics.

Kronecker rejected irrational numbers as bona fide entities since they involve the mathematical infinite. For example, the algebraic number field $Q(\sqrt{2})$ was defined by Kronecker as the quotient field of the polynomial ring $Q[x]$ relative to the ideal generated by $x^2 - 2$, though he would have put it in terms of congruences rather than quotient rings. These ideas contain the germ of what came to be known as Kronecker's Theorem, namely that every polynomial over a field has a root in some extension field [9], [13].

It is interesting to compare this definition of $Q(\sqrt{2})$ with Cauchy's definition in the 1840s of the complex numbers as polynomials over the reals modulo $x^2 + 1$ (and compare the latter with Gauss's integers modulo p). Cauchy's rationale was to give an "algebraic" definition of complex numbers that would avoid the use of $\sqrt{-1}$.

Dedekind vs. Kronecker. Dedekind and Kronecker were great contemporary algebraists. Both published pathbreaking works on algebraic number theory. But their approaches to the subject were very different. Both were guided in their works by their "philosophies" of mathematics, and these too were very different [13]. Kronecker was perhaps the first preintuitionist, Dedekind likely the first preformalist (cf. Kronecker's "God made the [positive] integers, all the rest is the work of man" with Dedekind's "[The natural] numbers are a free creation of the human mind"). To Kronecker mathematics had to be constructive and finitary. Dedekind did not hesitate to use axiomatic notions and the infinite. While Kronecker made frequent pronouncements on these topics, Dedekind made few; his views became known mainly from his works—conceptual and abstract. Some examples:

(i) Since Kronecker's domains of rationality had to be generated by *finitely* many elements (the $R', R'', R''', \ldots$), his definition would not admit the totality of algebraic numbers as a field. Dedekind had no problem in considering the set of all complex numbers that are roots of polynomial equations with integer coefficients (viz. the set of all algebraic numbers) as a bona fide mathematical object.

(ii) On the other hand, Kronecker put no restriction on the nature of the entities $R', R'', R''', \ldots$—they could, for example, be indeterminates or roots of algebraic equations. So $Q(x)$ was a legitimate field to Kronecker. In fact, the adjunction of indeterminates to a field was a cornerstone of his approach to algebraic number theory. Dedekind, recall, defined his fields to be subsets of the complex numbers (but see Section 4).

(iii) Since Kronecker did not accept π (say) as a legitimate number, he identified $Q(\pi)$ with $Q(x)$ (x an indeterminate), thus claiming that transcendental numbers are indeterminate! To Dedekind $Q(\pi)$ was a perfectly legitimate entity not requiring any assistance from $Q(x)$.

4. ALGEBRAIC GEOMETRY.
The examples of fields we have come across so far have been mainly fields of numbers. Here we encounter principally fields of

functions, in particular, algebraic functions and rational functions. The ideas are due mainly to Kronecker and Dedekind-Weber.

Fields of algebraic functions. Algebraic geometry is the study of algebraic curves and their generalizations to higher dimensions, algebraic varieties. An *algebraic curve* is the set of roots of an algebraic function, that is, a function $y = f(x)$ defined implicitly by a polynomial equation $P(x, y) = 0$.

Several approaches were used in the study of algebraic curves, notably the analytic, the geometric-algebraic, and the algebraic-arithmetic. In the analytic approach, to which Riemann (in the 1850s) was the major contributor, the main objects of study were algebraic functions $f(w, z) = 0$ of a complex variable and their integrals, the so-called abelian integrals. It was in this connection that Riemann introduced the fundamental notion of a Riemann surface, on which algebraic functions become single-valued. Riemann's methods, however, were nonrigorous, relying heavily on the physically obvious but mathematically questionable Dirichlet Principle [3], [11].

Dedekind and Weber, in their important 1882 paper "Theorie der algebraischen Funktionen einer Veränderlichen," set for themselves the task of making Riemann's ideas rigorous, or, as they put it [11, p. 154]:

> The purpose of the[se] investigations ... is to justify the theory of algebraic functions of a single variable, which is one of the main achievements of Riemann's creative work, from a simple as well as rigorous and completely general viewpoint.

To accomplish this, they carried over to algebraic functions the ideas that Dedekind had earlier introduced for algebraic numbers. Specifically, just as an algebraic number field is a finite extension $Q(a)$ of the field Q of rational numbers, so Dedekind and Weber defined an *algebraic function field* as a finite extension $K = C(z)(w)$ of the field $C(z)$ of rational functions (in the indeterminate z). That is, w is a root of a polynomial $p(t) = a_0 + a_1 t + a_2 t^2 + \cdots + a_n t^n$, where $a_i \in C(z)$ (we can take $a_i \in C[z]$). Thus $w = f(z)$ is an algebraic function defined implicitly by the polynomial equation $P(z, w) = a_0 + a_1 w + a_2 w^2 + \cdots + a_n w^n = 0$. In fact, all the elements of $K = C(z)(w) = C(z, w)$ are algebraic functions.

Now let A be "the integers of K"; that is, A consists of the elements of $K = C(z)(w)$ that are roots of monic polynomials over $C[z]$ (cf. "the integers of $Q(a)$," Section 3). By analogy with the case of algebraic numbers, here too A is an integral domain and every nonzero ideal of A is a unique product of prime ideals [1], [3]. Incidentally, the meromorphic functions on a Riemann surface form a field of algebraic functions, with the entire functions as their "integers."

Dedekind and Weber were now ready to give a rigorous, algebraic definition of a Riemann surface S of the algebraic function field K: It is (in our terminology) the set of nontrivial discrete valuations on K. The finite points of S correspond to the ideals of A; to deal with points at infinity of S, they introduced the notions of "place" and "divisor" [3]. They developed many of Riemann's ideas on algebraic functions algebraically and rigorously. In particular, they gave a rigorous algebraic proof of the important Riemann-Roch Theorem [1], [3], [11].

Dedekind and Weber were at heart algebraists. They felt that algebraic function theory is intrinsically an algebraic subject, hence it ought to be developed algebraically. As they put it: "In this way, a well-delimited and relatively comprehensive part of the theory of algebraic functions is treated solely by means belonging to its own domain" [**11**, p. 156].

Beyond their technical achievements in putting major parts of Riemann's algebraic function theory on solid ground, the conceptual breakthrough by Dedekind and Weber lay in pointing to the strong analogy between algebraic number fields and algebraic function fields, hence between algebraic number theory and algebraic geometry. This analogy proved most fruitful for both theories. Another noteworthy aspect of their work was its generality, in particular its applicability to arbitrary fields; see [**6**], [**15**].

Fields of rational functions. As noted earlier, algebraic geometry is the study of algebraic varieties. An algebraic variety is the set of points in R^n (or C^n) satisfying a system of polynomial equations $f_i(x_1, x_2, \ldots, x_n) = 0$, $i = 1, 2, \ldots, k$; the Hilbert basis theorem implies that finitely many equations will do. The ideal structure of the ring $R[x_1, \ldots, x_n]$ (or $C[x_1, \ldots, x_n]$) to which the polynomials $f_i(x_1, x_2, \ldots, x_n)$ belong is fundamental for the understanding of the algebraic variety, as is the "natural habitat" of that ring—its field of quotients $R(x_1, \ldots, x_n)$ (or $C(x_1, \ldots, x_n)$). These are the fields of (formally) *rational functions*. We have seen that such fields were also introduced by Kronecker in connection with his work in algebraic number theory [**6**], [**13**].

5. CONGRUENCES. Gauss introduced the congruence notation in the *Disquisitiones Arithmeticae* of 1801 and showed (among other things) that one can add, subtract, multiply, and divide congruences modulo a prime p, in effect that the integers modulo p form a field—a *finite* field of p elements. Inspired by Gauss's work on congruences, Galois introduced finite fields with p^n elements in an 1830 paper entitled "Sur la theorie des nombres."

Galois's aim was to study the congruence $F(x) \equiv 0 \pmod{p}$ as a generalization of Gauss's quadratic congruences (cf. Gauss's quadratic reciprocity law). Here $F(x)$ is a polynomial of degree n that is irreducible mod p, i.e., $F(x)$ is irreducible over the field Z_p. Galois showed that $F(x)$ has no integral roots [mod p]. His conclusion was that [**7**, pp. 277–278]:

> One should therefore regard the roots of this congruence as some kind of imaginary symbols..., symbols whose employment in calculation will often prove as useful as that of the imaginary $\sqrt{-1}$ in ordinary analysis.

He continues:

> Let i [an arbitrary symbol, *not* the complex number i] denote one of the roots of the congruence $F(x) \equiv 0$, which can be supposed to have degree n. Consider the general expression
> $$a + a_1 i + a_2 i^2 + \cdots + a_{n-1} i^{n-1}, \quad (**)$$
> where $a, a_1, a_2, \ldots, a_{n-1}$ represent integers [mod p]. When these numbers are assigned all their possible values, expression (**) takes on p^n values, which possess, as I shall demonstrate, the same properties as the natural numbers in the *theory of residues of powers*.

Galois did, indeed, show that the expressions (**) form a field, now called a *Galois field*. He also showed that (in our terminology) the multiplicative group of that field is cyclic [1], [7], [13]. In an 1893 paper entitled "A doubly-infinite system of simple groups," E. H. Moore characterized the finite fields [12].

6. SYMBOLICAL ALGEBRA. In the third and fourth decades of the 19th century British mathematicians, notably Peacock, Gregory, and De Morgan, created what came to be known as symbolical algebra. Their aim was to set algebra—to them this meant the laws of operation with numbers, negative numbers especially—on an equal footing with geometry by providing it with logical justification. They did this by distinguishing between *arithmetical algebra*—laws of operation with positive numbers, and *symbolical algebra*—a subject newly created by Peacock, which dealt with laws of operation with numbers in general.

Although the laws were carried over verbatim from those of arithmetical algebra, in accordance with the so-called Principle of Permanence of Equivalent Forms, the point of view was remarkably modern. Witness Peacock's definition of symbolical algebra, given in his *Treatise of Algebra* of 1830 [**14**, p. 35]:

> The science which treats of the combinations of arbitrary signs and symbols by means of defined though arbitrary laws.

Quite a statement for the early 19th century! Such sentiments were about a century ahead of their time. And of course one did have to wait about a century to have what Peacock had preached put fully into practice. Nevertheless, the creation of symbolical algebra was a significant development, even if not directly related to fields, signalling (according to some) the birth of abstract algebra [2].

REFERENCES

1. I. G. Bashmakova and E. I. Slavutin, Algebra and algebraic number theory, in *Mathematics of the 19th Century*, ed. by A. N. Kolmogorov and A. P. Yushkevich, Birkhäuser, 1992, pp. 35–135.
2. G. Birkhoff, Current trends in algebra, *Amer. Math. Monthly* **80** (1973) 760–782, and corrections in **81** (1974) 746.
3. N. Bourbaki, *Elements of the History of Mathematics*, Springer-Verlag, 1984.
4. L. Corry, *Modern Algebra and the Rise of Mathematical Structures*, Birkhäuser, 1996.
5. H. M. Edwards, *Fermat's Last Theorem: A Genetic Introduction to Algebraic Number Theory*, Springer-Verlag, 1977.
6. D. Eisenbud, *Commutative Algebra with a View Toward Algebraic Geometry*, Springer-Verlag, 1995.
7. E. Galois, Sur la théorie des nombres, English translation in S. Stahl, *Introductory Modern Algebra: A Historical Approach*, Wiley, 1997, pp. 277–284.
8. K. Ireland and M. Rosen, *A Classical Introduction to Modern Number Theory*, 2nd ed., Springer-Verlag, 1982.
9. B. M. Kiernan, The development of Galois theory from Lagrange to Artin, *Arch. Hist. Exact Sci.* **8** (1971/72) 40–54.
10. I. Kleiner, The roots of commutative algebra in algebraic number theory, *Math. Mag.* **68** (1995) 3–15.
11. D. Laugwitz, *Bernhard Riemann, 1826–1866*, Birkhäuser, 1999. Translated from the German by A. Shenitzer.
12. E. H. Moore, A doubly-infinite system of simple groups, *New York Math. Soc. Bull.* **3** (1893) 73–78.
13. W. Purkert, Zur Genesis des abstrakten Körperbegriffs I, II, *Naturwiss., Techn. u. Med.* **8** (1971) 23–37 and **10** (1973) 8–20. Unpublished English translation by A. Shenitzer.

14. H. M. Pycior, George Peacock and the British origins of symbolical algebra, *Historia Math.* **8** (1981) 23–45.
15. J. H. Silverman and J. Tate, *Rational Points on Elliptic Curves*, Springer-Verlag, 1992.
16. J.-P. Tignol, *Galois' Theory of Algebraic Equations*, Wiley, 1988.

Part II

7. THE ABSTRACT DEFINITION OF A FIELD. The developments we have been describing thus far lasted close to a century. They gave rise to important "concrete" theories—Galois theory, algebraic number theory, algebraic geometry—in which the (at times implicit) field concept played a central role. At the end of the 19th century abstraction and axiomatics were "in the air." For example, Pasch (1882) gave axioms for projective geometry, stressing for the first time the importance of undefined notions, Cantor (1883) defined the real numbers essentially as equivalence classes of Cauchy sequences of rationals, and Peano (1889) gave his axioms for the natural numbers. In algebra, von Dyck (1882) gave an abstract definition of a group that encompassed both finite and infinite groups (about thirty years earlier Cayley had defined a *finite* group), and Peano (1888) gave a definition of a finite-dimensional vector space, though this was largely ignored by his contemporaries. The time was propitious for the abstract field concept to emerge. Emerge it did in 1893 in the hands of Weber (of Dedekind-Weber fame).

Weber's definition of a field appeared in his 1893 paper "Die allgemeinen Grundlagen der Galois'schen Gleichungstheorie" [**15**], in which he aimed to give an abstract formulation of Galois theory [**8**, p. 136]:

> In the following an attempt is made to present the Galois theory of algebraic equations in a way which will include equally well all cases in which this theory might be used. Thus we present it here as a direct consequence of the group concept illuminated by the field concept, as a formal structure completely without reference to any numerical interpretation of the elements used.

Weber's presentation of Galois theory is indeed very close to the way the subject is taught today. His definition of a field, preceded by that of a group, is as follows [**15**, pp. 526–527]:

> A group becomes a field if two types of composition are possible in it, the first of which may be called *addition*, the second *multiplication*. The general determination must be somewhat restricted, however.
> 1. We assume that both types of composition are commutative.
> 2. Addition shall generally satisfy the conditions which define a group.
> 3. Multiplication is such that
>
> $a(-b) = -(ab)$
>
> $a(b + c) = ab + ac$
>
> $ab = ac$ implies $b = c$, unless $a = 0$
>
> Given b and c, $ab = c$ determines a, unless $b = 0$.

Although the associative law under multiplication is missing, and the axioms are not independent, they are of course very much in the modern spirit. As examples of his newly defined concept Weber included the number fields and function fields of algebraic number theory and algebraic geometry, respectively, but also Galois's finite fields and Kronecker's "congruence fields" $K[x]/(p(x))$, K a field, $p(x)$ irreducible over K.

Weber proved (often reproved, after Dedekind) various theorems about fields, which later became useful in Artin's formulation of Galois theory, and which are today recognized as basic results of the theory. Among them are [8], [10]:

(i) Every finite algebraic extension of a field is simple (that is, it is generated by a single element).
(ii) Every polynomial over a field has a splitting field.
(iii) If $F \subseteq F(a) \subseteq F(b)$, then $(F(a):F)$ divides $(F(b):F)$, where for fields K and E with $E \subseteq K$, $(K:E)$ denotes the dimension of K as a vector space over E.

It should be emphasized that it was not Weber's aim to study fields as such, but rather to develop enough of field theory to give an abstract formulation of Galois theory [11]. In this he succeeded admirably. His paper, and somewhat later his two-volume *Lehrbuch der Algebra*, exerted considerable influence on the development of abstract algebra [3].

8. HENSEL'S P-ADIC NUMBERS. In an 1899 article entitled "New foundations of the theory of algebraic numbers," Hensel began a life-long study of p-adic numbers. Inspired by the work of Dedekind-Weber, Hensel took as his point of departure the analogy between function fields and number fields. Just as power series are useful for a study of the former, Hensel introduced p-adic numbers to aid in the study of the latter [**10**, II, p. 19]:

> The analogy between the results of the theory of algebraic functions of one variable and those of the theory of algebraic numbers suggested to me many years ago the idea of replacing the decomposition of algebraic numbers, with the help of ideal prime factors, by a more convenient procedure that fully corresponds to the expansion of an algebraic function in power series in the neighborhood of an arbitrary point.

Indeed, in the neighborhood of a given point α every algebraic function of a complex variable can be represented as an infinite series of integral and rational powers of $z - \alpha$, as Weierstrass had shown. The elements of Hensel's *field of p-adic numbers* are formal power series $\sum_n^\infty a_k p^k$, where $a_k \in Z_p$ and $n \in Z$. And just as every element of an algebraic function field can be identified with the set of its expansions at all points of the Riemann surface on which it is defined, so every element of an algebraic number field is identified with the set of its representations in the field of p-adic numbers $\sum_n^\infty a_k p^k$ for every prime p [**2**, p. 111].

In a 1907 book, Hensel introduced topological notions in his p-adic fields and applied the resulting p-adic analysis in algebraic number theory. The p-adic numbers proved extremely useful not only there but also in algebraic geometry [4], [7]. They were also influential in motivating the abstract study of rings and fields [3].

9. STEINITZ. The last major event in the evolution of field theory that we describe is Steinitz's great work of 1910 [**13**]. But first some background.

Algebra in the 19th century was by our standards concrete. It was connected in one way or another with the real or complex numbers. For example, some of the great contributors to 19th-century algebra, mathematicians whose ideas shaped the algebra of the 20th century, were Gauss, Galois, Jordan, Kronecker, Dedekind, and Hilbert, and their algebraic work dealt with quadratic forms, cyclotomy, permutation groups, ideals in rings of algebraic number fields and algebraic function fields, and invariant theory. All of these subjects were related in one way or another to the real or complex numbers.

At the turn of the 20th century the axiomatic method began to take hold as an important mathematical tool. Hilbert's *Foundations of Geometry* of 1899 was very influential in this respect (see also Section 7). Noteworthy also is the American school of axiomatic analysis, as exemplified in the works of Dickson, Huntington, E. H. Moore, and Veblen. In the first decade of the 20th century these mathematicians began to examine various axiom systems for groups, fields, associative algebras, projective geometry, and the algebra of logic. Their principal aim was to study the independence, consistency, and completeness of the axioms defining any one of these systems. Also relevant were Hilbert's axiomatic characterization in 1900 of the field of real numbers and Huntington's like characterization in 1905 of the field of complex numbers [**1**], [**3**].

Steinitz's groundbreaking 150-page paper "Algebraische Theorie der Körper" of 1910 initiated the abstract study of fields as an independent subject [**13**]. While Weber *defined* fields abstractly, Steinitz *studied* them abstractly.

Steinitz's immediate source of inspiration was Hensel's p-adic numbers [**3**, p. 194]:

> I was led into this general research especially by Hensel's *Theory of Algebraic Numbers*, whose starting point is the field of p-adic numbers, a field which counts neither as a field of functions nor as a field of numbers in the usual sense of the word.

More generally, Steinitz's work arose out of a desire to delineate the abstract notions common to the various contemporary theories of fields: fields in algebraic number theory, in algebraic geometry, and in Galois theory, p-adic fields, and finite fields. His goal was a comprehensive study of *all* fields, starting from the field axioms [**3**, p. 195]:

> The aim of the present work is to advance an overview of all the possible types of fields and to establish the basic elements of their interrelations.

Quite a task! Steinitz's plan was to start from the simplest fields and to build up all fields from these. The basic concept that he identified to study the former is the *characteristic* of the field. Here are several of his fundamental results, nowadays staples of field theory [**10**], [**13**]:

(i) Classification of fields into those of characteristic zero and those of characteristic p. The *prime fields*—the "simplest" fields—are Q and Z_p; one or the other is a subfield of every field.

(ii) Development of a theory of *transcendental extensions*, which became indispensable in algebraic geometry.

(iii) Recognition that it is precisely the *finite, normal, separable extensions* to which Galois theory applies.

(iv) Proof of the existence and uniqueness (up to isomorphism) of the *algebraic closure* of any field.

A description of all fields followed [**11**, p. 754]:

> Starting with an arbitrary prime field, by taking an arbitrary, purely transcendental extension followed by an arbitrary algebraic extension, we have a method of arriving at any field.

The notions of *transcendency base* and *degree of transcendence* of an extension field, both of which Steinitz introduced, played a crucial role here. Also important was the axiom of choice, whose use he acknowledged [**10**, II, p. 20]:

> Many mathematicians continue to reject the axiom of choice. The growing realization that there are questions in mathematics that cannot be decided without this principle is likely to result in the gradual disappearance of the resistance to it.

Steinitz's work was very influential in the development of abstract algebra in the 1920s and 1930s, as the following testimonials prove:

> Steinitz's paper was the basis for all [algebraic] investigations in the school of Emmy Noether (van der Waerden, [**14**, p. 162]).

> [Steinitz's work]... is not only a landmark in the development of algebra, but also... an excellent, in fact indispensable, introduction to a serious study of the new [modern] algebra (Baer and Hasse, [**13**, Preface]).

> Steinitz's work marks a methodological turning-point in algebra leading to ...'modern' or abstract algebra (Purkert and Wussing, [**11**, p. 754]).

> [Steinitz's work] can be considered as having given birth to the actual concept of Algebra (Bourbaki, [**2**, p. 83]).

10. A GLANCE AHEAD. We now list several major developments in field theory and related areas in the decades following Steinitz's fundamental work.

(a) *Valuation theory*. In 1913 Kürschak abstracted Hensel's ideas on p-adic fields by introducing the notion of a *valuation field*. He proved the existence of the completion of a field with respect to a valuation. In 1918 Ostrowski determined all valuations of the field Q of rational numbers. Valuation theory, which "forms a solid link between number theory, algebra, and analysis" [**7**, vol. II, p. 537], played fundamental roles in both algebraic number theory and algebraic geometry; see [**2**], [**4**], [**7**], [**14**].

(b) *Formally real fields*. In 1927 Artin and Schreier defined the notion of a *formally real field*, namely a field in which -1 is not a sum of squares. "One of [the] remarkable results [of the Artin-Schreier theory] is no doubt the discovery that the existence of an order relation on a field is linked to purely algebraic properties of the field" [**2**, p. 92]: A field can be ordered if and only if it is formally real. The theory of formally real fields enabled Artin in the same year to solve *Hilbert's* 17*th Problem* on the resolution of positive definite rational functions into sums of squares [**7**, vol. II, p. 640].

(c) *Class field theory*. This is the study of finite extensions of an algebraic number field having an abelian Galois group. It is a beautiful synthesis of algebraic, number-theoretic, and analytic ideas, in which *Artin's Reciprocity Law* has a central place. Major strides were already made by Hilbert in his "Zahlbericht" (Report on Number Theory) of 1897. More modern aspects of the theory were developed by Artin, Chevalley, Hasse, Tagaki, and others; see [**5**].

(d) *Galois theory*. Artin set out his now-famous abstract formulation of Galois theory in lectures given in 1926 (but published only in 1938). In a 1950 talk he said [**8**, p. 144]:

> Since my mathematical youth I have been under the spell of the classical theory of Galois. This charm has forced me to return to it again and again, and try to find new ways to prove its fundamental theorems.

Extensions of the classical theory were given in various directions. For example, in 1927 Krull developed a Galois theory of *infinite field extensions*, establishing a one-one correspondence between subfields and "closed" subgroups, and thereby introducing topological notions into the theory. There is also a Galois theory for *inseparable field extensions*, in which the notion of derivation of a field plays a central role, and a Galois theory for *division rings*, developed independently by H. Cartan and Jacobson in the 1940s; see [**7**], [**16**].

(e) *Finite fields*. Finite field theory is a thriving subject of investigation in its own right, but it also has important uses in number theory, coding theory, geometry, and combinatorics; see [**6**], [**9**].

REFERENCES

1. G. Birkhoff, Current trends in algebra, *Amer. Math. Monthly* **80** (1973) 760–782, and corrections in **81** (1974) 746.
2. N. Bourbaki, *Elements of the History of Mathematics*, Springer-Verlag, 1984.
3. L. Corry, *Modern Algebra and the Rise of Mathematical Structures*, Birkhäuser, 1996.
4. D. Eisenbud, *Commutative Algebra with a View Toward Algebraic Geometry*, Springer-Verlag, 1995.
5. H. Hasse, History of class field theory, in *Algebraic Number Theory, Proceedings of an Instructional Conference*, ed. by J. Cassels and A. Fröhlich, Thompson Book Co., 1967, pp. 266–279.
6. K. Ireland and M. Rosen, *A Classical Introduction to Modern Number Theory*, 2nd ed., Springer-Verlag, 1982.
7. N. Jacobson, *Basic Algebra I, II*, W. H. Freeman, 1974 and 1980.
8. B. M. Kiernan, The development of Galois theory from Lagrange to Artin, *Arch. Hist. Exact Sci.* **8** (1971/72) 40–154.
9. R. Lidl and H. Niederreiter, *Introduction to Finite Fields and their Applications*, Cambridge University Press, 1986.
10. W. Purkert, Zur Genesis des abstrakten Körperbegriffs I, II, *Naturwiss., Techn. u. Med.* **8** (1971) 23–37 and **10** (1973) 8–20. Unpublished English translation by A. Shenitzer.
11. W. Purkert and H. Wussing, Abstract algebra, in *Companion Encyclopedia of the History and Philosophy of the Mathematical Sciences*, vol. I, ed. by I. Grattan-Guinness, Routledge, 1994, pp. 741–760.
12. J. H. Silverman and J. Tate, *Rational Points on Elliptic Curves*, Springer-Verlag, 1992.
13. E. Steinitz, *Algebraische Theorie der Körper*, 2nd ed., Chelsea, 1950.
14. B. L. van der Waerden, Die Algebra seit Galois, *Jahresber. Deutsch. Math. Verein.* **68** (1966) 155–165.
15. H. Weber, Die allgemeinen Grundlagen der Galois'schen Gleichungstheorie, *Math. Ann.* **43** (1893) 521–549.
16. D. Winter, *The Structure of Fields*, Springer-Verlag, 1974.

Elliptic Curves

John Stillwell

In recent years, elliptic curves have played a leading role in number theory, most famously in Wiles' program to prove Fermat's last theorem. However, since these developments are highly technical, it may be useful to look back to earlier times, when elliptic curves led a simpler life. For about 1500 years, from the time of Diophantus to Newton, elliptic curves were known only as curves defined by certain cubic equations. This put them just a step beyond the conic sections, and some of their geometric and arithmetic properties can in fact be viewed as generalisations of properties of conics. In particular, it is possible to find rational solutions of both quadratic and cubic equations by simple geometric constructions.

It was only with the development of calculus, in the 17th century, that sharp differences between conics and elliptic curves began to emerge. Conic sections can be parametrised by rational functions. For example, the circle $x^2 + y^2 = 1$ is parametrised by
$$x = \frac{1-t^2}{t+t^2}, \quad y = \frac{2t}{1+t^2}$$
but the elliptic curves cannot. Their simplest parametrising functions are *elliptic functions*, which arise in calculus as the inverses of elliptic integrals, so-called because a typical example is the integral for the arc length of the ellipse. It is for this fairly accidental reason that they are called elliptic curves—an unfortunate accident since the ellipse itself is *not* an elliptic curve.

The difference between conics and elliptic curves was "felt" in the 17th century in the apparent intractability of elliptic integrals, though the parametrisation of cubic curves was not known at that time. The idea of inverting elliptic integrals to create elliptic functions had to wait until the early 19th century. The nonrationality of elliptic curves was not fully understood until the mid-19th century, when the introduction of complex coordinates revealed a *topological* difference between them and conics. This brings us within sight of the modern view of elliptic curves—a remarkable synthesis of number theory, geometry, algebra, analysis and topology. In what follows I shall attempt to describe what led up to this state of affairs.

Diophantus. Very little is known about Diophantus except that he lived sometime between 150 AD and 350 AD and was a wizard at finding rational solutions to polynomial equations in two or more variables. His *Arithmetica* (available in the English edition of Heath [4]), contains the solutions of hundreds of equations, among them the following instructive examples.

1. A rational solution of $x^2 + y^2 = 16$, other than an obvious one such as $x = 0$, $y = 4$, is found by solving the simultaneous equations
$$x^2 + y^2 = 16,$$
$$y = 2x - 4,$$
which yield the solution $x = 16/5$, $y = 12/5$ (Heath [4], p. 145).

2. A rational solution of $x^3 - 3x^2 + 3x + 1 = y^2$, other than the obvious one $x = 0, y = 1$, is found by solving the simultaneous equations

$$x^3 - 3x^2 + 3x + 1 = y^2,$$
$$y = \tfrac{3}{2}x + 1,$$

which yield the solution $x = 21/4, y = 71/8$ (Heath [3], p. 242).

How did Diophantus choose the linear equations in these two examples? The simplest explanation is geometric, although he makes no mention of geometry.

In the first example the linear equation represents a line through the "obvious" rational point $(0, -4)$. Its slope is not important, since any line through $(0, -4)$ with rational slope t will meet the circle at a second rational point $(8t/(1 + t^2), (4t^2 - 4)/(1 + t^2))$. Conversely, all rational points on the circle are obtainable in this way, so Diophantus has essentially *parametrised* the rational points on the circle by rational functions of a rational parameter t.

The linear equation in the second example has an even stronger geometric smell. It is the *tangent* to $x^3 - 3x^2 + 3x + 1 = y^2$ at the "obvious" rational point $(0, 1)$. Here there is no option about the slope because a line has to meet a cubic curve in *two* rational points for its third intersection to be rational. When only one rational point is known, this forces us to use the tangent, which is the line through two "coincident" points.

It is possible, of course, that Diophantus discovered these facts purely algebraically, and did not notice their geometric interpretation. However, that would be a truly amazing departure from the Greek mathematical culture of his time. Even in the more algebraic culture of the 17th century. Fermat and Newton immediately recognised Diophantus' work as geometry, with Newton [6] explicitly interpreting Diophantus' solutions as chord and tangent constructions. Later discoveries added more weight to the geometric interpretation, as we shall see below.

Fermat and Newton. Fermat was the first mathematician to make significant progress in number theory beyond Diophantus. Among his many discoveries were methods for proving *non*existence of integer or rational solutions for certain equations. For example, he proved that there are no positive rationals a, b, c such that

$$a^4 \pm b^4 = c^2$$

This implies in particular that no positive integer fourth powers sum to a fourth power (the $n = 4$ case of Fermat's last theorem), but it is also a statement about an elliptic curve. It says that there are no nontrivial rational points on the curve

$$y^2 = 1 - x^4,$$

since a rational point $(p/r, q/r)$ with $p, q \neq 0$ and

$$\frac{p^2}{r^2} = 1 - \frac{q^4}{r^4}$$

gives nonzero integers $a = r, b = q, c = pr$ with $a^4 - b^4 = c^2$.

Now I know I said that elliptic curves are cubics, but they are cubic *in a suitable coordinate system*. Any quartic curve of the form

$$y^2 = (x - \alpha)(x - \beta)(x - \gamma)(x - \delta)$$

can be rewritten

$$\left(\frac{y}{(x-\alpha)^2}\right)^2 = \left(1 - \frac{\beta - \alpha}{x - \alpha}\right)\left(1 - \frac{\gamma - \alpha}{x - \alpha}\right)\left(1 - \frac{\delta - \alpha}{x - \alpha}\right)$$

and hence it is cubic in the coordinates

$$X = \frac{1}{x - \alpha}, \quad Y = \frac{y}{(x - \alpha)^2}.$$

In particular, $y^2 = 1 - x^4$ is a cubic $Y^2 = 4X^3 - 6X^2 + 4X - 1$ in the coordinates $X = 1/(1-x), Y = y/(1-x)^2$. Notice that this is an appropriate coordinate change from the point of view of number theory, because it makes the rational points (x, y) on one curve correspond to the rational points (X, Y) on the other. Such a coordinate change is called *birational*.

Newton made the surprising discovery that all cubic equations in x and y can be reduced to the form

$$Y^2 = X^3 + aX + b$$

by a birational coordinate transformation. In fact, the transformations he used were simply projections. He called this "genesis of curves by shadows." His result can be viewed as an analogue of the well known theorem that second degree curves are conic sections and hence, in nondegenerate cases, projections of the circle. The degenerate cubic curves are those for which the right hand side $X^3 + aX + b$ has a repeated factor. The corresponding repeated root $X = \alpha$ is either a double point (Fig. 1) or cusp (Fig. 2) of the curve, and by drawing a line of slope t through this point we obtain the coordinates of the general point on the curve as rational functions of t.

The curves for which $X^3 + aX + b$ has no repeated factor cannot be parametrised by rational functions, and are what we now call elliptic curves (Fig. 3).

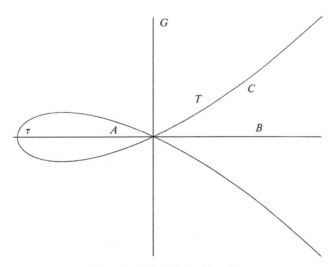

Figure 1. Cubic with double point.

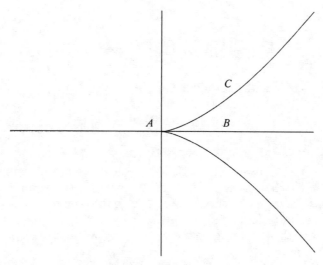

Figure 2. Cubic with cusp.

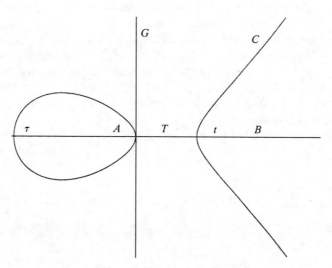

Figure 3. Nonsingular cubic.

Elliptic integrals. Early in the development of integral calculus, mathematicians encountered the problem of "rationalising" square roots of polynomials. For example, to find the area or arc length of a circle one finds an integral involving $\sqrt{1-x^2}$. This can be rationalised by the "Diophantine" substitution $x = (1-t^2)/(1+t^2)$, and fact Jakob Bernoulli [1], in a similar situation, actually attributed the substitution to Diophantus. He used it to obtain the expression

$$\frac{\pi}{4} = \int_0^1 \frac{dt}{1+t^2},$$

whence he obtained the famous series

$$\frac{\pi}{4} = 1 - \frac{1}{3} + \frac{1}{5} - \frac{1}{7} + \cdots$$

by expanding $1/(1 + t^2)$ in a geometric series and integrating term by term.

Integrals involving square roots of cubic or quartic polynomials proved more intractable. They were called *elliptic integrals* because one of them expresses the arc length of the ellipse. Cubics and quartics were lumped together because of birational equivalences between them, as noted above for $y^2 = 1 - x^4$ and $Y^2 = 4X^3 - 6X^2 + 4X - 1$. Such integrals arise from a great number of natural geometric and mechanical problems, so a lot of effort was expended on them, but without success.

Perhaps the first to see why rationalisation might be impossible was Jakob Bernoulli [2], who noted that a rationalisation of $\sqrt{1 - x^4}$, at least by a rational function $x = f(t)$ with *rational* coefficients, would violate Fermat's theorem on the nonexistence of positive integer solutions of $a^4 \pm b^4 = c^2$. In fact, it can be shown that $\sqrt{1 - x^4}$ cannot be rationalised by *any* rational function $x = f(t)$, by repeating Fermat's argument with polynomials in place of integers, so Jakob Bernoulli was on the right track. However, this type of argument was not used until the 19th century, so the nature of elliptic integrals remained unclear until then (when ideas not only from number theory, but also from analysis and topology, were directed at the problem).

Elliptic functions. In the 1820s, Abel and Jacobi finally saw what to do with elliptic integrals—*Invert* them. Instead of studying the integral

$$u = g^{-1}(x) = \int_0^x \frac{dt}{\sqrt{t^3 + at + b}},$$

say, study its inverse function $x = g(u)$. The gain in simplicity is comparable to studying the function $x = \sin u$ instead of the integral $\sin^{-1} x = \int_0^x (dt/\sqrt{1 - t^2})$. In particular, instead of a multi-valued integral $g^{-1}(x)$, one has a *periodic function* $x = g(u)$.

The difference between $\sin u$ and $g(u)$ is that the periodicity of $g(u)$ cannot be properly seen until complex values of the variables are admitted, at which stage it emerges that $g(u)$ has *two* periods. That is, there are nonzero $\omega_1, \omega_2 \in \mathbf{C}$, with $\omega_1/\omega_2 \notin \mathbf{R}$, such that

$$g(u) = g(u + \omega_1) = g(u + \omega_2).$$

The two periods can be brought to light in various ways. One method, originating with Eisenstein [1847] and commonly used today, is to write down a function that obviously has periods ω_1 and ω_2, namely

$$g(u) = \sum_{m,n \in \mathbf{Z}} \frac{1}{(u + m\omega_1 + n\omega_2)^2},$$

and derive its properties by manipulation of infinite series. Eventually one finds that $g^{-1}(x)$ is an integral of the type we started with.

A more insightful approach, though harder to make rigorous, is to study the behaviour of the integrand $1/\sqrt{t^3 + at + b}$ as t varies over the complex plane. Following Riemann [7], and viewing the 2-valued "function" $1/\sqrt{t^3 + at + b}$ as a 2-sheeted surface over $\mathbf{C}$, one finds that there are two independent closed paths of integration, over which the integrals are ω_1 and ω_2. This accounts for the periods ω_1 and ω_2 of the inverse function $g(u)$.

Since $g(u) = x$, it follows by basic calculus that

$$g'(u) = \frac{dx}{du} = \frac{1}{du/dx} = \frac{1}{1/\sqrt{x^3 + ax + b}} = \sqrt{x^3 + ax + b} = y,$$

so $x = g(u), y = g'(u)$ gives a parametrisation of the curve $y^2 = x^3 + ax + b$. With a little more work it can be shown that $u \mapsto (g(u), g'(u))$ is in fact a continuous one-to-one correspondence between $\mathbf{C}/\langle \omega_1, \omega_2 \rangle$ and the curve. $\mathbf{C}/\langle \omega_1, \omega_2 \rangle$ is the quotient of $\mathbf{C}$ by the subgroup generated by ω_1 and ω_2 and is topologically a *torus*, hence so is the curve $y^2 = x^3 + ax + b$. This is the deeper reason why elliptic curves are not rationally parametrisable—a curve parametrised by rational functions $x = p(u), y = q(u)$ is the topological image of the completed plane $\mathbf{C} \cup \{\infty\}$ of u values, and $\mathbf{C} \cup \{\infty\}$ is topologically a *sphere*.

Another consequence of the parametrisation $x = g(u), y = g'(u)$ is that the curve $y^2 = x^3 + ax + b$ is an abelian group. The "sum" of points with parameter values u_1, u_2 is simply the point with parameter value $u_1 + u_2$. Under this definition of sum, the curve is isomorphic to the group $\mathbf{C}/\langle \omega_1, \omega_2 \rangle$. Amazingly, there is an equivalent definition of the sum that Diophantus would have under-

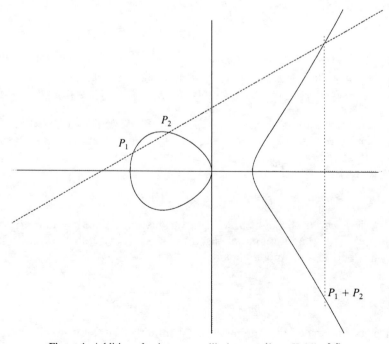

Figure 4. Addition of points on an elliptic curve (from Koblitz [5]).

stood (and which helps to explain why elliptic functions are useful in number theory): the sum of the points P_1 and P_2 is simply the reflection, in the x-axis, of the third point on the curve collinear with P_1 and P_2 (Fig. 4). For an explanation of this face we must refer the reader to a recent book on elliptic curves, such as Koblitz [5]. In the same book you will find many beautiful modern results on elliptic curves, motivated by ancient problems in number theory and geometry.

REFERENCES

1. Bernoulli, Jakob (1696) Positionum de seriebus infinitis pars tertia. *Werke*, **4**, 85–106.
2. _____ (1704) Positionum de seriebus infinitis... pars quinta. *Werke*, **4**, 127–147.
3. Eisenstein, G. (1847) Beiträge zur Theorie der elliptischen Functionen. *J. reine angew. Math.* **35**, 137–274.
4. Heath, T. L. (1910) *Diophantus of Alexandria*, Cambridge University Press.
5. Koblitz, N. (1985) *Introduction to Elliptic Curves and Modular Forms*, Springer-Verlag, New York.
6. Newton, I. (late 1670s) De resolutione quaestionum circa numeros. *Math. Papers* **4**, 110–115.
7. Riemann, G. B. H. (1851) Grundlagen für eine allgemeine Theorie der Functionen einer veränderlichen complexen Grösse. *Werke*, 2nd ed., 3–48.

Modular Miracles

John Stillwell

Over the last 20 years, the modular function has become widely known through its miraculous intervention in two great mathematical achievements: the proof of Fermat's last theorem and the "moonshine" of the monster simple group. In both cases, the modular function appears where no one expected it, and it bridges a chasm between seemingly unrelated fields. It is probably fair to say that, in these two cases, we do not yet fully understand how the modular magic works.

However, it can at least be said that these are not the first modular miracles. Ever since its discovery, in the early 19th century, the modular function has been an engine for spectacular and unexpected results. Now that things modular are back in the news, it is a good time to recall some of the modular miracles of the 19th century. They help us see the recent results in some perspective, and encourage us to believe that there is a lot more to be learned.

THE MODULAR FUNCTION j. Modular functions may be defined as meromorphic functions on the upper half plane with the periodicity of the *modular tessellation* shown in Figure 1. When the half plane is interpreted as the hyperbolic

Figure 1. The modular tessellation.

plane, the black and white tiles of the tessellation are congruent triangles with one vertex at infinity, and the whole tessellation is generated by reflections in the sides of any one of them.

It follows that a modular function is determined by its values on any one tile of the tessellation, the other values being obtained by reflection in the sides of the tile. The values on a tile can be defined by mapping the tile conformally onto the upper half plane, and they in turn are completely determined by the images of the three vertices.

This idea was used by Dedekind (1877) to define the classical modular function j by the unique conformal map

$$\text{white region} \to \text{half plane}$$

which sends i, $e^{\pi i/3}$, ∞ to $0, 1, \infty$ respectively [3].

The periodicity of j can be described algebraically by saying that

$$j(\tau) = j\left(\frac{a\tau + b}{c\tau + d}\right)$$

for any $a, b, c, d \in \mathbb{Z}$ with $ad - bc = 1$. The transformations

$$\tau \mapsto \frac{a\tau + b}{c\tau + d}$$

carry any particular black and white region to any other. These transformations are generated by the two simple transformations $\tau \mapsto \tau + 1$ and $\tau \mapsto -1/\tau$, so the latter transformations also define the periodicity of j.

Because of its periodicity under $\tau \mapsto \tau + 1$, j has a Fourier series, that is, an expansion in powers of $q = e^{2i\pi\tau}$. This expansion happens to be

$$j(\tau) = q^{-1} + 744 + 196884q + 21493760q^2 + \cdots$$

Neither the tessellation definition nor the expansion in powers of q is close to the original definition of j, which comes from the theory of elliptic functions, as we explain below. We began with the geometric definition because it is probably the simplest to grasp, though it hides some difficulties (among them the Riemann mapping theorem, which ensures the *existence* of a conformal map of any simply-connected region onto the half plane). The definition of j as a mapping also yields the most famous application of the modular function to analysis—Picard's proof that any entire function omits at most one complex value. We do not go further into Picard's theorem, because it can be found in most complex analysis books; Picard's beautiful proof is presented in [**1**, p. 307].

For a thorough treatment of j and its history, including most of the topics discussed in this article, McKean and Moll's book [**8**] is warmly recommended.

THE QUINTIC MIRACLE. *The general quintic equation can be solved by j.* This result was proved by Hermite in 1858 [**5**]. It was not completely out of the blue, because Galois had pointed out a quintic equation related to j in 1832, and Kronecker had similar ideas about the same time as Hermite. Nevertheless, it is a startling result, and it remains so even when its antecedents are pointed out.

Hermite compared his solution of the quintic by j to the solution of the cubic equation that takes advantage of the "angle-tripling" equation

$$4\cos^3\theta - 3\cos\theta = \cos 3\theta$$

satisfied by the cosine function. One transforms the general cubic equation into the special form

$$4x^3 - 3x = c,$$

and then sets $x = \cos\theta$, where $c = \cos 3\theta$.

There are analogous *modular equations* satisfied by j, and it turns out that the general quintic equation can be transformed to the quintic modular equation.

WHERE DO MODULAR EQUATIONS COME FROM? The function j is not the only function with the periodicity of the modular tessellation, but it is simplest in the sense that all other such functions are rational functions of j. The first of them to be encountered, and the origin of the name "modular," was *modulus k^2* in the elliptic integral

$$\int \frac{dt}{\sqrt{(1-t^2)(1-k^2t^2)}}.$$

A thought-provoking result about such integrals was Fagnano's 1718 formula for doubling the arc length of the lemniscate:

$$2\int_0^x \frac{dt}{\sqrt{1-t^4}} = \int_0^y \frac{dt}{\sqrt{1-t^4}}, \quad \text{where } y = \frac{2x\sqrt{1-x^4}}{1+x^4}$$

which gives a polynomial equation between x and y:

$$y^2(1+x^4)^2 = 4x^2(1-x^4).$$

This is analogous to the doubling formula for the arcsine integral,

$$2\int_0^x \frac{dt}{\sqrt{1-t^2}} = \int_0^y \frac{dt}{\sqrt{1-t^2}}, \quad \text{where } y = 2x\sqrt{1-x^2},$$

which in turn is just a restatement of the double angle formula

$$\sin 2\theta = 2\sin\theta\cos\theta = 2\sin\theta\sqrt{1-\sin^2\theta},$$

and the polynomial relation

$$y^2 = 4x^2(1-x^2)$$

between $y = \sin 2\theta$ and $x = \sin\theta$. This analogy with circular functions led to great interest in n-tupling (and later, multiplication by complex numbers, or "complex multiplication") of elliptic integrals and to computation of the corresponding polynomial equations. When the value of the integral is regarded as a function of the modulus, the equations obtained are called *modular equations*.

Modular equations were a popular topic with many leading mathematicians of the early 19th century—Legendre, Gauss, Abel, Jacobi, Galois—and the results of Galois were particularly tantalising. Galois left only some cryptic remarks about the equations for multiplication by 5, 7, and 11 (implying that they yield equations of degrees 5, 7, and 11) in the letter he wrote to Chevalier just before his death. It was several decades before these remarks were really understood, and Hermite's 1858 paper was both a step towards understanding Galois, and a step beyond him.

THE QUADRATIC MIRACLE. Kronecker (1857) discovered that *j detects the class number of* $\mathbb{Q}(\sqrt{-D})$ *for an imaginary quadratic integer* $\sqrt{-D}$ [7]. This result is to my mind even more startling than the solution of the quintic, because class numbers are a deeper topic, which mathematicians did not begin to grasp until the 1830s.

In 1832 Gauss studied the *Gaussian integers* $a + ib$, where $a, b \in \mathbb{Z}$ and $i = \sqrt{-1}$, and showed that they have unique prime factorisation or *class number 1*. (The terminology goes back to the older language of quadratic forms, where the equivalent fact in this case is that all forms $ax^2 + bxy + cy^2$ with $b^2 - 4ac = -4$ are in the same "class" as $x^2 + y^2$.) Soon afterwards, mathematicians noticed examples, such as the quadratic integers $a + b\sqrt{-5}$, where prime factorisation is *not* unique because the class number is > 1. (In this case the class number is 2, and the two classes of forms are represented by $x^2 + 5y^2$ and $2x^2 + 2xy + 3y^2$.) In 1839, Dirichlet introduced the powerful analytic method of Dirichlet series to determine the class number of the integers of a quadratic field $\mathbb{Q}(\sqrt{-D})$, but it was a complete surprise when Kronecker showed that j could do the same job.

He showed that, for any integer τ in the quadratic field $\mathbb{Q}(\sqrt{-D})$, $j(\tau)$ *is an algebraic integer whose degree is the class number of* $\mathbb{Q}(\sqrt{-D})$.

For example, the Gaussian integers are the integers of the field $\mathbb{Q}(i)$ with $D = 1$, and it turns out that $j(i) = 12^3$—an ordinary integer, as we expect because $\mathbb{Q}(i)$ has class number 1. A second example, which happens to be the largest D for which $\mathbb{Q}(\sqrt{-D})$ has class number 1, is where $D = 163$. Gauss also found this example, and the class number 1 is confirmed by the ordinary integer value

$$j((1 + \sqrt{-163})/2) = (-640320)^3.$$

Finally, an example with class number 2 is $\mathbb{Q}(\sqrt{-15})$, and indeed

$$j((1 + \sqrt{-15})/2) = (-191025 + 85995\sqrt{5})/2,$$

which is an integer of degree 2. (The existence of "integers" with denominator 2 is a quirk of certain quadratic fields that the reader may take on trust here.)

Kronecker's result is difficult to explain in a short article, but we can give the following hint. What the quadratic integers have in common with elliptic functions is an underlying *lattice L* in the plane $\mathbb{C}$—a set of points at the corners of a tessellation of the plane by identical parallelograms. An elliptic function f has two *periods* ω_1 and ω_2, which have different directions in $\mathbb{C}$ and hence generate the *lattice of periods* $\{m\omega_1 + n\omega_2 : m, n \in \mathbb{Z}\}$, at each point of which f takes the same value. In a quadratic field $\mathbb{Q}(\sqrt{-D})$ the set $\mathscr{O}$ of *integers* is a lattice, either

$$\{m + n\sqrt{-D} : m, n \in \mathbb{Z}\} \quad \text{or} \quad \left\{\frac{m}{2} + \frac{n\sqrt{-D}}{2} : m, n \in \mathbb{Z} \text{ with same parity}\right\},$$

and more generally so is any *ideal* of $\mathscr{O}$—a set of integers closed under addition and under multiplication by any $\alpha \in \mathscr{O}$. The algebraic significance of ideals is that $\mathscr{O}$ has unique prime factorisation if and only if every ideal of $\mathscr{O}$ is *principal*—that is, equal to $\alpha\mathscr{O}$ for some $\alpha \in \mathscr{O}$—and a principal ideal $\alpha\mathscr{O}$ is geometrically

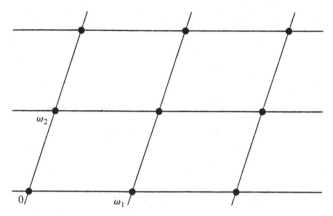

Figure 2. Parallelograms generated by ω_1 and ω_2.

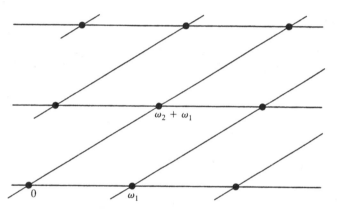

Figure 3. Parallelograms generated by ω_1 and $\omega_2 + \omega_1$.

significant because it has the *same shape* as $\mathscr{O}$ (being the result of magnifying $\mathscr{O}$ by $|\alpha|$ and rotating by arg α).

The modular function is pertinent to both elliptic functions and quadratic integers because *j is really a function of lattice shapes*. The idea of lattice shape may be illustrated by the lattice of periods $\{m\omega_1 + n\omega_2 : m, n \in \mathbb{Z}\}$. The points of this lattice occur at the corners of the tessellation of the plane by parallelograms shown in Figure 2.

The "shape" of a parallelogram is captured by the number $\omega = \omega_2/\omega_1$, because $|\omega|$ is the ratio $|\omega_2|/|\omega_1|$ of the side lengths and arg ω = arg ω_2 − arg ω_1 is the angle between the sides. However, this parallelogram is just one of infinitely many that define the same lattice. Another is shown in Figure 3.

The shape of the basic parallelogram is now

$$\frac{\omega_2 + \omega_1}{\omega_1} = \omega + 1.$$

so the lattice shape is represented equally well by $\omega + 1$. It turns out that, for each ω representing the shape of a lattice L, the number $(a\omega + b)/(c\omega + d)$ also represents the shape of L, provided $a, b, c, d \in \mathbb{Z}$ and $ad - bc = 1$. A *lattice shape* is therefore a whole *class* of numbers, of the form

$$\frac{a\omega + b}{c\omega + d} \quad \text{for some } \omega.$$

where $a, b, c, d \in \mathbb{Z}$ and $ad - bc = 1$.

This brings us to the reason for saying that j is a function of lattice shapes: as mentioned at the beginning, j has the property that

$$j(\omega) = j\left(\frac{a\omega + b}{c\omega + d}\right)$$

for any $a, b, c, d \in \mathbb{Z}$ with $ad - bc = 1$. Thus *j takes the same value at each number in a lattice shape*. In other words, j is *well defined* on lattice shapes.

Now the properties of an elliptic function are largely controlled by the shape of its period lattice, and the properties of a quadratic field $\mathbb{Q}(\sqrt{-D})$ are controlled by the shapes of its ideals. In particular, it turns out that *$\mathbb{Q}(\sqrt{-D})$ has unique prime factorisation if and only if all its ideals have the same shape*. This is why j can have something to say about unique prime factorisation in quadratic fields—it is an echo of what j says about the so-called "complex multiplication" of period lattices—though the way j says it is still pretty amazing.

The equation satisfied by $j(\sqrt{-D})$ happens to be another modular equation, which factorises into terms like $x - j(\sqrt{-D})$, and the number of factors is the number of different lattice shapes in the integers of $\mathbb{Q}(\sqrt{-D})$.

THE NUMERICAL MIRACLE. Hermite (1859) noticed a curious numerical consequence of Kronecker's theorem on the values of $j(\tau)$:

$$e^{\pi\sqrt{163}} = 262537412640768744,$$

(an integer!) correct to 12 decimal places [6].

This little known discovery of Hermite was exploited by Martin Gardner in an amusing hoax edition of his column in *Scientific American*. On 1 April 1975 Gardner announced—among several other "sensational discoveries that have somehow or another escaped public attention"—that

$$e^{\pi\sqrt{163}} = 262537412640768744 \text{ exactly}. \tag{1}$$

He gave the announcement an extra coat of varnish by claiming that it settled a conjecture of Ramanujan, supposedly made in a paper of 1914. The paper cited by Gardner does indeed discuss near integers of the form $e^{\pi\sqrt{n}}$, but without claiming that they could be integers, and without mentioning $e^{\pi\sqrt{163}}$. Still, in the pocket calculator days of 1975, it was pretty hard to decide whether $e^{\pi\sqrt{163}}$ is an integer or not.

Its true value. as Hermite and Ramanujan knew, is the integer in (1) *minus a very tiny number* ($< 10^{-12}$).

In fact, putting $\tau = (1 + \sqrt{-163})/2$ in $q = e^{2i\pi\tau}$ gives the tiny

$$q = e^{i\pi - \pi\sqrt{163}} = -e^{-\pi\sqrt{163}},$$

and putting this q in
$$j(\tau) = q^{-1} + 744 + 196884q + 21493760q^2 + \cdots$$
gives
$$j\bigl((1 + \sqrt{-163}\,/2\bigr) = -e^{\pi\sqrt{163}} + 744 - \text{tiny number},$$
and therefore
$$e^{\pi\sqrt{163}} = \text{integer} - \text{tiny number}.$$

THE ORIGIN OF MOONSHINE. "Moonshine" is a theory linking j to the monster simple group $\mathbb{M}$, and it has its origin an an apparent coincidence observed by McKay in 1977: *The coefficient* 196884 *in the Fourier expansion of j is* 1 *plus the dimension of the smallest nontrivial representation of* $\mathbb{M}$.

Actually, several other coincidences were discovered around the same time, and are listed in [2]. But nonetheless moonshine could hardly have been discovered without knowing the Fourier expansion of j, so one would like to know who discovered the coefficient 196884. Hermite 1859 actually has the *incorrect* expansion
$$j(\tau) = q^{-1} + 744 + 196880q + \cdots,$$
though the error does not affect his result that $e^{\pi\sqrt{163}}$ is an integer to 12 decimal places.

As far as I know, the first correct expansion as far as the coefficient 196884 was given by Weber in 1891 [**9**, p. 248]. Was this the first glimpse of moonshine? Or did Hermite also see 196884, but write it down incorrectly? I am inclined to vote for Hermite because his 1859 paper contains another series that later became part of moonshine [**2**, p. 334]:
$$q^{-1} + 104 + 4372q + 96256q^2 + \cdots$$
and in this series Hermite got all the digits right.

REFERENCES

1. L. V. Ahlfors, *Complex Analysis*, McGraw-Hill Kogakusha, Tokyo, 1979.
2. J. H. Conway and S. P. Norton, Monstrous moonshine, *Bull. London Math. Soc.* 11 (1979) 308–339.
3. R. Dedekind, Schreiben an Herrn Borchardt über die Theorie der elliptischen Modulfunktionen, *J. Reine Angew. Math.* 83 (1877) 265–292.
4. M. Gardner, Mathematical Games. *Scientific American* 232 (April 1975), p. 126.
5. C. Hermite, Sur la résolution de l'équation du cinquième degré, *C. R. Acad. Sci. Paris Sér. I Math.* XLVI (1858) 508. Also in *Oeuvres de Charles Hermite*, Paris, Gauthier-Villars, 1905–17, vol. 2, pp. 5–12.
6. _____, Sur la théorie des équations modulaires, *C. R. Acad. Sci. Paris Sér. I Math.* XLVIII (1859) 940. Also in *Oeuvres de Charles Hermite*, Paris, Gauthier-Villars, 1905–17, vol. 2, pp. 38–82.
7. L. Kronecker, Über die elliptischen Functionen für welche complexe Multiplication stattfinder, *Leopold Kroncker's Werke*, Chelsea Pub. Co., New York, 1968, vol. 4, pp. 179–183.
8. H. P. McKean and V. Moll, *Elliptic Curves*, Cambridge University Press, Cambridge, 1997.
9. Weber, *Elliptische Functionen und algebraischen Zahlen*, Vieweg, Braunschweig, 1891.

Topology and Abstract Algebra as Two Roads of Mathematical Comprehension*

Unterrichtsblätter für Mathematik und Naturwissenschaften 38, 177–188 (1932). (A lecture in the summer course of the Swiss Society of Gymnasium Teachers, given in Bern, in October 1931.)

Hermann Weyl

We are not very pleased when we are forced to accept a mathematical truth by virtue of a complicated chain of formal conclusions and computations, which we traverse blindly, link by link, feeling our way by touch. We want first an overview of the aim and of the road; we want to understand the *idea* of the proof, the deeper context. A modern mathematical proof is not very different from a modern machine, or a modern test setup: the simple fundamental principles are hidden and almost invisible under a mass of technical details. When discussing Riemann in his lectures on the history of mathematics in the 19th century, Felix Klein said:

> Undoubtedly, the capstone of every mathematical theory is a convincing proof of all of its assertions. Undoubtedly, mathematics inculpates itself when it foregoes convincing proofs. But the mystery of brilliant productivity will always be the posing of new questions, the anticipation of new theorems that make accessible valuable results and connections. Without the creation of new viewpoints, without the statement of new aims, mathematics would soon exhaust itself in the rigor of its logical proofs and begin to stagnate as its substance vanishes. Thus, in a sense, mathematics has been most advanced by those who distinguished themselves by intuition rather than by rigorous proofs.

The key element of Klein's own method was an intuitive perception of inner connections and relations whose foundations are scattered. To some extent, he failed when it came to a concentrated and pointed logical effort. In his commemorative address for Dirichlet, Minkowski contrasted the minimum principle that Germans tend to name for Dirichlet (and that was actually applied most comprehensively by William Thomson) with the true Dirichlet principle: to conquer problems with a minimum of blind computation and a maximum of insightful thoughts. It was Dirichlet, said Minkowski, who ushered in the new era in the history of mathematics.

What is the secret of such an understanding of mathematical matters, what does it consist in? Recently, there have been attempts in the philosophy of science to contrast understanding, the art of interpretation as the basis of the humanities, with scientific explanation, and the words intuition and understanding have been

---

*The original German version of this article is found in vol. 3, pp. 348–358, of the four-volume edition of Hermann Weyl's collected works published by Springer-Verlag in 1968. The translation is by Abe Shenitzer.

invested in this philosophy with a certain mystical halo, an intrinsic depth and immediacy. In mathematics, we prefer to look at things somewhat more soberly. I cannot enter into these matters here, and it strikes me as very difficult to give a precise analysis of the relevant mental acts. But at least I can single out, from the many characteristics of the process of understanding, one that is of decisive importance. One separates in a natural way the different aspects of a subject of mathematical investigation, makes each accessible through its own relatively narrow and easily surveyable group of assumptions, and returns to the complex whole by combining the appropriately specialized partial results. This last synthetic step is purely mechanical. The great art is in the first, analytic, step of appropriate separation and generalization. The mathematics of the last few decades has revelled in generalizations and formalizations. But to think that mathematics pursues generality for the sake of generality is to misunderstand the sound truth that a natural generalization *simplifies* by reducing the number of assumptions and by thus letting us understand certain aspects of a disarranged whole. Of course, it can happen that different directions of generalization enable us to understand different aspects of a particular concrete issue. Then it is subjective and dogmatic arbitrariness to speak of the true ground, the true source of an issue. Perhaps the only criterion of the naturalness of a severance and an associated generalization is their fruitfulness. If this process is systematized according to subject matter by a researcher with a measure of skill and "sensitive fingertips" who relies on all the analogies derived from his experience, then we arrive at axiomatics, which today is an instrument of concrete mathematical investigation rather than a method for the clarification and "deep-laying" of foundations.

In recent years mathematicians have had to focus on the general and on formalization to such an extent that, predictably, there have turned up many instances of cheap and easy generalizing for its own sake. Pólya has called it generalizing by dilution. It does not increase the essential mathematical substance. It is much like stretching a meal by thinning the soup. It is deterioration rather than improvement. The aged Klein said: "Mathematics looks to me like a store that sells weapons in peacetime. Its windows are replete with luxury items whose ingenious, artful and eyecatching execution delights the connoisseur. The true origin and purpose of these objects—the strike that defeats the enemy—have receded into the background and have been all but forgotten." There is perhaps more than a grain of truth in this indictment, but, on the whole, our generation regards this evaluation of its efforts as unjust.

There are two modes of understanding that have proved, in our time, to be especially penetrating and fruitful. The two are topology and abstract algebra. A large part of mathematics bears the imprint of these two modes of thought. What this is attributable to can be made plausible at the outset by considering the central concept of real number. The system of real numbers is like a Janus head with two oppositely directed faces. In one respect it is the domain of the operations + and × and their inverses, in another it is a continuous manifold, and the two are continuously related. One is the algebraic and the other is the topological face of numbers. Since modern axiomatics is simpleminded and (unlike modern politics) dislikes such ambiguous mixtures of peace and war, it made a clean break between the two. The notion of size of number, expressed in the relations $<$ and $>$, occupies a kind of intermediate relation between algebra and topology.

Investigations of continua are purely topological if they are restricted to just those properties and differences that are unchanged by arbitrary continuous deformations, by arbitrary continuous mappings. The mappings in question need only be faithful to the extent to which they don't collapse what is distinct. Thus it is a topological property of a surface to be closed like the surface of a sphere or open like the ordinary plane. A piece of the plane is said to be simply connected if, like the interior of a circle, it is partitioned by every crosscut. On the other hand, an annulus is doubly connected because there exists a crosscut that does not partition it but every subsequent crosscut does. Every closed curve on the surface of a sphere can be shrunk to a point by means of a continuous deformation, but this is not the case for a torus. Two closed curves in space can be intertwined or not. These are examples of topological properties or dispositions. They involve the primitive differences that underlie all finer differentiations of geometric figures. They are based on the single idea of continuous connection. References to a particular structure of a continuous manifold, such as a metric, are foreign to them. Other relevant concepts are limit, convergence of a sequence of points to a point, neighborhood and continuous line.

After this preliminary sketch of topology I want to tell you briefly about the motives that have led to the development of abstract algebra. Then I will use a simple example to show how the same issue can be looked at from a topological and from an abstract-algebraic viewpoint.

All a pure algebraist can do with numbers is apply to them the four operations of addition, subtraction, multiplication and division. If a system of numbers is a field, that is, if it is closed under these operations, then the algebraist has no means of going beyond it. The simplest field is the field of rationals. Another example is the field of numbers of the form $a + b\sqrt{2}$, a, b rational. The well-known concept of irreducibility of polynomials is relative and depends on the field of coefficients of the polynomials, namely a polynomial $f(x)$ with coefficients in a field K is said to be irreducible over K if it cannot be written as a product $f_1(x) \cdot f_2(x)$ of two non-constant polynomials with coefficients in K. The solution of linear equations and the determination of the greatest common divisor of two polynomials by means of the Euclidean algorithm are carried out within the field of the coefficients of the equations and of the polynomials respectively. The classical problem of algebra is the solution of an algebraic equation $f(x) = 0$ with coefficients in a field K, say the field of rationals. If we know a root ϑ of the equation, then we know the numbers obtained by applying to ϑ and to the (presumably known) numbers in K the four algebraic operations. The resulting numbers form a field $K(\vartheta)$ that contains K. In $K(\vartheta)$, ϑ plays a role of a determining number from which all other numbers in $K(\vartheta)$ are rationally derivable. But many—virtually all—numbers in $K(\vartheta)$ can play the same role as ϑ. It is therefore a breakthrough if we replace the study of the equation $f(x) = 0$ by the study of the field $K(\vartheta)$. By doing this we eliminate all manner of trivia and consider at the same time all equations that can be obtained from $f(x) = 0$ by means of Tschirnhausen transformations. The algebraic, and above all the arithmetical, theory of number fields is one of the sublime creations of mathematics.

From the viewpoint of the richness and depth of its results it is the most perfect such creation.

There are domains in algebra whose elements are not numbers. The polynomials in one variable, or indeterminate, x, [with coefficients in a field], are closed under addition, subtraction and multiplication but not under division. Such a system of magnitudes is called an integral domain. The idea that the argument x is a variable that traverses continuously its values is foreign to algebra; it is just an indeterminate, an empty symbol that binds the coefficients of the polynomial into a uniform expression that makes it easier to remember the rules for addition and multiplication. 0 is the polynomial all of whose coefficients are 0 (not the polynomial which takes on the value 0 for all values of the variable x). It can be shown that the product of two nonzero polynomials is $\neq 0$. The algebraic viewpoint does not rule out the substitution for x of a number a taken from the field in which we operate. But we can also substitute for x a polynomial in one or more indeterminates $y, z \ldots$. Such substitution is a formal process which effects a faithful projection of the integral domain $K[x]$ of polynomials in x onto K or onto the integral domain of polynomials $K[y, z, \ldots]$; here "faithful" means subject to the preservation of the relations established by addition and multiplication. It is this formal operating with polynomials that we are required to teach students studying algebra in school. If we form quotients of polynomials, then we obtain a field of rational functions which must be treated in the same formal manner. This, then, is a field whose elements are functions rather than numbers. Similarly, the polynomials and rational functions in two or three variables, x, y or x, y, z with coefficients in K form an integral domain and field respectively.

Compare the following three integral domains: the integers, the polynomials in x with rational coefficients, and the polynomials in x and y with rational coefficients. The Euclidean algorithm holds in the first two of these domains, and so we have the theorem: If a, b are two relatively prime elements, then there are elements p, q in the appropriate domain such that

$$(*) \qquad 1 = p \cdot a + q \cdot b.$$

This implies that the two domains in question are unique factorization domains. The theorem $(*)$ fails for polynomials in two variables. For example, $x - y$ and $x + y$ are relatively prime polynomials such that for every choice of polynomials $p(x, y)$ and $q(x, y)$ the constant term of the polynomial $p(x, y)(x - y) + q(x, y)(x + y)$ is 0 rather than 1. Nevertheless polynomials in two variables with coefficients in a field form a unique factorization domain. This example points to interesting similarities and differences.

There is yet another way of making fields in algebra. It involves neither numbers nor functions but congruences. Let p be a prime integer. Identify two integers if their difference is divisible by p, or, briefly, if they are congruent mod p. (To "see" what this means wrap the real line around a circle of circumference p.) The result is a field with p elements. This representation is extremely useful in all of number theory. Consider, for example, the following theorem of Gauss that has numerous applications: If $f(x)$ and $g(x)$ are two polynomials with integer coefficients such that all coefficients of the product $f(x) \cdot g(x)$ are divisible by a prime p, then all coefficients of $f(x)$ or all coefficients of $g(x)$ are divisible by p. This is just the trivial theorem that the product of two polynomials can be 0 only if one of its factors is 0, applied to the field just described as the field of coefficients. This integral domain contains polynomials that are not 0 but vanish for all values of the argument; one such polynomial is $x^p - x$. In fact, by Fermat's theorem, we have

$$a^p - a \equiv 0 \pmod{p}.$$

Cauchy uses a similar approach to construct the complex numbers. He regards the imaginary unit i as an indeterminate and studies polynomials in i over the reals modulo $i^2 + 1$, that is he regards two polynomials as equal if their difference is divisible by $i^2 + 1$. In this way, the actually unsolvable equation $i^2 + 1 = 0$ is rendered, in some measure, solvable. Note that the polynomial $i^2 + 1$ is prime over the reals. Kronecker generalized Cauchy's construction as follows. Let K be a field and $p(x)$ a polynomial prime over K. Viewed modulo $p(x)$, the polynomials $f(x)$ with coefficients in K form a field (and not just an integral domain). From an algebraic viewpoint, this process is fully equivalent to the one described previously, and can be thought of as the process of extending K to $K(\vartheta)$ by adjoining to K a root of the equation $p(\vartheta) = 0$. But it has the advantage that it takes place within pure algebra and gets around the demand for solving an equation that is actually unsolvable over K.

It is quite natural that these developments should have prompted a purely axiomatic buildup of algebra. A field is a system of objects, called numbers, closed under two operations, called addition and multiplication, that satisfy the usual axioms: both operations are associative and commutative, multiplication is distributive over addition, and both operations are uniquely invertible yielding subtraction and division respectively. If the axiom of invertibility of multiplication is left out, then the resulting system is called a ring. Now "field" no longer denotes, as before, a kind of sector of the continuum of real or complex numbers but a self-contained universe. One can apply the field operations to elements of the same field but not to elements of different fields. In this process we need not resort to artificial abstracting from the size relations $<$ and $>$. These relations are irrelevant for algebra and the "numbers" of an abstract "number field" are not subject to such relations. In place of the uniform number continuum of analysis we now have the infinite multiplicity of structurally different fields. The previously described processes, namely adjunction of an indeterminate and identification of elements that are congruent with respect to a fixed prime element, are now seen as two modes of construction that lead from rings and fields to other rings and fields respectively.

The elementary axiomatic grounding of geometry also leads to this abstract number concept. Take the case of plane projective geometry. The incidence axioms alone lead to a "number field" that is naturally associated with it. Its elements, the "numbers," are purely geometric entities, namely dilations. A point and a straight line are ratios of triples of "numbers" in that field, $x_1 : x_2 : x_3$ and $u_1 : u_2 : u_3$ respectively, such that incidence of the point $x_1 : x_2 : x_3$ on the line $u_1 : u_2 : u_3$ is represented by the equation

$$x_1 u_1 + x_2 u_2 + x_3 u_3 = 0.$$

Conversely, if one uses these algebraic expressions to define the geometric terms, then every abstract field leads to an associated projective plane that satisfies the incidence axioms. It follows that a restriction involving the number field associated with the projective plane cannot be read off from the incidence axioms. Here the preexisting harmony between geometry and algebra comes to light in the most impressive manner. For the geometric number system to coincide with the continuum of ordinary real numbers one must introduce axioms of order and continuity, very different in kind from the incidence axioms. We thus arrive at a reversal of the development that has dominated mathematics for centuries and seems to have arisen originally in India and to have been transmitted to the West by Arab scholars: Up till now, we have regarded the number concept as the logical antecedent of geometry, and have therefore approached every realm of magnitudes with a universal and systematically developed number concept independent of the applications involved. Now, however, we revert to the Greek viewpoint that every subject has an associated intrinsic number realm that must be derived from within it. We experience this reversal not only in geometry but also in the new quantum physics. According to quantum physics, the physical magnitudes associated with a particular physical setup (*not* the numerical values that they may take on depending on its different states) admit of an addition and a non-commutative multiplication, and thus give rise to a system of algebraic magnitudes intrinsic to it that cannot be viewed as a sector of the system of real numbers.

And now, as promised, I will present a simple example that illustrates the mutual relation between the topological and abstract-algebraic modes of analysis. I consider the theory of algebraic functions of a single variable x. Let $K(x)$ be the field of rational functions of x with arbitrary complex coefficients. Let $f(z)$, more precisely $f(z; x)$, be an n-th degree polynomial in z with coefficients in $K(x)$. We explained earlier when such a polynomial is said to be irreducible over $K(x)$. This is a purely algebraic concept. Now construct the Riemann surface of the n-valued algebraic function $z(x)$ determined by the equation $f(z; x) = 0$. Its n sheets extend over the x-plane. For easier transformation of the x-plane into the x-sphere by means of a stereographic projection we add to the x-plane a point at infinity. Like the sphere, our Riemann surface is now closed. The irreducibility of the polynomial f is reflected in a very simple topological property of the Riemann surface of $z(x)$, namely its connectedness: if we shake a paper model of that surface it does not break into distinct pieces. Here you witness the coincidence of a purely algebraic and a purely topological concept. Each suggests generalization in a different direction. The algebraic concept of irreducibility depends only on the fact that the coefficients of the polynomial are in a field. In particular, $K(x)$ can be replaced by the field of rational functions of x with coefficients in a preassigned field k which takes the place of the continuum of all complex numbers. On the other hand, from the viewpoint of topology it is irrelevant that the surface in question is a Riemann surface, that it is equipped with a conformal structure, and that it consists of a finite number of sheets that extend over the x-plane. Each of the two antagonists can accuse the other of admitting side issues and of neglecting essential features. Who is right? Questions such as these, involving not facts but ways of looking at facts, can lead to hatred and bloodshed when they touch human emotions. In mathematics, the consequences are not so serious. Nevertheless, the contrast between Riemann's topological theory of algebraic functions and Weierstrass' more algebraically directed school led to a split in the ranks of mathematicians that lasted for almost a generation.

Weierstrass himself wrote to his faithful pupil H. A. Schwarz: "The more I reflect on the principles of function theory—and I do this all the time—the stronger is my conviction that this theory must be established on the foundation of algebraic truths, and that it is therefore not the right way when, contrariwise, the 'transcendent' (to put it briefly) is invoked to establish simple and fundamental algebraic theorems—this is so no matter how attractive are, at a first glance, say, the considerations by means of which Riemann discovered so many of the most important properties of algebraic functions." This strikes us now as onesided; neither one of the two ways of understanding, the topological or the algebraic, can be acknowledged to have unconditional advantage over the other. And we cannot spare Weierstrass the reproach that he stopped midway. True, he explicitly constructed the functions as algebraic, but he also used as coefficients the algebraically unanalyzed, and in a sense unfathomable for algebraists, continuum of complex numbers. The dominant general theory in the direction followed by Weierstrass is the theory of an abstract number field and its extensions determined by means of algebraic equations. Then the theory of algebraic functions moves in the direction of a shared axiomatic basis with the theory of algebraic numbers. In fact, what suggested to Hilbert his approaches in the theory of number fields was the analogy [between the latter] and the state of things in the realm of algebraic functions discovered by Riemann by his topological methods. (Of course, when it came to proofs, the analogy was useless.)

Our example "irreducible-connected" is typical also in another respect. How visually simple and understandable is the topological criterion (shake the paper model and see if it falls apart) in comparison with the algebraic! The visual primality of the continuum (I think that in this respect it is superior to the 1 and the natural numbers) makes the topological method particularly suitable for both discovery and synopsis in mathematical areas, but is also the cause of difficulties when it comes to rigorous proofs. While it is close to the visual, it is also refractory to logical approaches. That is why Weierstrass, M. Noether and others preferred the laborious, but more solid-feeling, procedure of direct algebraic construction to Riemann's transcendental-topological justification. Now, step by step, abstract algebra tidies up the clumsy computational apparatus. The generality of the assumptions and axiomatization force one to abandon the path of blind computation and to break the complex state of affairs into simple parts that can be handled by means of simple reasoning. Thus algebra turns out to be the El Dorado of axiomatics.

I must add a few words about the method of topology to prevent the picture from becoming altogether vague. If a continuum, say, a two-dimensional closed manifold, a surface, is to be the subject of mathematical investigation, then we must think of it as being subdivided into finitely many "elementary pieces" whose topological nature is that of a circular disk. These pieces are further fragmented by repeated subdivision in accordance with a fixed scheme, and thus a particular spot in the continuum is ever more precisely intercepted by an infinite sequence of nested fragments that arise in the course of successive subdivisions. In the one-dimensional case, the repeated "normal subdivision" of an elementary segment is its bipartition. In the two-dimensional case, each edge is first bipartitioned, then each piece of surface is divided into triangles by means of lines in the surface that lead from an arbitrary center to the (old and new) vertices. What proves that a piece is elementary is that it can be broken into arbitrarily small pieces by repetition of this division process. The scheme of the initial subdivision into elementary pieces—to be referred to briefly in what follows as the "skeleton"—is best described by labelling the surface pieces, edges and vertices by means of symbols, and thus prescribing the mutual bounding relations of these elements. Following the successive subdivisions, the manifold may be said to be spanned by an increasingly dense net of coordinates which makes it possible to determine a particular point by means of an infinite sequence of symbols that play a role comparable to that of numbers. The reals appear here in the particular form of dyadic fractions, and serve to describe the subdivision of an open one-dimensional continuum. Other than that, we can say that each continuum has its own arithmetical scheme; the introduction of numerical coordinates by reference to the special division scheme of an open one-dimensional continuum violates the nature of things, and its sole justification is the practical one of the extraordinary convenience of the calculational manipulation of the continuum of numbers with its four operations. In the case of an actual continuum, the subdivisions can be realized only with a measure of imprecision; one must imagine that, as the process of subdivision progresses step by step, the boundaries set by the earlier subdivisions are ever more sharply fixed. Also, in the case of an actual continuum, the process of subdivision that runs virtually ad infinitum can reach only a certain definite stage. But in distinction to concrete realization, the localization in an actual continuum, the combinatorial scheme, the arithmetical nullform, is a priori determined ad infinitum; and mathematics deals with this combinatorial scheme alone. Since the continued subdivision of the initial topological skeleton progresses in accordance with a fixed scheme, it must be possible to read off all the topological properties of the nascent manifold from that skeleton. This means that, in principle, it must be possible to pursue topology as finite combinatorics. For topology, the ultimate elements, the atoms, are, in a sense, the elementary parts of the skeleton and not the points of the relevant continuous manifold. In particular, given two such skeletons, it must be possible to decide if they lead to concurrent manifolds. Put differently, it must be possible to decide if we can view them as subdivisions of one and the same manifold.

The algebraic counterpart of the transition from the algebraic equation $f(z; x) = 0$ to the Riemann surface is the transition from the latter equation to the field determined by the function $z(x)$; this is so because the Riemann surface is uniquely occupied not only by the function $z(x)$ but also by all algebraic functions in this field. What is characteristic for Riemann's function theory is the converse problem: given a Riemann surface construct its field of algebraic functions. The problem has always just one solution. Since every point p of the Riemann surface lies over a definite point of the x-plane, the Riemann surface, as presently constituted, is embedded in the x-plane. The next step is to abstract from the embedding relation $p \to x$. As a result, the Riemann surface becomes, so to say, a free-floating surface equipped with a conformal structure and an angle measure. Note that in ordinary surface theory we must learn to distinguish between the surface as a continuous structure made up of elements of a specific kind, its points, and the embedding in 3-space that associates with each point p of the surface, in a continuous manner, the point P in space at which p is located. In the case of a

Riemann surface, the only difference is that the Riemann surface and the embedding plane have the same dimension. To abstraction from the embedding there corresponds, on the algebraic side, the viewpoint of invariance under arbitrary birational transformations. To enter the realm of topology we must ignore the conformal structure associated with the free-floating Riemann surface. Continuing the comparison, we can say that the conformal structure of the Riemann surface is the equivalent of the metric structure of an ordinary surface, controlled by the first fundamental form, or of the affine and projective structures associated with surfaces in affine and projective differential geometry respectively. In the continuum of real numbers, it is the algebraic operations of + and · that reflect its structural aspect, and in a continuous group the law that associates with an ordered pair of elements their product plays an analogous role. These comments may have increased our appreciation of the relation of the methods. It is a question of rank, of what is viewed as primary. In topology we begin with the notion of continuous connection, and in the course of specialization we add, step by step, relevant structural features. In algebra this order is, in a sense, reversed. Algebra views the operations as the beginning of all mathematical thinking and admits continuity, or some algebraic surrogate of continuity, at the last step of specialization. The two methods follow opposite directions. Little wonder that they don't get on well together. What is most easily accessible to one is often most deeply hidden to the other. In the last few years, in the theory of representation of continuous groups by means of linear substitutions, I have experienced most poignantly how difficult it is to serve these two masters at the same time. Such classical theories as that of algebraic functions can be made to fit both viewpoints. But viewed from these two viewpoints they present completely different sights.

After all these general remarks I want to use two simple examples that illustrate the different kinds of concept building in algebra and in topology. The classical example of the fruitfulness of the topological method is Riemann's theory of algebraic functions and their integrals. Viewed as a topological surface, a Riemann surface has just one characteristic, namely its connectivity number or genus p. For the sphere $p = 0$ and for the torus $p = 1$. How sensible it is to place topology ahead of function theory follows from the decisive role of the topological number p in function theory on a Riemann surface. I quote a few dazzling theorems: The number of linearly independent everywhere regular differentials on the surface is p. The total order (that is, the difference between the number of zeros and the number of poles) of a differential on the surface is $2p - 2$. If we prescribe more than p arbitrary points on the surface, then there exists just one single-valued function on it that may have simple poles at these points but is otherwise regular; if the number of prescribed poles is exactly p, then, if the points are in general position, this is no longer true. The precise answer to this question is given by the Riemann-Roch theorem in which the Riemann surface enters only through the number p. If we consider all functions on the surface that are everywhere regular except for a single place $\wp$ at which they have a pole, then its possible orders are all numbers $1, 2, 3, \ldots$ except for certain powers of p (the Weierstrass gap theorem). It is easy to give many more such examples. The genus p permeates the whole theory of functions on a Riemann surface. We encounter it at every step, and its role is direct, without complicated computations, understandable from its topological meaning (provided that we include, once and for all, the Thomson-Dirichlet principle as a fundamental function-theoretic principle).

The Cauchy integral theorem gives topology the first opportunity to enter function theory. The integral of an analytic function over a closed path is 0 only if the domain that contains the path and is also the domain of definition of the analytic function is simply connected. Let me use this example to show how one "topologizes" a function-theoretic state of affairs. If $f(z)$ is analytic, then the integral $\int_\gamma f(z)\,dz$ associates with every curve a number $F(\gamma)$ such that

(†) $$F(\gamma_1 + \gamma_2) = F(\gamma_1) + F(\gamma_2).$$

$\gamma_1 + \gamma_2$ stands for the curve such that the beginning of γ_2 coincides with the end of γ_1. The functional equation (†) marks the integral $F(\gamma)$ as an additive path function. Also, each point has a neighborhood such that $F(\gamma) = 0$ for each closed path γ in that neighborhood. I will call a path function with these properties a topological integral, or briefly, an integral. In fact, all this concept assumes is that there is given a continuous manifold on which one can draw curves; it is the topological essence of the analytic notion of an integral. Integrals can be added and multiplied by numbers. The topological part of the Cauchy integral theorem states that on a simply connected manifold every integral is homologous to 0 (not only in the small but in the large), that is, $F(\gamma) = 0$ for every closed curve γ on the manifold. In this we can spot the definition of "simply connected." The function-theoretic part states that the integral of an analytic function is a topological integral in our sense of the term. The definition of the order of connectivity [that we are about to state] fits in here quite readily. Integrals $F_1, F_2, \ldots, F_n$ on a closed surface are said to be linearly independent if they are not connected by a homology relation

$$c_1 F_1 + c_2 F_2 + \cdots + c_n F_n \sim 0$$

with constant coefficients c_i other than the trivial one, when all the c_i vanish. The order of connectivity of a surface is the maximal number of linearly independent integrals. For a closed two-sided surface the order of connectivity h is always an even number $2p$, where p is the genus. From a homology between integrals we can go over to a homology between closed paths. The path homology

$$n_1 \gamma_1 + n_2 \gamma_2 + \cdots + n_r \gamma_r \sim 0$$

states that for every integral F we have the equality

$$n_1 F(\gamma_1) + n_2 F(\gamma_2) + \cdots + n_r F(\gamma_r) = 0.$$

If we go back to the topological skeleton that decomposes the surface into elementary pieces and replace the continuous point-chains of paths by the discrete chains constructed out of elementary pieces, then we obtain an expression for the order of connectivity h in terms of the numbers s, k and e of pieces, edges and vertices. The expression in question is the well-known Euler polyhedral formula $h = k - (e + s) + 2$. Conversely, if we start with the topological skeleton, then our reasoning yields the result that this combination h of the number of pieces, edges and vertices is a topological invariant, namely it has the same value for "equivalent" skeletons which represent the same manifold in different subdivisions.

When it comes to application to function theory, it is possible, using the Thomson-Dirichlet principle, to "realize" the topological integrals as actual integrals of everywhere regular-analytic differentials on a Riemann surface. One can say that all of the constructive work is done on the topological side, and that the topological results are realized in a function-theoretic manner with the help of a universal transfer principle, namely the Dirichlet principle. This is, in a sense, analogous to analytic geometry, where all the constructive work is carried out in the realm of numbers, and then the results are geometrically "realized" with the help of the transfer principle lodged in the coordinate concept.

All this is seen more perfectly in uniformization theory, which plays a central role in all of function theory. But at this point, I prefer to point to another application which is probably close to many of you. I have in mind enumerative geometry, which deals with the determination of the number of points of intersection, singularities, and so on, of algebraic relational structures, which was made into a general but very poorly justified, system by Schubert and Zeuthen. Here, in the hands of Lefschetz and v.d. Waerden, topology achieved a decisive success in that it led to definitions of multiplicity valid without exception, as well as to laws likewise valid without exception. Of two curves on a two-sided surface one can cross the other at a point of intersection from left to right or from right to left. These points of intersection must enter every setup with opposite weights $+1$ and -1. Then the total of the weights of the intersections (which can be positive or negative) is invariant under arbitrary continuous deformations of the curves; in fact, it remains unchanged if the curves are replaced by homologous curves. Hence it is possible to master this number through finite combinatorial means of topology and obtain transparent general formulas. Two algebraic curves are, actually, two closed Riemann surfaces embedded in a space of four real dimensions by means of an analytic mapping. But in algebraic geometry a point of intersection is counted with positive multiplicity, whereas in topology one takes into consideration the sense of the crossing. This being so, it is surprising that one can resolve the algebraic question by topological means. The explanation is that in the case of an analytic manifold, crossing always takes place with the same sense. If the two curves are represented in the x_1, x_2-plane in the vicinity of their point of intersection by the functions $x_1 = x_1(s), x_2 = x_2(s)$, and $x_1 = x_1^*(t), x_2 = x_2^*(t)$, then the sense ± 1 with which the first curve intersects the second is given by the sign of the Jacobian

$$\begin{vmatrix} \dfrac{\partial x_1}{\partial s} & \dfrac{\partial x_2}{\partial s} \\ \dfrac{\partial x_1^*}{\partial t} & \dfrac{\partial x_2^*}{\partial t} \end{vmatrix} = \dfrac{\partial(x_1, x_2)}{\partial(x, t)},$$

evaluated at the point of intersection. In the case of complex-algebraic "curves" this criterion always yields the value $+1$. Indeed, let z_1, z_2 be complex coordinates in the plane and let s and t be the respective complex parameters on the two "curves." The real and imaginary parts of z_1 and z_2 play the role of real coordinates in the plane. In their place we can take $z_1, \bar{z}_1, z_2, \bar{z}_2$. But then the determinant whose sign determines the sense of the crossing is

$$\dfrac{\partial(z_1, \bar{z}_1, z_2, \bar{z}_2)}{\partial(s, \bar{s}, t, \bar{t})} = \dfrac{\partial(z_1, z_2)}{\partial(s, t)} \cdot \dfrac{\partial(\bar{z}_1, \bar{z}_2)}{\partial(\bar{s}, \bar{t})} = \left| \dfrac{\partial(z_1, z_2)}{\partial(s, t)} \right|^2,$$

and thus invariably positive. Note that the Hurwitz theory of correspondence between algebraic curves can likewise be reduced to a purely topological core.

On the side of abstract algebra, I will emphasize just one fundamental concept, namely the concept of an ideal. If we use the algebraic method, then an algebraic manifold is given in 3-dimensional space with complex cartesian coordinates x, y, z by means of a number of simultaneous equations

$$f_1(x, y, z) = 0, \ldots, f_n(x, y, z) = 0.$$

The f_i are polynomials. In the case of a curve it is not at all true that two equations suffice. Not only do the polynomials f_i vanish at points of the manifold but also every polynomial f of the form

$$(**) \qquad f = A_1 f_1 + \cdots + A_n f_n \ (A_i \text{ are polynomials}).$$

Such polynomials f form an "ideal" in the ring of polynomials. Dedekind defined an ideal in a given ring as a system of ring elements closed under addition and subtraction as well as under multiplication by ring elements. This concept is not too broad for our purposes. The reason is that, according to the Hilbert basis theorem, every ideal in the polynomial ring has a finite basis; there are finitely many polynomials $f_1, \ldots, f_n$ in the ideal such that every polynomial in the ideal can be written in the form $(**)$. Hence the study of algebraic manifolds reduces to the study of ideals. On an algebraic surface there are points and algebraic curves. The latter are represented by ideals that are divisors of the ideal under consideration. The fundamental theorem of M. Noether deals with ideals whose manifold of zeros consists of finitely many points, and makes membership of a polynomial in such an ideal dependent on its behavior at these points. This theorem follows readily from the decomposition of an ideal into prime ideals. The investigations of E. Noether show that the concept of an ideal, first introduced by Dedekind in the theory of algebraic number fields, runs through all of algebra and arithmetic like Ariadne's thread. v.d. Waerden was able to justify the enumerative calculus by means of the algebraic resources of ideal theory.

If one operates in an arbitrary abstract number field rather than in the continuum of complex numbers, then the fundamental theorem of algebra, which asserts that every complex polynomial in one variable can be [uniquely] decomposed into linear factors, need not hold. Hence the general prescription in algebraic work: See if a proof makes use of the fundamental theorem or not. In every algebraic theory there is a more elementary part that is independent of the fundamental theorem, and therefore valid in every field, and a more advanced part for which the fundamental theorem is indispensable. The latter part calls for the algebraic closure of the field. In most cases the fundamental theorem marks a crucial split; its use should be avoided as long as possible. To establish theorems that hold in an arbitrary field it is often useful to embed the given field in a larger field. In particular, it is possible to embed any field in an algebraically closed field. A well-known example is the proof of the fact that a real polynomial can be decomposed over the reals into linear and quadratic factors. To prove this, we adjoin i to the reals and thus embed the latter in the algebraically closed field of complex numbers. This procedure has an analogue in topology which is used in the study and characterization of manifolds; in the case of a surface, this analogue consists in the use of its covering surfaces.

At the center of today's interest is noncommutative algebra in which one does not insist on the commutativity of multiplication. Its rise is dictated by concrete needs of mathematics. Composition of operations is a kind of noncommutative operation. Here is a specific example. We consider the symmetry properties of functions $f(x_1, x_2, \ldots, x_n)$ of a number of arguments. The latter can be subjected to an arbitrary permutation s. A symmetry property is expressed in one or more equations of the form

$$\sum_s a(s) \cdot sf = 0.$$

Here $a(s)$ stands for the numerical coefficients associated with the permutation. These coefficients belong to a given field K. $\sum_s a(s) \cdot s$ is a "symmetry operator." These operators can be multiplied by numbers, added and multiplied, that is, applied in succession. The result of the latter operation depends on the order of the "factors." Since all formal rules of computation hold for addition and multiplication of symmetry operators, they form a "noncommutative ring" (hypercomplex number system). The dominant role of the concept of an ideal persists in the noncommutative realm. In recent years, the study of groups and their representations by linear substitutions has been almost completely absorbed by the theory of noncommutative rings. Our example shows how the multiplicative group of $n!$ permutations s is extended to the associated ring of magnitudes $\sum_s a(s) \cdot s$ that admit, in addition to multiplication, addition and multiplication by numbers. Quantum physics has given noncommutative algebra a powerful boost.

Unfortunately, I cannot here produce an example of the art of building an abstract-algebraic theory. It consists in setting up the right general concepts, such as fields, ideals, and so on, in decomposing an assertion to be proved into steps (for example, an assertion "A implies B," or $A \to B$, may be decomposed into steps $A \to C, C \to D, D \to B$), and in the appropriate generalization of these partial assertions in terms of general concepts. Once the main assertion has been subdivided in this way and the inessential elements have been set aside, the proofs of the individual steps do not, as a rule, present serious difficulties.

Whenever applicable, the topological method appears, thus far, to be more effective than the algebraic one. Abstract algebra has not yet produced successes comparable to the successes of the topological method in the hands of Riemann. Nor has anyone reached by an algebraic route the peak of uniformization scaled topologically by Klein, Poincaré and Koebe. Here are questions to be answered in the future. But I do not want to conceal from you the growing feeling among mathematicians that the fruitfulness of the abstracting method is close to exhaustion. It is a fact that beautiful general concepts do not drop out of the sky. The truth is that, to begin with, there are definite concrete problems, with all their undivided complexity, and these must be conquered by individuals relying on brute force. Only then come the axiomatizers and conclude that instead of straining to break in the door and bloodying one's hands one should have first constructed a magic key of such an such shape and then the door would have opened quietly, as if by itself. But they can construct the key only because the successful breakthrough enables them to study the lock front and back, from the outside and from the inside. Before we can generalize, formalize and axiomatize there must be mathematical substance. I think that the mathematical substance on which we have practiced formalization in the last few decades is near exhaustion and I predict that the next generation will face in mathematics a tough time.

[The sole purpose of this lecture was to give the audience a feeling for the intellectual atmosphere in which a substantial part of modern mathematical research is carried out. For those who wish to penetrate more deeply I give a few bibliographical suggestions. The true pioneers of abstract axiomatic algebra are Dedekind and Kronecker. In our own time, this orientation has been decisively advanced by Steinitz, by E. Noether and her school, and by E. Artin. The first great advance in topology came in the middle of the 19th century and was due to Riemann's function theory. The more recent developments are linked primarily to a few works of H. Poincaré devoted to analysis situs (1895–1904). I mention the following books:

1. *On algebra*: Steinitz, *Algebraic Theory of Fields*, appeared first in Crelle's *Journal* in 1910. It was issued as a paperback by R. Baer and H. Hasse and published by Verlag W. de Gruyter, 1930.
   H. Hasse, *Higher algebra I, II*. Sammlung Göschen 1926/27.
   B. v.d. Waerden, *Moderne algebra I, II*. Springer 1930/31.
2. *On topology*: H. Weyl, *The Idea of a Riemann Surface*, second ed. Teubner 1923.
   O. Veblen, *Analysis Situs*, second ed., and S. Lefschetz, *Topology*. Both of these books are in the series Colloquium Publications of the American Mathematical Society, New York 1931 and 1930 respectively.
3. Volume 1 of F. Klein, *History of Mathematics in the 19th Century*, Springer 1926.

Symmetry

Jacques Tits
Translated by John Stillwell

The word "symmetry" is used in mathematics quite differently from the way it is used in ordinary speech. In everyday life one applies it mainly to two-sided, right-left symmetry; but not so in mathematics. Admittedly, the word sometimes has a more general meaning in everyday speech. For example, everyone recognises that Figure 1 is highly symmetric, although it has no two-sided symmetry. However, this is really an exception. (The example of Figure 1 calls for some comment: in preparing this lecture it struck me that one can easily run into political or religious symbols when seeking examples of highly symmetric figures. This shows that symmetry has always had a powerful effect on people.)

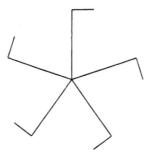

Figure 1

A second difference between symmetry in mathematics and symmetry in ordinary speech lies in the fact that *perfect* symmetry exists only in mathematics and not in real life. Here I need only allude to the fact emphasised by Hermann Weyl in his book *Symmetry*, that in western art the artist avoids the possibility of perfect symmetry and always breaks it slightly. There are beautiful examples of this, such as the famous Etruscan riders on the triclinic tomb in Corneto (Figure 2).

The picture is almost symmetric, but not quite. Perfect symmetry in art is often a little boring! In mathematics it is not so (even though in recent times mathematicians have also been interested in "near symmetries"). However, the assertion that complete symmetry never appears in reality is more fundamental than that. Look at Figure 3, for example.

Symmetries appear at first glance—and we shall come back to them—but they vanish when one looks more closely. This is immediately clear when one observes the symbols attached to the vertices—the 30 vertices have 30 different names—but even when one overlooks this, one easily finds small irregularities in the drawing that destroy all visible symmetries.

Figure 2

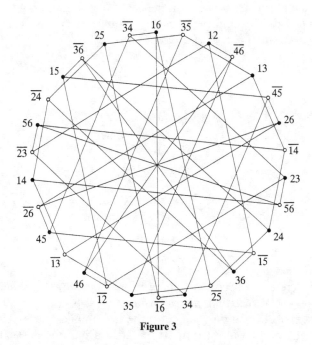

Figure 3

If one disregards the symbols and the small irregularities, however, then one immediately recognises a rotation symmetry of order 5, indeed a symmetry of order 10 if one makes no distinction between the "white" and "black" points. But much more symmetry is hidden in this figure, and it becomes apparent when one regards the figure simply as a *graph*. This means that one notices only the 30 points (lying on the boundary) and pairs of these that are connected. One may think of the 30 points, for example, as 30 people, with connections between acquaintances. The lengths and angles of the connecting lines (called edges) play no role. The hidden symmetries may then be observed as follows.

Each white point is connected to three black points, and the pairs of symbols attached to the latter three points form a so-called *partition* of the set {1, 2, 3, 4, 5, 6}.

For example, the point $\overline{12}$ is connected to 46, 15, and 23, so we obtain the partition (46)(15)(23) of the set $\{1, 2, 3, 4, 5, 6\}$. Conversely, if one takes an arbitrary partition, say (12)(34)(56), one finds that the corresponding three black points 12, 34, and 56 are connected to the same white point, in this case $\overline{23}$. Thus the white points are associated with the 15 partitions of $\{1, 2, 3, 4, 5, 6\}$ into pairs.

Now each permutation σ of the set $\{1, 2, 3, 4, 5, 6\}$ yields a permutation of the black points (since these correspond to pairs of elements of that set), as well as a permutation of the white points (regarded as partitions), which together represent a symmetry of the whole figure (regarded as a graph). For example, the permutation

$$1 \to 2 \to 3 \to 4 \to 5 \to 6 \to 1 \tag{1}$$

yields the symmetry

$$12 \to 23 \to 34 \to 45 \to 56 \to 16 \to 12,$$
$$13 \to 24 \to 35 \to 46 \to 15 \to 26 \to 13,$$
$$14 \to 25 \to 36 \to 14,$$
$$\overline{12} = (15)(23)(46)$$
$$\to (26)(34)(15) = \overline{45}$$
$$\to (13)(45)(26) = \overline{26}$$
$$\to (24)(56)(13) = \overline{14}$$
$$\to (35)(16)(24) = \overline{25}$$
$$\to (46)(12)(35) = \overline{46}$$
$$\to (15)(23)(46) = \overline{12},$$

and similarly

$$\overline{13} \to \overline{35} \to \overline{36} \to \overline{13}$$
$$\overline{15} \to \overline{56} \to \overline{16} \to \overline{15}$$
$$\overline{23} \leftrightarrow \overline{34}$$
$$\overline{24} \to \overline{24}.$$

We remark that the resulting permutation of the white points may be simply described as being induced by the permutation

$$\overline{\sigma}: \ \overline{1} \to \overline{5} \to \overline{6} \to \overline{1}, \ \overline{2} \leftrightarrow \overline{4}, \ \overline{3} \to \overline{3}. \tag{2}$$

Things are similar for any permutation σ of $\{1, 2, 3, 4, 5, 6\}$, so each permutation σ becomes associated with a permutation $\overline{\sigma}$ of $\{\overline{1}, \overline{2}, \overline{3}, \overline{4}, \overline{5}, \overline{6}\}$, which usually seems very different from σ. (One notices, for example, that the σ in (1) permutes the six symbols 1, 2, 3, 4, 5, 6 cyclically, whereas $\overline{\sigma}$ in (2) has a "fixed point", namely $\overline{3}$.)

The 720 permutations of 1, 2, 3, 4, 5, 6 (and also those of $\overline{1}, \overline{2}, \overline{3}, \overline{4}, \overline{5}, \overline{6}$ form a *group*, the so-called *symmetric group* S_6. The correspondence $\sigma \leftrightarrow \overline{\sigma}$ is a symmetry —or, as one says in mathematical language, an *automorphism*—of the group S_6. The existence of this so-called outer automorphism of S_6 is a well-known and remarkable phenomenon, which has no analogue when 6 is replaced by another integer. By means of the above method (inducing symmetries by permutations) we have obtained 720 symmetries of the graph in Figure 3; there are 1440 when one also combines them with the central symmetry $12 \leftrightarrow \overline{12}, 13 \leftrightarrow \overline{13}, \ldots, 56 \leftrightarrow \overline{56}$, which indeed is a symmetry only when white and black points are not distinguished.

One sees that the symmetry properties of a figure depend very much on how one visualises the figure: in Figure 3, whether one pays attention to irregularities in drawing, lengths of lines, the difference between white and black points etc. This leads in a natural way to the concept of a *mathematical object,* namely a thing for which the properties one intends to consider are prescribed at the beginning. Such objects can have proper symmetries. In daily life, on the other hand, it is usual and often necessary to consider all aspects of a thing, as far as possible, and this naturally destroys all symmetry.

In mathematics and physics, remarkable symmetries are often hidden. Figure 3 showed us two examples of this: on the one hand there are 720 symmetries of the graph, which come from permutations of $\{1, 2, 3, 4, 5, 6\}$ and are not at all apparent, apart from the 72 degree rotation and its multiples; on the other hand there is the outer automorphism of S_6. One of the most interesting tasks of the mathematician is to discover such hidden symmetries. We shall give further examples.

In order to introduce the next example, we present two apparently elementary problems. The first is: in how many ways may a natural number N be decomposed into sums of odd numbers? One must frank a letter with N cents, say, using stamps of denominations 1 cent, 3 cents, 5 cents, etc., and we ask in how many ways this is possible. With $N = 6$, for example, there are four solutions. The number M of solutions grows with N according to a law that is not at first easy to ascertain (Figure 4).

The second problem formulated in Figure 4 is not so easy to explain in terms of stamps, but let us try. The stamps are of two kinds: "normal" stamps whose values are even numbers, and "special" stamps whose values are the so-called triangular

---

PROBLEM 1. In how many ways may a given positive integer N be expressed as a monotonically decreasing series of positive odd integers?

Example.
$$\begin{aligned} 6 &= 1 + 1 + 1 + 1 + 1 + 1 \\ &= 3 + 1 + 1 + 1 \\ &= 3 + 3 \\ &= 5 + 1 \end{aligned}$$

PROBLEM 2. In how many ways may the number N be expressed as an ordered sum whose first term is of the form $n(n + 1)/2$, while the other terms form a monotonic increasing sequence of positive even integers?

Example.
$$\begin{aligned} 6 &= 0 + 2 + 2 + 2 \\ &= 0 + 2 + 4 \\ &= 0 + 6 \\ &= 6 \end{aligned}$$

Both problems have the same answer M:

N	1	2	3	4	5	6	7	8	9	10	11	12	13	...
M	1	1	2	2	3	4	5	6	8	10	12	15	18	...

Figure 4

> The equivalence of Problems 1 and 2 corresponds to the formula of Gauss
>
> $$\frac{\eta(q^2)^2}{\eta(q)} = \sum_{n=-\infty}^{+\infty} q^{2(n+\frac{1}{4})^2}$$
>
> where
>
> $$\eta(q) = q^{1/24} \cdot \prod_{n=1}^{\infty} (1 - q^n)$$

Figure 5

numbers $\{1, 3, 6, 10, \ldots\}$, and they are subject to the condition that at most one special stamp is used. Again we ask for the number of combinations of such stamps that add up to N cents.

It is remarkable that these two very different problems have the same answer: for every N the numbers of combinations of the two types are equal (Figure 4). This corresponds to a well-known and deep formula of Gauss (Figure 5).

Should one now say that the whole situation is completely understood as soon as a proof of the Gauss formula is produced? I do not think so. One gains a deeper insight by constructing a mathematical object that reflects the two problems, and in such a way that their equivalence (that is, the equality of their number of solutions) corresponds to a symmetry (no doubt hidden) of the object. The existence of this symmetry then explains not only the equivalence of the problems but also the formula of Gauss. Such an object was found by Frenkel, Kac, Lepowsky, et al. in the representation theory of certain Kac-Moody-Lie algebras.

To introduce our last example of a hidden symmetry, consider the lattices shown in Figures 6 and 7, whose symmetries we investigate briefly. Both have the so-called translation symmetries, which we wish to disregard here. To exclude them we could, for example, fix a point of the lattice. The symmetry group then becomes finite, and indeed of order 12 for Figure 6—one finds 6 rotations and 6 reflections that leave the lattice invariant—and order 8 for the lattice in Figure 7 (which is therefore somewhat less symmetric than the first).

Symmetry groups of lattices have excited great interest in recent times, for number theory reasons among others. If one investigates lattices in three-dimensional space, four-dimensional space, etc. from the standpoint of symmetry, there

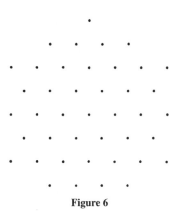

Figure 6

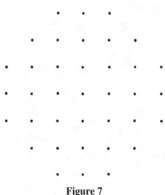

Figure 7

suddenly appears in twenty-four-dimensional space a quite special lattice, which is highly symmetric. It is the so-called Leech lattice (unknown until 30 years ago, which seems scarcely conceivable to many mathematicians today). When J. Leech discovered that lattice he did not know that it had extraordinary symmetry properties. He was interested in quite a different problem, namely, dense packings of spheres. Looking at the lattice as he did, with the construction he gave, the lattice does not appear to be particularly symmetrical. The construction may be roughly described as follows: one begins with a nice symmetric lattice, the rectangular one, removes some points, and replaces them with others. Both modifications are asymmetric and would appear to partially destroy the original symmetry. However, the process produces new symmetries that are not immediately visible: they are "hidden symmetries". J. H. Conway was the first to notice that the Leech lattice has an enormous symmetry group, a group of order 8 315 553 513 086 720 000. I know of no explicit construction of the Leech lattice that allows the full symmetry to be immediately seen: there always remain hidden symmetries that are difficult to find.

A last, famous, example is the following. It is known that in the space of one hundred and ninety six thousand eight hundred and eighty three dimensions there is a wonderful lattice whose symmetry group has order

808 017 424 794 512 875 886 459 904 961 710 757 005 754 368 000 000 000
$$= 2^{46} \cdot 3^{20} \cdot 5^9 \cdot 7^6 \cdot 11^2 \cdot 13^3 \cdot 17 \cdot 19 \cdot 23 \cdot 29 \cdot 31 \cdot 41 \cdot 47 \cdot 59 \cdot 71$$

Of course, no one has ever really seen this lattice: we know that it exists, but an explicit construction is lacking. Nevertheless, one can construct the symmetry group of the lattice, the so-called *monster group M* of R. Griess and B. Fischer. Here again, finding a hidden symmetry is an essential step in the construction. The group M has a certain subgroup of order

$$2^{46} \cdot 3^9 \cdot 5^4 \cdot 7^2 \cdot 11 \cdot 13 \cdot 23,$$

which is well understood. To generate M, Griess constructs in the space of 196 883 dimensions a certain object, a so-called algebra, which has this smaller group as (part of) its symmetry group. Then with greater difficulty he determines another symmetry, which is truly hidden. Together with the known subgroup, the hidden symmetry generates the group M. The author showed later that M is the *full* symmetry group of the Griess algebra.

Now it is natural to ask: why is one particularly interested in this monster group? Is it more than a beautiful game? I would like to show that the answer is decidedly positive.

Since Galois, the question of finding all possible symmetry types, and hence all existing groups (and here I always mean *finite* groups) has had a clear meaning. It is indeed a natural question, even a fundamental one, but it turns out not to be a reasonable problem, as I now briefly explain.

It is well known that each natural number is a product of *prime numbers*, which are therefore the "atoms" of number theory. Finite group theory also has its "atoms", which are called *simple groups*: each group is composed from such atoms in a certain way. But while the composition process in number theory is nothing else but the ordinary multiplication, in group theory the corresponding process—called "extension"—is considerably more complicated and diverse: here, given atoms (simple groups) may be combined in many ways, and when the number of constituents and their mutual "reactivities" is large, the totality of combinations becomes completely beyond apprehension.

Still, there remains the obviously natural question of enumerating at least all the finite simple groups. Until 40 years ago even this problem was considered unrealistic, yet it has recently been solved completely. The proof has not been completely written down yet, despite the production of thousands of pages (the combined work of many specialists coordinated by D. Gorenstein), but the result is astounding. As may be expected, there are infinitely many finite simple groups, but they can all be described in a concise and unified way apart from 26 exceptions, which lie outside this nice framework and are called *sporadic groups*. Here the monster group plays a special role: it is the largest sporadic simple group and was the last discovered. In addition, it has remarkable and mysterious number-theoretic properties. Understanding these properties is one of the most fascinating problems in finite group theory today. It is worth mentioning that, in a recent work of I. Frenkel, J. Lepowsky, and A. Meurman on this theme, the hidden symmetry between the two problems in Figure 4 plays an essential role.

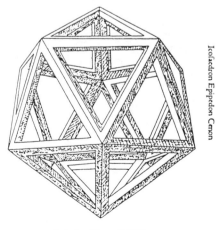

Figure 8

To conclude, I show another beautiful and well-known figure, an *icosahedron*. The psychological effect of highly symmetric objects, mentioned at the beginning, is reflected in the increasingly frequent appearance of the icosahedron in our publicity-oriented world. However, the model shown here (Figure 8) is particularly worthy of respect, since it is is supposedly due to Leonardo da Vinci.

It is an easy exercise to determine all the symmetries of the icosahedron: one finds 60 rotations and the 60 products of these with the central symmetry. Now imagine the space of 196 883 dimensions, containing a crystal somewhat like the icosahedron, except that it has

808 017 424 794 512 875 886 459 904 961 710 757 005 754 368 *billion*

rotational symmetries. These symmetries make up the monster group, which the reader can now begin to imagine—at the same time, perhaps, gaining some impression of the beauty of symmetries in mathematics.

ACKNOWLEDGMENT. This is an adaptation of the text of a lecture given on 28 November 1986 in Bonn (see *Mathematische Betrachtungen*, Bouvier Verlag, Bonn, 1988, pp. 32–44. I heartily thank Bouvier Verlag for permission to reproduce part of this text). The main alterations are the replacement of Figures 3 and 4 by a new Figure 3, essentially richer in symmetry properties, which I particularly dedicate to Herr Dr. H. Götze.

TRANSLATOR'S NOTE. The adaptation was published in *Miscellanea Mathematica*, edited by P. Hilton, F. Hirzebruch, and R. Remmert, Springer-Verlag, Heidelberg, 1991; this translation is published with the permission of the copyright holder, Springer-Verlag. Following an approach from the MONTHLY, Professor Tits has kindly agreed to allow this translation to be published.

REFERENCES

1. Robert L. Griess, Jr., *Twelve Sporadic Groups*, Springer-Verlag, New York, 1998.
2. Thomas M. Thompson, *From Error-correcting Codes through Sphere Packings to Simple Groups*, Mathematical Association of America, Washington, DC, 1983.

Exceptional Objects

John Stillwell

1. INTRODUCTION. In the mind of every mathematician, there is tension between the general rule and exceptional cases. Our conscience tells us we should strive for general theorems, yet we are fascinated and seduced by beautiful exceptions. We can't help loving the regular polyhedra, for example, but arbitrary polyhedra leave us cold. (Apart from the Euler polyhedron formula, what theorem do you know about *all* polyhedra?) The dream solution to this dilemma would be to find a general theory of exceptions—a complete description of their structure and relations—but of course it is still only a dream. A more feasible project than mathematical unification of the exceptional objects is historical unification: a description of some (conveniently chosen) objects, their evolution, and the way they influenced each other and the development of mathematics as a whole.

It so happens that patterns in the world of exceptional objects have often been discovered through an awareness of history, so a historical perspective is worthwhile even for experts. For the rest of us, it gives an easy armchair tour of a world that is otherwise hard to approach. A rigorous understanding of the exceptional Lie groups and the sporadic simple groups, for example, might take a lifetime. They are some of the least accessible objects in mathematics, and from most viewpoints they are way over the horizon. The historical perspective at least puts these objects in the picture, and it also shows lines that lead naturally to them, thus paving the way for a deeper understanding. I hope to show that the exceptional objects do have a certain unity and generality, but at the same time they are important *because* they are exceptional.

Some of the ideas in this article arose from discussions with David Young, in connection with his Monash honours project on octonions. I also received inspiration from some of the Internet postings of John Baez, and help on technical points from Terry Gannon.

2. REGULAR POLYHEDRA. The first exceptional objects to emerge in mathematics were the five regular polyhedra, known to the Greeks and the Etruscans around 500 BC.

These are exceptional because there are *only* five of them, whereas there are infinitely many regular polygons.

The Pythagoreans may have known a proof, by considering angles, that only five regular polyhedra exist. Many more of their properties were worked out by Theaetetus, around 375 BC, and by 300 BC they were integrated into the general theory of numbers and geometry in Euclid's *Elements*. The construction of the regular polyhedra in Book XIII, and the proof that there are only five of them, is the climax of the *Elements*. And the elaborate theory of irrational magnitudes in Book X is probably motivated by the magnitudes arising from the regular polyhedra, such as $((5 + \sqrt{5})/2)^{1/2}$, the diagonal of the icosahedron with unit edge.

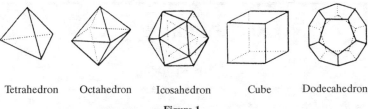

Figure 1

Thus it is probable that the regular polyhedra were the inspiration for the *Elements*, and hence for most of the later development of mathematics. If I wished to demonstrate the influence of exceptional objects on mathematics, I could rest my case right there. But there is much more. The most interesting cases of influence have been comparatively recent, but before discussing them we should recall a famous example from 400 years ago.

In Kepler's *Mysterium Cosmographicum* of 1596 the regular polyhedra made a spectacular, though premature, appearance in mathematical physics. Kepler "explained" the distances from the sun of the six known planets by a model of six spheres inscribing and circumscribing the five regular polyhedra. Alas, while geometry could not permit more regular polyhedra, physics could permit more planets, and the regular polyhedra were blown out of the sky by the discovery of Uranus in 1781.

Kepler of course never knew the fatal flaw in his model, and to the end of his days it was his favourite creation. Indeed it was not a complete waste of time, because it also led him to mathematical results of lasting importance about polyhedra.

3. REGULAR POLYTOPES. It is a measure of the slow progress of mathematics until recent times that the place of the regular polyhedra did not essentially change until the 19th century. Around 1850, they suddenly became part of an infinite panorama of exceptional objects, the n-dimensional analogues of polyhedra, called *polytopes*. Among other things, this led to the realisation that the dodecahedron and icosahedron are *more exceptional* than the other polyhedra, because the tetrahedron, cube, and octahedron have analogues in all dimensions.

The 4-dimensional analogues of the tetrahedron, cube, and octahedron are known as the 5-cell (because it is bounded by 5 tetrahedra), the 8-cell (bounded by 8 cubes), and the 16-cell (bounded by 16 tetrahedra), respectively.

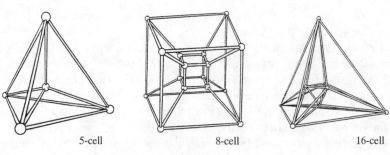

Figure 2

The other 4-dimensional regular polytopes were discovered by Schläfli in 1852: the 24-cell (bounded by 24 octahedra), 120-cell (bounded by 120 dodecahedra), and 600-cell (bounded by 600 tetrahedra).

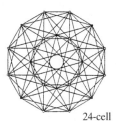
24-cell
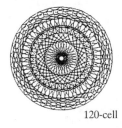
120-cell
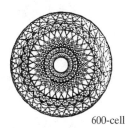
600-cell

Figure 3

Schläfli also discovered that in five or more dimensions the only regular polytopes are the analogues of the tetrahedron, cube, and octahedron. Thus from the n-dimensional perspective the dodecahedron, icosahedron, 24-cell, 120-cell, and 600-cell are the genuine exceptions. This also suggests that 3 and 4 are exceptional dimensions, a fact that just might explain the dimension of the space (or spacetime) in which we live, though after Kepler one should be cautious about such speculations!

4. INFINITE FAMILIES PLUS EXCEPTIONS. Schläfli's discovery may be summarised by saying that the regular polytopes may be classified into three infinite families plus five exceptions.

The classification of other structures follows a similar pattern:

- Regular tessellations of $\mathbb{R}^n$: one infinite family and four exceptional tessellations. These were enumerated by Schläfli in 1852. The infinite family is the tessellation of $\mathbb{R}^n$ by "n-cubes", which generalises the tessellation of the plane by squares. Two of the exceptions are the dual tessellations of $\mathbb{R}^2$ by equilateral triangles and regular hexagons. The other two, discovered by Schläfli, are dual tessellations of $\mathbb{R}^4$, by 16-cells and 24-cells.

- Simple Lie groups: four infinite families A_n, B_n, C_n, D_n, plus five exceptional groups G_2, F_4, E_6, E_7, E_8. The infinite families were discovered by Lie, and the exceptions by Killing in 1888 and Cartan in 1894. The process of classification resembles that of regular polyhedra, because it reduces these continuous groups (in a far from obvious way) to discrete arrangements of line segments in Euclidean space (root systems), with certain constraints on angles and relative lengths.

- Finite reflection groups: four infinite families plus seven exceptional groups, discovered by Coxeter in 1934. These generalise the symmetry groups of polyhedra, and are linked to the simple Lie groups via root systems. It emerged from Coxeter's work, and also the work of Weyl in 1925, Cartan in 1926, and Stiefel in 1942, that each simple Lie group is determined by a reflection group, now called its *Weyl group*.

- Finite simple groups: 18 infinite families plus 26 exceptional (sporadic) groups, discovered by the collective work of many mathematicians between

1830 and 1980. These are also linked to the simple Lie groups by passage from the continuous to the discrete. Galois, around 1830, first noticed that finite simple groups arise from groups of transformations when complex coefficients are replaced by elements of a finite field. The idea was extended by Jordan and Dickson, and reached full generality with Chevalley in 1955.

However, even the exceptional Lie groups yield infinite families of finite simple groups, so the sporadic simple groups appear to be the height of exceptionality. Still, they are not completely unrelated to the other exceptional objects. Other links between exceptions appear when we pick up other threads of the story.

5. SUMS OF SQUARES. Entirely different from the story of polyhedra, but at least as old, is the story of sums of squares. Sums of two squares have been studied since the Babylonian discovery of "Pythagorean triples" around 1800 BC. Around 200 AD, Diophantus made the striking discovery that sums of two squares can be multiplied, in a certain sense. His *Arithmetica*, Book III, Problem 19 says

> 65 is "naturally" divided into two squares in two ways... due to the fact that 65 is the product of 13 and 5, each of which numbers is the sum of two squares.

In 950 AD, al-Khazin interpreted this as a reference to the *two square identity*

$$(a_1^2 + b_1^2)(a_2^2 + b_2^2) = (a_1a_2 \mp b_1b_2)^2 + (b_1a_2 \pm a_1b_2)^2,$$

as did Fibonacci in his *Liber Quadratorum* of 1225. Fibonacci also gave a proof, which is not trivial in his algebra—it takes five pages!

There is no similar identity for sums of three squares, as Diophantus probably realised. 15 is not a sum of three integer squares, yet $15 = 5 \times 3$, and $3 = 1^2 + 1^2 + 1^2$, $5 = 0^2 + 1^2 + 2^2$. Thus there can be no identity, with integer coefficients, expressing the product of sums of three squares as a sum of three squares. It is also true that 15 is not the sum of three *rational* squares. Diophantus may have known this too. He stated that 15 is not the sum of two rational squares, and the proof for three squares is similar (involving congruence mod 8 instead of mod 4).

But claims about sums of four squares are conspicuously absent from the *Arithmetica*. This led Bachet to conjecture, in his 1621 edition of the book, that *every natural number is the sum of four (natural number) squares*.

Fermat claimed a proof of Bachet's conjecture, but the first documented step towards a proof was Euler's *four square identity*, given in a letter to Goldbach, 4 May 1748:

$$(a^2 + b^2 + c^2 + d^2)(p^2 + q^2 + r^2 + s^2)$$
$$= (ap + bq + cr + ds)^2 + (aq - bp - cs + dr)^2$$
$$+ (ar + bs - cp - dq)^2 + (as - br + cq - dp)^2.$$

Using this, Lagrange completed the proof of Bachet's conjecture in 1770.

In 1818, Degen discovered an *eight square identity*, which turned out to be the last in the series, so sums of 2, 4, and 8 squares are exceptional. This was not proved until 1898, by Hurwitz. In fact, Degen's identity was virtually unknown until the discovery of...

6. THE DIVISION ALGEBRAS $\mathbb{R}$, $\mathbb{C}$, $\mathbb{H}$, $\mathbb{O}$. A division algebra of dimension n over $\mathbb{R}$ consists of n-tuples of real numbers under vector addition, together with a "multiplication" that distributes over addition and admits "division". The idea is to make n-tuples add and multiply as "n-dimensional numbers". Addition is no problem, but a decent multiplication is exceptional—apart from the obvious case $n = 1$ it exists only in dimensions 2, 4, and 8. The first clue how to multiply in each of these dimensions was the identity for sums of 2, 4, or 8 squares.

Diophantus associated $a^2 + b^2$ with the pair (a, b), identified with the right-angled triangle with sides a and b. From triangles (a_1, b_1) and (a_2, b_2) he formed the "product" triangle $(a_1 a_2 - b_1 b_2, b_1 a_2 + a_1 b_2)$. His identity

$$(a_1^2 + b_1^2)(a_2^2 + b_2^2) = (a_1 a_2 - b_1 b_2)^2 + (b_1 a_2 + a_1 b_2)^2$$

shows that the hypotenuse of the "product" is the product of the hypotenuses. In modern notation his identity is

$$|z_1|^2 |z_2|^2 = |z_1 z_2|^2 \quad \text{where} \quad z_1 = a_1 + ib_1, \quad z_2 = a_2 + ib_2,$$

and it expresses the *multiplicative property of the norm* $|z|^2 = a^2 + b^2$ of $z = a + ib$: the norm of a product is the product of the norms. A multiplicative norm makes division possible, because it guarantees that the product of nonzero elements is nonzero.

The same product of pairs (without the identification with triangles) was rediscovered by Hamilton in 1835 as a *definition* of the product of complex numbers:

$$(a_1, b_1)(a_2, b_2) = (a_1 a_2 - b_1 b_2, b_1 a_2 + a_1 b_2).$$

Hamilton had been trying since 1830 to define multiplication of n-tuples and retain the basic properties of multiplication on $\mathbb{R}$ and $\mathbb{C}$:

- multiplication is commutative and associative
- multiplication is distributive over vector addition
- the norm $x_1^2 + x_2^2 + \cdots + x_n^2$ of $(x_1, x_2, \ldots, x_n)$ is multiplicative.

He got stuck on triples for 13 years, not knowing that a multiplicative norm was ruled out by elementary results on sums of three squares.

If he had known this earlier, would he have given up the whole idea, without trying quadruples? van der Waerden [**7**, p. 185] thought so, implying that Hamilton tried quadruples in October 1843 only to salvage something from his long and fruitless commitment to triples. He had already given up on commutative multiplication, but with quadruples he saved the other properties, in his algebra $\mathbb{H}$ of *quaternions*. It was lucky, in van der Waerden's opinion, that Hamilton didn't know about sums of three squares.

But what if he had known about sums of four squares? He would then have seen a multiplicative norm of quadruples, and would perhaps have discovered quaternions in 1830. This is not pure speculation. Gauss, who knew Euler's four square identity, discovered both a "quaternion form" of it and also a "quaternion representation" of rotations of the sphere, the latter around 1819.

However, Gauss did not publish these discoveries, so Hamilton was lucky after all—he was first to see the full structure of the quaternions, and he received all the credit for them.

John Graves, who had been in correspondence with Hamilton for years on the problem of multiplying n-tuples, was galvanised by the discovery of quaternions and the associated four square identity (which Hamilton and he at that time believed to be new). In December 1843 Graves rediscovered Degen's eight square identity, and immediately constructed the 8-dimensional division algebra $\mathbb{O}$ of *octonions*.

Dickson [4, p. 159] condensed all these results into one formula, showing that each algebra in the sequence $\mathbb{R}, \mathbb{C}, \mathbb{H}, \mathbb{O}$ comes from the one before by a simple generalisation of Diophantus' rule for multiplying pairs:

$$(a_1, b_1)(a_2, b_2) = \left(a_1 a_2 - \overline{b_2} b_1, b_2 a_1 + b_1 \overline{a_2}\right) \quad \text{where} \quad \overline{(a, b)} = (\bar{a}, -b).$$

From this it is easily proved that the quaternions are associative but not commutative, and that the octonions are not associative. The next algebra in the sequence, consisting of pairs (a, b) of octonions, is not a division algebra. Theorems by Frobenius, Zorn, and others confirm that $\mathbb{R}, \mathbb{C}, \mathbb{H},$ and $\mathbb{O}$ are indeed exceptional—they are the only finite-dimensional division algebras over $\mathbb{R}$. For further information on them see [5].

7. LATTICES. The polyhedron thread of our story intertwines with the division algebra thread when we reconsider tessellations of $\mathbb{R}^n$. The two exceptional tessellations of $\mathbb{R}^2 = \mathbb{C}$, by equilateral triangles and hexagons, are both based on the lattice of *Eisenstein integers*

$$m + n \frac{1 + \sqrt{-3}}{2} \quad \text{for} \quad m, n \in \mathbb{Z}.$$

The triangle tessellation has lattice points at vertices; the hexagon tessellation has them at face centres.

Similarly, the two exceptional tessellations of $\mathbb{R}^4 = \mathbb{H}$ are based on the *Hurwitz integers*

$$p \frac{1 + \mathbf{i} + \mathbf{j} + \mathbf{k}}{2} + q\mathbf{i} + r\mathbf{j} + s\mathbf{k} \quad \text{for} \quad p, q, r, s \in \mathbb{Z},$$

called the "integer quaternions," and were used by Hurwitz in 1896 to give a new proof that every natural number is a sum of four squares.

The analogous "integer octonions" form a lattice in $\mathbb{R}^8$, in which the neighbours of each lattice point form a polytope discovered by Gosset in 1897. It is not regular, but is nevertheless highly symmetrical. Its symmetry group is none other than the Weyl group of the exceptional Lie group E_8. Gosset's story, and the history of regular polytopes in general, is told in [3].

8. PROJECTIVE CONFIGURATIONS. Other exceptional structures in geometry are the projective configurations discovered by Pappus, around 300 AD, and by Desargues in 1639.

Theorem of Pappus. *If the vertices of a hexagon lie alternately on two straight lines, then the intersections of the opposite sides of the hexagon lie in a straight line.*

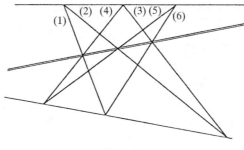

Figure 4

Theorem of Desargues. *If two triangles are in perspective, then the intersections of their corresponding sides lie in a line.*

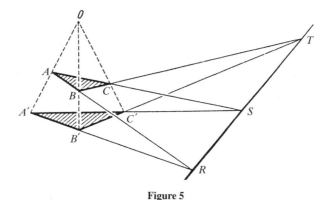

Figure 5

These statements are "projective" because they involve only incidence: whether or not points meet lines. Moreover, Desargues' theorem in space has a *proof* from incidence properties—the properties that

- two planes meet in a line,
- two lines in the same plane meet in a point.

The diagram of the Desargues configuration hints at this, by suggesting that the triangles lie in three dimensions, even though the diagram itself necessarily lies in the plane. Yet, strangely, the Desargues and Pappus theorems in the plane do *not* have proofs from obvious incidence properties; their proofs involve the concept of distance or coordinates.

Why is this so? In 1847 von Staudt gave geometric constructions of $+$ and $\times$, thus "coordinatising" each projective plane by a division ring. Then in 1899 Hilbert made the wonderful discovery that the geometry of the plane is tied to the algebra of the ring:

- *Pappus' theorem holds* ⇔ *the division ring is commutative*
- *Desargues' theorem holds* ⇔ *the division ring is associative*.

Conversely, any division ring R yields a projective plane RP^2. So by Hilbert's theorem,

- $\mathbb{R}P^2$ and $\mathbb{C}P^2$ satisfy Pappus,
- $\mathbb{H}P^2$ satisfies Desargues but not Pappus, and
- $\mathbb{O}P^2$ satisfies neither.

In 1933 Ruth Moufang completed Hilbert's results with a theorem satisfied by $\mathbb{O}P^2$, the "little Desargues' theorem", which states uniqueness of the construction of the fourth harmonic point D of points A, B, C. She showed that "little Desargues" holds if and only if the division ring is *alternative*, that is, for all a and b, $a(ab) = (aa)b$ and $b(aa) = (ba)a$.

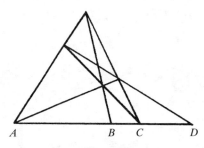

Figure 6

Alternativity is a characteristic property of $\mathbb{O}$ by the following result of Zorn from 1930: a finite-dimensional alternative division algebra over $\mathbb{R}$ is $\cong \mathbb{R}, \mathbb{C}, \mathbb{H}$ or $\mathbb{O}$. Thus the octonion projective plane $\mathbb{O}P^2$ is exceptional, simply because $\mathbb{O}$ is. However, $\mathbb{O}P^2$ is *more* exceptional than $\mathbb{R}P^2$, $\mathbb{C}P^2$, and $\mathbb{H}P^2$, because each of the latter belongs to an infinite family of projective spaces.

The method of homogeneous coordinates can be used to construct a projective space RP^n of each dimension n for $R = \mathbb{R}, \mathbb{C}, \mathbb{H}$. But there is no $\mathbb{O}P^3$, because

existence of $\mathbb{O}P^3 \Rightarrow$ Desargues' theorem holds $\Rightarrow \mathbb{O}$ is associative.

Thus $\mathbb{O}P^1$ and $\mathbb{O}P^2$ are *exceptional projective spaces*, compared with the infinite families $\mathbb{R}P^n$, $\mathbb{C}P^n$, $\mathbb{H}P^n$.

9. $\mathbb{O}$: THE MOTHER OF ALL EXCEPTIONS? We have already observed that many objects inherit their classification from that of the simple Lie groups A_n, B_n, C_n, D_n, G_2, F_4, E_6, E_7, E_8. In particular, classifications based on the A, D, and E series arise so often that explaining them has become a flourishing industry. For a recent survey of this field of "ADE classifications" see [6].

To find the source of such classifications, we should try to understand where the simple Lie groups come from. The groups A_n, B_n, C_n, D_n are automorphism groups of n-dimensional projective spaces (or subgroups), so the infinite families of simple Lie groups arise from the infinite families of projective spaces $\mathbb{R}P^n$, $\mathbb{C}P^n$, $\mathbb{H}P^n$.

Since there is no $\mathbb{O}P^n$ for $n > 2$, one does not expect many Lie groups to come from the octonions, but in fact all five exceptional Lie groups are related to $\mathbb{O}$. It seems that, in some sense, the exceptional Lie groups inherit their exceptionality from $\mathbb{O}$.[1]

This astonishing relationship began to emerge when Cartan discovered that $G_2 \cong \text{Aut}(\mathbb{O})$. He mentioned this casually in a 1908 article [2] on the history of hypercomplex number systems, saying only that the automorphism group of the octonions is a simple Lie group with 14 parameters. However, he knew from his 1894 classification of the simple Lie groups that G_2 is the only compact simple Lie group of dimension 14.

In the 1950s several constructions of F_4, E_6, E_7, and E_8 from $\mathbb{O}$ were discovered by Freudenthal, Tits, and Rosenfeld. For example, they found that F_4 is the isometry group of $\mathbb{O}P^2$, and E_6 is its collineation group. Their success in finding links between these exceptional objects raises an interesting question, though perhaps one that will never be completely answered: How many exceptional objects inherit their exceptionality from $\mathbb{O}$?

Some connections between $\mathbb{O}$ and sporadic simple groups are known, but the latter groups remain the most mysterious exceptions to date. Naturally, the sheer mystery of these groups has only intensified the search for a broad theory of exceptional objects. This has led to some surprising developments, among them a revival of the near-dead subject of "foundations of geometry". Foundations have now grown into "buildings" (a concept due to Tits), which are just part of a huge field of "incidence geometry" spanning most of the results we have discussed. A comprehensive survey of this new field with ancient roots may be found in [1].

REFERENCES

1. F. Buekenhout, *Handbook of Incidence Geometry*, Elsevier 1995.
2. É. Cartan, Nombres Complexes, *Œuvres Complètes* II, 1, pp. 107–246.
3. H. S. M. Coxeter, *Regular Polytopes*, Macmillan, 1963.
4.. L. E. Dickson, On quaternions and their generalisation and the history of the eight square theorem, *Ann. of Math.* **20** (1919), 155–171.
5. H.-D. Ebbinghaus et al., *Numbers*, Springer-Verlag, 1991.
6. I. Reiten, Dynkin diagrams and the representation theory of algebras, *Notices Amer. Math. Soc.* **44** (1997), 546–556.
7. B. L. van der Waerden, *A History of Algebra*, Springer-Verlag, 1985.

---

[1]The five exceptional Lie groups could be of truly cosmic importance. Recent work in theoretical physics seeks to reconcile general relativity with quantum theory by means of "string theories" in which atomic particles arise as modes of vibration. The possible string theories correspond to the exceptional Lie groups, hence string theory allows five "possible worlds". This point was raised by Ed Witten in his Gibbs Lecture at the AMS-MAA Joint Meetings in Baltimore 1998, along with the question: if there are five possible worlds, who lives in the other four?

The Problem of Squarable Lunes

M. M. Postnikov
translated from the Russian by Abe Shenitzer

Translator's note. Hippocrates of Chios (second half of the fifth century BCE) seems to have been the first mathematician to square a figure with curved boundary. The figure in question was a *lune*. (For details about Hippocrates and his work see pp. 131–136 in B. L. van der Waerden, *Science Awakening*, P. Noordhoff, Groningen, 1954.)

A lune is a figure bounded by two circular arcs with a common chord. Figure 1 shows a concave-convex lune and Figure 2 shows a convex lune.

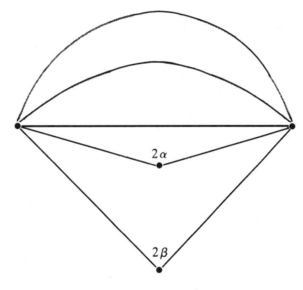

concave-convex lune

Figure 1

The problem of squarable lunes is usually referred to as *The problem of the lunes of Hippocrates*. The lunes in this problem are *constructible*, i.e., they can be constructed by ruler and compass, and *squarable*, i.e., one can construct for such a lune (by ruler and compass) a square of the same area. Using these terms we can state the problem of the lunes of Hippocrates as follows:

Find all constructible squarable lunes.

Over the ages, the problem of squarable lunes was dealt with by many mathematicians (see the index of *The Dictionary of Scientific Biography* under Quadrature, of lunes), but was solved only in the 20th century by two Russian algebraists.

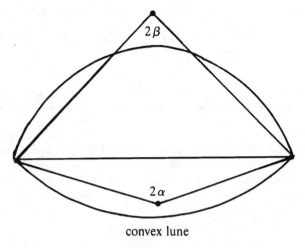

convex lune

Figure 2

The two were Chebotarev, who came close to the complete solution in 1934, and his student Dorodnov, who completed his teacher's work in 1947. Their principal tool was Galois theory.

The solution of the problem of squarable lunes presented in the following essay is taken from M. M. Postnikov's book *Galois Theory*, published by Fizmatgiz, Moscow, in 1963. This is an enlarged version of Postnikov's earlier book with the same title. The earlier version was translated by L. F. Boron and published by Noordhoff in 1962. It does not contain the material on lunes.

* * *

Note. In what follows, the remarks in brackets [ ] are due to the translator.)

[Figures 1 and 2 make it clear that] A lune is given by the length of the common chord that subtends its bounding arcs and by the central angles 2α and 2β that are their measures (we assume for definiteness that $\alpha > \beta$). We limit ourselves to lunes with *commensurable* angles α and β, i.e., angles for which there exists an angle θ and coprime (positive) integers m and n ($m > n$) such that $\alpha = m\theta$ and $\beta = n\theta$. With these restrictions, the construction of a lune reduces to the construction of the angle θ.

We make use of the following result: *If $\theta \neq 0$, then θ and $\sin \theta$ cannot both be algebraic*. [For a proof see p. 131 in I. Niven, *Irrational Numbers*, Carus Monograph No. 11, MAA, 1956.]

We consider a squarable lune with angles α and β. Without restriction of generality we may assume that the length of the chord that subtends the bounding arcs of the lune is 1. Then it is not difficult to see that the area of the lune is given by the formula

$$S = \frac{\alpha}{\sin^2 \alpha} - \frac{\beta}{\sin^2 \beta} + \frac{\cot \beta}{4} - \frac{\cot \alpha}{4}$$

if both bounding arcs are on the same side of the common chord (this is a so-called concave–convex lune) and by the formula

$$S = \frac{\alpha}{\sin^2 \alpha} + \frac{\beta}{\sin^2 \beta} + \frac{\cot \beta}{4} + \frac{\cot \alpha}{4} \tag{1}$$

otherwise (a convex lune). We consider the latter case first. Since $\alpha = m\theta$ and $\beta = n\theta$, it follows from (1) that

$$\theta = \frac{\left(S - \frac{\cot n\theta}{4} - \frac{\cot m\theta}{4}\right)\sin^2 m\theta \sin^2 n\theta}{m\sin^2 n\theta + n\sin^2 m\theta}. \qquad (*)$$

[We are dealing with constructible lunes of common chord length 1. The vertices of the central angles in Figures 1 and 2 are assumed to be constructible. Hence their distances from the common chord are algebraic. Specifically, these distances are $(\cot \alpha)/2$ and $(\cot \beta)/2$. It follows that $\sin \alpha = \sin m\theta, \sin \beta, \cot \alpha$, and $\cot \beta$ are algebraic. But then θ in $(*)$ is algebraic. Note also that if $\sin m\theta$ is algebraic then $\sin \theta$ is also algebraic.] Thus in this case θ and $\sin \theta$ are both algebraic. But this, we know, can happen only if $\theta = 0$. Hence

there are no squarable convex lunes.

A similar argument shows that

a squarable concave–convex lune with angles $\alpha = m\theta$ and $\beta = n\theta$ can exist only if

$$n\sin^2 m\theta - m\sin^2 n\theta = 0. \qquad (2)$$

Thus

the construction of squarable lunes (with commensurable angles α and β) reduces to the construction of an angle θ satisfying (2).

If in (2) we express $\sin m\theta$ and $\sin n\theta$ in terms of $\cos \theta$, then we obtain for $\cos \theta$ an equation that we must investigate. A lune with angles $m\theta$ and $n\theta$ can be constructed with ruler and compass if and only if this equation has a real root whose absolute value does not exceed 1 and whose computation reduces to the solution of a chain of quadratic equations.

Incidentally, from a computational standpoint it is more convenient to consider not the number $\cos \theta$ but the number $\xi = \cos 2\theta + i\sin 2\theta$. Of course, this change is of no major significance (ξ is constructible if and only if $\cos \theta$ is constructible). Since

$$-\sin^2(k\theta) = \frac{(\xi^k - 1)^2}{4\xi^k}, \qquad k = m, n,$$

the equation for ξ has the form

$$n(x^m - 1)^2 - mx^{m-n}(x^n - 1)^2 = 0. \qquad (3)$$

Thus we have arrived at the following, purely algebraic, problem:

for what coprime positive integers m and n (with $m > n$) does the solution of (3) *reduce to the solution of quadratic equations?*

After finding all equations (3) that are reducible to quadratic equations, we can then find among them the equations that correspond to "real" lunes, i.e., equations with root ξ with absolute value 1.

It is possible to prove that

if m is composite, then, except for the case $m = 9, n = 1$, the solution of (3) *cannot be reduced to the solution of quadratic equations.*

The proof of this statement is beyond the level of this book and we leave it out.

Now let m be a prime. It turns out that

if the solution of (3) *for a prime $m = p$ reduces to the solution of quadratic equations, then $p = 2$ or p is a Fermat prime.*

To prove this proposition we make the substitution $x = y + 1$. We obtain the equation

$$n\left(\frac{(y+1)^p - 1}{y}\right)^2 = p(y+1)^{p-n}\left(\frac{(y+1)^n - 1}{n}\right)^2,$$

i.e., the equation

$$n\left(y^{p-1} + \binom{1}{p}y^{p-2} + \binom{2}{p}y^{p-3} + \cdots + \binom{p-1}{p}\right)^2$$
$$- p(y+1)^{p-n}\left(y^{n-1} + \binom{1}{n}y^{n-2} + \binom{2}{n}y^{n-3} + \cdots + \binom{n-1}{n}\right)^2 = 0. \quad (4)$$

This yields the equation

$$ny^{2(p-1)} + a_1 y^{2p-3} + \cdots + a_{2p-3} y + a_{2p-2} = 0,$$

where $a_1, \ldots, a_{2p-2}$ are certain integers.

Now we note that

If $k \neq 0, p$ then the binomial coefficient $\binom{k}{p}$ is divisible by p.

Indeed,

$$\binom{k}{p} = \frac{p(p-1)\cdots(p-k+1)}{1 \cdot 2 \cdots k},$$

and the prime p in the numerator cannot be reduced (for all factors in the denominator are less than p).

It follows that we can write (4) as

$$n\left(y^{p-1} + pf_1(y)\right)^2 + pf_2(y) = 0,$$

where $f_1(y)$ and $f_2(y)$ are certain polynomials with integer coefficients, and thus as

$$ny^{2(p-1)} + pf(y) = 0, \quad (5)$$

where $f(y)$ is a polynomial with integer coefficients. This means that all the coefficients $a_1, \ldots, a_{2p-2}$ are divisible by p. Since the leading coefficient n is not divisible by p and the constant term a_{2p-2} (which, clearly, is equal to $np^2 - pn = pn(n - p)$) is not divisible by p^2, (5) satisfies the Eisenstein criterion [see A. Clark, *Elements of Abstract Algebra*, Wadsworth, 1971, pp. 86–87] and thus is irreducible (over Q). It follows that its solution is reducible to quadratic equations only if its degree $2(p - 1)$ is equal to two or is a Fermat prime. This completes the proof.

It turns out that the necessary condition just established is not sufficient. Specifically, it can be shown that

if $p > 5$ then (3) (for $m = p$) cannot be reduced to the solution of quadratic equations.

We will not prove this result.

It remains to consider the cases $m = 2, m = 3, m = 5$, and the case $m = 9$ and $n = 1$.

Let $m = 2$. Then, after division by $(x - 1)^2$, (3) becomes

$$x^2 + 1 = 0.$$

Hence in this case $2\theta = 90°$ and we see that

A lune with angles $2\alpha = 180°$ and $2\beta = 180°$ is squarable.

This is the well-known lune of Hippocrates.

Let $m = 3$. Then $n = 1$ or $n = 2$. In the first case, after division by $(x - 1)^2$, (3) becomes

$$(x^2 + x + 1)^2 - 3x^2 = 0.$$

This equation has two real roots (which are of no interest to us) and two complex roots

$$\frac{\sqrt{3} - 1}{2} \pm \sqrt{-\frac{\sqrt{3}}{2}}.$$

Hence $\cos 2\theta = (\sqrt{3} - 1)/2$, whence $2\theta \approx 68°.5$. It follows that

a lune with angles $\arccos(\sqrt{3} - 1)/2 \approx 68°.5$ and $3\arccos(\sqrt{3} - 1)/2 \approx 205°.6$ is squarable.

For $n = 2$ we obtain the equation

$$2(x^2 + x + 1)^2 - 3x(x + 1)^2 = 0.$$

Putting $x = y^2$ we obtain an equation that splits over the field $Q(\sqrt{2}, \sqrt{3})$ into two reciprocal quartic equations that reduce to quadratic equations. [An equation $a_0 x^n + a_1 x^{n-1} + \cdots + a_{n-1} x + a_n = 0$ is said to be *reciprocal* if $a_i = a_{n-i}$ for all $i = 0, 1, \ldots, n$.] After carrying out the necessary computations we find that $\cos 2\theta = (\sqrt{33} - 1)/8$, whence $2\theta \approx 53°.6$. It follows that

a lune whose angles are $\arccos(\sqrt{33} - 1)/8 \approx 107°.2$ and $3\arccos(\sqrt{33} - 1)/8 \approx 160°.9$ is squarable.

These two lunes were also constructed by Hippocrates.

Lastly, let $m = 5$. Then $n = 1, 2, 3$, or 4. For $n = 1$ we obtain the equation

$$(x^4 + x^3 + x^2 + x + 1)^2 - 5x^4 = 0,$$

which splits into two reciprocal quartic equations

$$x^4 + x^3 + (1 \pm \sqrt{5})x^2 + x + 1 = 0,$$

and hence is reducible to quadratic equations. In this case $\cos 2\theta = (\sqrt{33} - 1)/8$, whence $2\theta \approx 46°.9$. It follows that

a lune with angles $\arccos(\sqrt{5 + 4\sqrt{5}} - 1)/4 \approx 46°.9$ and $5\arccos(\sqrt{5 + 4\sqrt{5}} - 1)/4 \approx 234°.4$ is squarable.

This lune was found in 1840 by Clausen.

For $n = 2$ we obtain the equation

$$2(x^4 + x^3 + x^2 + x + 1)^2 - 5x^3(x + 1)^2 = 0.$$

Putting $x = y^2$ we reduce it to two reciprocal equations of degree eight

$$y^8 + y^6 + y^4 + y^2 + 1 \pm \sqrt{\frac{5}{2}} y^3 (y^2 + 1) = 0.$$

Solving these equations by the well-known method (using the substitution $y + y^{-1} = z$) we obtain two quartic equations

$$4z^4 - 3z^2 \pm \sqrt{\frac{5}{2}} z + 1 = 0. \tag{6}$$

When we solve these equations by Ferrari's method [see A. Clark, op. cit., pp. 141–143] we obtain for the auxiliary unknown u the cubic equation

$$16u^3 - 24u^2 + 20u - 5 = 0. \tag{7}$$

It is easy to see that this equation is irreducible (it has no rational roots). Hence the initial equation is not reducible to quadratic equations (for the roots of (7) are rationally expressible in terms of the roots of (6)). It follows that

there is no squarable lune corresponding to the values $m = 5$ and $n = 2$.

For $n = 3$ we obtain the equation

$$3(x^4 + x^3 + x^2 + x + 1)^2 - 5x^2(x^2 + x + 1)^2 = 0,$$

which splits over the field $Q(\sqrt{3}, \sqrt{5})$ into two reciprocal quartic equations that reduce to quadratic equations. After carrying out the necessary computations we find that $\cos 2\theta = t$, where

$$t = \frac{\sqrt{\frac{5}{3}} - 1 + \sqrt{\frac{20}{3} + \sqrt{\frac{20}{3}}}}{4},$$

whence $2\theta \approx 33°.6$. It follows that

a lune with angles $3 \arccos t \approx 100°.8$ and $5 \arccos t \approx 168°.0$ is squarable.

This lune was also found by Clausen.

For $n = 4$ we obtain the equation

$$4(x^4 + x^3 + x^2 + x + 1)^2 - 5x(x^3 + x^2 + x + 1)^2 = 0.$$

Just as in the case $n = 2$, it is easy to show that the solution of this equation is not reducible to the solution of quadratic equations. It follows that

there is no squarable lune corresponding to the values $m = 5$ and $n = 4$.

The last possibility to be considered is that of $m = 9$ and $n = 1$. In this case we obtain the reducible equation

$$(x^8 + x^7 + x^6 + x^5 + x^4 + x^3 + x^2 + x + 1)^2 - 9x^8 = 0,$$

which splits into two reciprocal equations

$$\begin{cases} x^8 + x^7 + x^6 + x^5 - 2x^4 + x^3 + x^2 + x + 1 = 0, \\ x^8 + x^7 + x^6 + x^5 + 4x^4 + x^3 + x^2 + x + 1 = 0. \end{cases} \tag{8}$$

Now the substitution $x + x^{-1} = y$ yields the two quartic equations

$$y^4 + y^3 - 3y^2 - 2y - 2 = 0,$$
$$y^4 + 4y^3 - 3y^2 - 2y + 4 = 0.$$

The solution of the first of these equations does not reduce to the solution of quadratic equations, for the cubic equation obtained by Ferrari's method is irreducible. On the other hand, the second equation splits over $Q(i\sqrt{3})$ into the two quadratic equations

$$\begin{cases} y^2 + \eta y - 2 = 0, \\ y^2 + \bar{\eta} y - 2 = 0. \end{cases} \tag{9}$$

where $\eta = (1 + i\sqrt{3})/2$, so that its solution reduces to the solution of quadratic equations. Hence

for $m = 9$ and $n = 1$ the solution of (3) (more correctly, of one of its irreducible factors of degree eight) reduces to the solution of quadratic equations.

We will try to obtain the corresponding lune. It is easy to see that the equations in (9) have no real roots. It follows that for any root ξ of (3) that is of interest to us (i.e., for any root of the second one of the equations in (8)), the quantity $\xi + \xi^{-1}$ is not real. If a lune with angle θ would correspond to this root, then, since $\xi = \cos 2\theta + i\sin 2\theta$, the quantity $\xi + \xi^{-1} = 2\cos 2\theta$ would be real. This contradiction shows that

there is no squarable lune corresponding to the values $m = 9$ and $n = 1$.

The final result of our investigation can be stated in the form of the following theorem:

Squarable lunes exist only in the following five cases:

$$\begin{array}{lllll} m = 2, & m = 3, & m = 3, & m = 5, & m = 5 \\ n = 1, & n = 1, & n = 2, & n = 1, & n = 3. \end{array}$$

How Hyperbolic Geometry Became Respectable

Abe Shenitzer

Many accounts of the evolution of hyperbolic geometry mention the names of Beltrami and Klein but give few details of their contributions in this area. The present paper focuses on some of these very details. The reader is assumed to have some familiarity with basic concepts of differential geometry.

HISTORICAL INTRODUCTION. The discovery of hyperbolic geometry* led to the realization—at the end of the 19th century—that the truths of mathematics are relative rather than absolute, and to the resolution of the millennial doubts about Euclid's parallel postulate, J. Coolidge describes its effect as follows:

> The point which I wish to insist on... is that it is to the doubts about Euclid's parallel postulate, and efforts of such thinkers as Saccheri, Lobachevskiĭ, Bolyai, Beltrami, Riemann and Pasch to settle these doubts, that we owe the whole modern abstract conception of mathematical science. ([1], p. 87.)

Another way of assigning to the discovery of hyperbolic geometry its due place is to note that it was one of the two greatest geometric discoveries (or inventions, if you are not a Platonist) of the 19th century, the other being the discovery (by Riemann) of the concept of an n-dimensional manifold. Of course, while the first of these discoveries marked a profound intellectual discontinuity, the second did not; in fact, one can safely say that an n-dimensional manifold and its geometry are direct "descendants" of the notions of a surface and of Gauss' notion of the intrinsic geometry of a surface.

The story of the discovery of hyperbolic geometry can be conveniently divided into three parts. The first was largely negative in the sense that it consisted of doomed attempts to deduce the Euclidean parallel postulate from the other postulates of Euclidean geometry. The first of these attempts was presumably due to Archimedes. Other such attempts were due to Arab mathematicians, and, more recently, to Saccheri (1733) and Legendre (1794).

The second part of our story involved attempts in the first third of the 19th century to deduce basic logical consequences of the axioms of Euclidean geometry with its parallel axiom replaced by the hyperbolic parallel axiom. The mathematicians involved were Schweikart, Lambert, and, more importantly, Gauss, Bolyai,

---

*The terms "hyperbolic geometry" and "hyperbolic plane" refer to the system of plane geometry based on Euclid's axioms with his parallel axiom replaced by the "hyperbolic parallel axiom" which asserts the existence of a line a and a point A not on a such that there are at least two lines passing through A that are parallel to a. An isometric replica of the hyperbolic plane is called a model of that plane.

and Lobachevskiĭ. Gauss and Lobachevskiĭ considered the possibility that the geometry of the universe is hyperbolic rather than Euclidean. Bolyai's chief concern was about the consistency of the new geometry. This is where matters stood in the 1830s. During the next 30 years the effect on mathematics of these investigations and insights was virtually nil.

The third stage of the evolution of hyperbolic geometry began in the 1860s and ended in the 1880s. Here the key contributions—the first models of the hyperbolic plane—were due to Beltrami and, in part, to Klein. Their work provided a proof of the relative consistency of hyperbolic geometry and put it logically on a par with Euclidean geometry.

TECHNICAL ACCOUNT. What follows is a description, in four parts, of the genesis and nature of the early models of the hyperbolic plane due to Beltrami and, in part, to Klein. The approach in the first part is based on an essay by A. M. Lopshits in volume II of [4].

(α) Beltrami studied mappings of surfaces in E^3 into the Euclidean plane that preserve geodesics, that is, map geodesics to (straight) lines in the plane.

A familiar example of such a mapping is the central projection of a sphere to the plane. We give a few highly relevant details of this mapping.

Consider (Figure 1) the projection of the lower hemisphere onto the plane. If $m(X, Y)$ is the plane image of M on the sphere, then we assign (X, Y) to M as its curvilinear coordinates. If the radius of the sphere is a, then its (Gaussian) curvature is $1/a^2$ and its length element ds^2 turns out to be

$$ds^2 = \frac{(1 + KY^2)\, dX^2 + (1 + KX^2)\, dY^2 - 2KXY\, dX\, dY}{\left[1 + K(X^2 + Y^2)\right]^2}, \quad (1)$$

with K the (Gaussian) curvature.

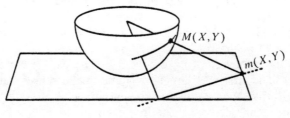

Figure 1

In 1866 Beltrami showed, quite generally, that if a surface in E^3 can be mapped in a one-one geodesic manner into the plane, then, after transfer of the coordinates (X, Y) of the image points to the points of the surface, its length element ds^2 can be reduced to the form (1) with K its constant curvature. Furthermore, the only surfaces that admit such mappings are surfaces with constant curvature. Since surfaces with the same ds^2 are locally isometric, it follows that a surface with constant curvature $K > 0$ is locally isometric to a sphere, and a surface with constant curvature $K = 0$ is locally isometric to a plane. The case of greatest interest to Beltrami was that of pseudospherical surfaces, that is, surfaces in E^3 with constant curvature $K < 0$. Beltrami studied such surfaces in "Saggio...," the first of the two papers he published in 1868. What follows is a brief summary of his findings.

In (1) put $K = -1/k^2$, $X = kx$, and $Y = ky$. Then the length element of a pseudospherical surface takes the form

$$ds^2 = k^2 \frac{(1-y^2) dx^2 + (1-x^2) dy^2 + 2xy\, dx\, dy}{(1-x^2-y^2)^2} \qquad (2)$$

where k is a positive constant. Since the necessarily positive discriminant of the form (2) is equal to

$$k^4/(1-x^2-y^2)^3, \qquad (3)$$

it follows that

$$x^2 + y^2 < 1.$$

This means that a geodesic mapping of a pseudospherical surface into a Euclidean plane carries each point of the surface to a point of the unit disk (see Figure 2).

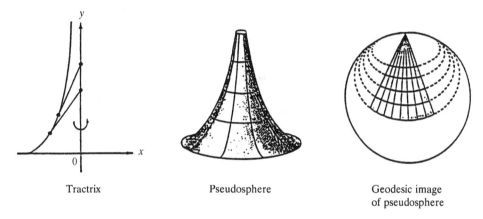

Tractrix   Pseudosphere   Geodesic image of pseudosphere

Figure 2

Now there is a 1-1 geodesic mapping of the hyperbolic plane onto the unit disk (see [**2**]. Also see Figure 3; the "dish" represents a unit disk that is part of a horosphere tangent to a hyperbolic plane. The curved lines represent hyperbolic straight lines belonging to the mapping pencil of parallel lines). If (as in the case of the stereographic projection of the hemisphere) we transfer the coordinates (x, y) of a point in the unit disk to its preimage in the hyperbolic plane, then the length element ds^2 of the latter takes the form (2). *The sameness of the length elements of*

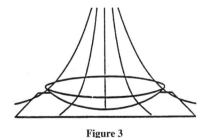

Figure 3

the pseudospherical surface and of the hyperbolic plane proves that these two surfaces are locally isometric.

Beltrami's realization of the geometry of (part of) the hyperbolic plane as the intrinsic geometry of a pseudospherical surface had a tremendous impact on his contemporaries. It was seen as changing what was thought to be a figment of the imagination into a mathematical fact.

If a pseudospherical surface is to be an isometric replica of the full hyperbolic plane, then it must have the basic property that the geodesics through any of its points are infinite in both directions. This turns out to be equivalent to the assumption that there exists a pseudospherical surface that admits a one-one geodesic mapping *onto* the unit disk. Beltrami thought that this is a defensible assumption. It is not, at least for smooth surfaces in E^3. In 1902 Hilbert showed that there is no such surface in E^3 that is an isometric replica of the hyperbolic plane.

(β) While the main emphasis in Beltrami's "Saggio..." is on showing that a pseudospherical surface is locally isometric to the hyperbolic plane, it also points to the possibility of turning the unit disk into a model of the hyperbolic plane by exploiting various features of the geodesic mapping linking it to a pseudospherical surface. Hence the term "the Beltrami disk model." The relevant definitions are: a hyperbolic (straight) line is a chord of the unit disk, hyperbolic parallel lines are chords that meet at a point of the unit circle, and hyperbolic diverging lines are chords that don't meet in the closed unit disk. The source of quantitative relations is the length element ds^2 given by equation (2). (Thus the length of a curve $x = x(t), y = y(t)$ is $\int ds$, with suitable limits in the integral.) The key missing element in Beltrami's disk model is the definition of a motion.

(γ) In 1868 Dedekind published Riemann's inaugural lecture of 1854. After reading it, Beltrami published the second of his 1868 papers in which he once more obtained his earlier disk model as well as *three conformal models* of the hyperbolic plane. The following description of these models is taken from the introduction in J. Stillwell's translation of the second of Beltrami's 1868 papers.

The starting point is a *hemisphere model* consisting of the open hemisphere

$$x_1^2 + x_2^2 + x^2 = a^2, \quad x > 0 \tag{4}$$

in 3-dimensional (x_1, x_2, x)-space, with the line element

$$ds = R \frac{\sqrt{dx_1^2 + dx_2^2 + dx^2}}{x}. \tag{5}$$

The geodesics for this metric are vertical sections of the hemisphere. Perpendicular projection of this model onto the plane $x = 0$ which we take to be horizontal (Figure 4) yields Beltrami's earlier disk model.

The third model is obtained by stereographic projection of the hemisphere onto its horizontal tangent plane (Figure 5), and the line element is computed to be

$$ds = \frac{\sqrt{d\xi_1^2 + d\xi_2^2}}{1 - \frac{1}{4R^2}(\xi_1^2 + \xi_2^2)}, \tag{6}$$

a metric which was stated by Riemann (1854) to be of constant curvature. We observe that (6) is a multiple of the Euclidean line element $\sqrt{d\xi_1^2 + d\xi_2^2}$ in the

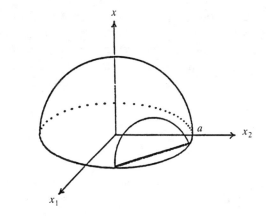

Figure 4

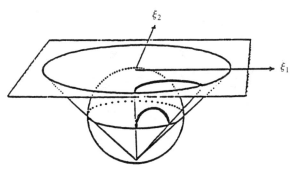

Figure 5

(ξ_1, ξ_2)-plane; the multiple of course varies with position but not with direction, hence angles are preserved. Thus we have the conformal, or *"Poincaré disk"* model, 14 years before its appearance in Poincaré (1882).

The fourth model is obtained by stereographic projection of the hemisphere onto (the upper half of) a vertical plane (Figure 6). This gives the line element

$$ds = R\frac{\sqrt{d\eta_1^2 + d\eta_2^2}}{\eta} \qquad (7)$$

for the "Poincaré half plane."

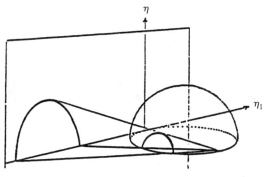

Figure 6

(δ) In 1871 Klein used the unit disk and its chords to create his so-called projective model of the hyperbolic plane. His points, lines, parallel lines, and diverging lines were the same as Beltrami's but he based his definitions of distance, angle, and motion on the ideas of Cayley. He defined the distance between points M_1 and M_2 (see Figure 7) as, essentially, the logarithm of the cross ratio $\{P_1P_2, M_2M_1\}$,

$$M_1M_2 = \frac{k}{2}\ln J = \frac{k}{2}\ln\left(\frac{P_1M_2}{P_1M_1} \bigg/ \frac{P_2M_2}{P_2M_1}\right),$$

defined the angle between two lines by, essentially, a projective dual of his definition of distance, and defined a motion of his model as a projective transformation of the plane of the disk that mapped its boundary onto itself. Klein's version of the unit-disk model of the hyperbolic plane was the first instance of a geometry in the sense of his famous Erlangen Program of 1872.

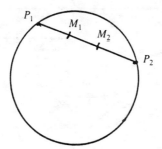

Figure 7

In sum we can say that the works of Beltrami and Klein that appeared between 1866 and 1871 put hyperbolic geometry on a par with Euclidean geometry. *From then on it became ever clearer that the truths of mathematics are relative rather than absolute.*

REFERENCES

1. J. L. Coolidge, *A history of geometrical methods*, Dover, 1963. Chapter IV, pp. 68–87. ("Nice summary.")
2. N. V. Efimov, *Higher geometry*, Mir, 1980. Chapter 8, pp. 474–528. (Useful beyond the needs of this essay.)
3. H. Eves, *Great moments in mathematics*, DME #7, MAA, 1983. Lectures 26 and 27. (Fine elementary account.)
4. V. F. Kagan, *Foundations of geometry*, GITTL, Moscow, Parts I and II. (In Russian.) Lectures 26 and 27. (Included because Part II contains the Lopshits essay mentioned on p. 465.)
5. J. Stillwell, *Mathematics and its history*, Springer, 1989. Chapter 17, pp. 255–274.
6. I. M. Yaglom, *Geometric transformations III*, NML #24, MAA, 1973. Introduction and Supplement, pp. 103–135. (Supplement contains detailed elementary account of Klein's disk model.)

Does Mathematics Distinguish Certain Dimensions of Spaces?

Zdzisław Pogoda and Leszek M. Sokołowski

translated by Abe Shenitzer with the editorial assistance of Hardy Grant

1. DIMENSION IN PHYSICS, OR: WHAT WORLD DO WE LIVE IN? The Greeks of, say, Euclid's time talked of *diastasis*—extension—of bodies rather than of the number of their dimensions.

It was Claudius Ptolemy who advanced the idea that the extension of a body should be measured by the number of mutually perpendicular straight lines (rigid rods) issuing from a point in the body and claimed that this number cannot exceed three. Ptolemy's idea is a precursor of the idea of a system of coordinate axes used to this day in mathematics (outside topology) and in physics. His claim—we call it Ptolemy's law—is deeply rooted in Western culture and remained unchallenged until the third decade of the 20th century. The remarkableness of this fact becomes clear if we recall that the totality of the physics of the ancients, with the exception of the Pythagorean thesis of the sphericity of the Earth and of the celestial bodies, of Archimedes' law, and of the elements of the theory of simple machines, was questioned and rejected at the beginning of the modern era. Ptolemy's law was first challenged in 1919 when Theodor Kaluza, then an instructor at Königsberg, sent a letter to Einstein in which he advanced the hypothesis of a 5-dimensional universe. Kaluza did not base his hypothesis on any experimental facts. His thinking was guided by the notion that in a 5-dimensional universe he could unify gravitation and electromagnetism.

In a sense, nothing has changed since 1919. Ptolemy's law is supported by countless experiments in all areas of physics. Moreover, there is not a single empirical fact that would suggest even indirectly the existence of additional dimensions of space. At the present time, the only reason for questioning Ptolemy's law is the hope that in a multidimensional universe all elementary interactions could be combined into one. Viewed by an observer in "our" 4-dimensional spacetime, this single interaction breaks down into gravitation, nuclear forces, and electroweak interactions. The only thing that has evolved between 1919 and today is the concrete version of the unification idea. First it was the classical 5-dimensional theory of Kaluza-Klein, then the multidimensional theory of Kaluza-Klein with arbitrary gauge fields, based largely on the conceptions of De Witt and Witten, and, as of 1984, the hopes of the physicists have focussed on the theory of superstrings.

These theories have in common the view that physical spacetime has $d = 4 + n$ dimensions of which four are directly observable and the remaining n are not—at least not with present experimental techniques. The two key questions to be answered here are: what is the value of n, i.e., the value of the dimension d? and why does the d-dimensional universe contain a distinguished subspace of dimension four? The second question is actually made up of the following two: why does the universe not have a structure analogous to that of Minkowskian spacetime, all of whose dimensions are equivalent in a well-known sense of the term, but decomposes into an "easy-to-see" component and a "difficult-to-see" component, and what is the significance of the dimensions of these components.

We have no satisfactory answer to any of these questions. The classical Kaluza-Klein theory, in which $d = 5$, was given up long ago. There is also a modern version of this theory based on supergravity. In 1981 Edward Witten, working with this version, offered quite convincing arguments in favor of the value $d = 11$. But this theory had to be abandoned three years later because of insurmountable difficulties. In its place came the theory of superstrings, which engendered great hopes and strong emotions. Some important scientists hailed it as the General Theory of Everything. By 1987 this enthusiasm decreased somewhat. Yet the theory of superstrings remains the most promising candidate for a theory that bids fair to quantize gravitation and unify all interactions. This would lead to a coherent theory of all elementary particles. Still, the theory of superstrings is only a candidate for such a theory. Also, it is, so to say, very much nonunique, in that it admits a great variety of models. The trouble is that we do not know enough to choose the "true" one of these models. Furthermore, the theory is nonunique when it comes to the dimension of spacetime. At first it seemed that if one considers bosonic strings then the theory can be internally consistent only for $d = 26$, while the theory of fermionic and supersymmetric strings, of greater interest for physicists, requires the value $d = 10$. Later, however, it was shown that what had previously been regarded as consistency was actually just simplicity: both dimensions are singled out by the theory for "pure" strings. By associating a number of physical fields with a string one can achieve a consistent string theory for any value of d. Thus the theory again gives no hint regarding the dimension of the physical world.

Suppose we accept $d = 10$ as a tentative answer to the first of our two questions. What about an answer to the second one? All we can say at this point is that we know neither the mechanism that brought into being the "easy-to-see" and the "difficult-to-see" components of the universe nor the factors that determine their respective dimensions.

In spite of the complete absence of relevant experimental data, the idea of a multidimensional universe has become a fixed part of theoretical physics and can be expected to influence the direction of its development for a long time to come. This being so, it is reasonable to accept tentatively its validity. Even if the idea turns out to be false, the question it gave rise to, namely the question of why Ptolemy's law holds, i.e., why $d = 4$, retains its relevance. In other words, a satisfactory future physical theory must not treat the dimension of physical spacetime as a free parameter whose value is determined experimentally but must be able to explain why its value is what it is.

2. DIMENSION IN MATHEMATICS: IS IT WEIGHTY? We have seen that, so far, physics provides no persuasive answer to the question of the dimension of physical space. Does mathematics? Does mathematics distinguish certain dimensions of space and are they close to 3, the only empirically distinguished dimension? More generally, is dimension a significant mathematical concept? In other words, will a mathematician agree that the question of the number of dimensions of spacetime is important or will he tell us that the number of dimensions of, say, Euclidean space conveys information that is as trivial as that conveyed by the fact that we describe the surface of the earth with respect to a rectangular coordinate system oriented in the east-west and north-south directions? If this were the viewpoint of mathematics, then we would have to admit that the empirically determined dimension of spacetime is something "local" and devoid of deeper physical significance, something akin to the empirically determined difference between the horizontal and vertical directions, something due to the local gravitational field of the earth and without a universal character. This would also imply that every physical theory in which dimension plays an important role is false in a definite sense. To illustrate what we are trying to say note that Aristotle's physics, with its division of the universe into sublunar and translunar regions, has no adequate mathematical apparatus; Euclidean geometry, known in antiquity, does not fit it in the least. We hope that these remarks make it clear that the determination of the status of the concept of dimension in mathematics is an issue of profound importance for physics.

While the mathematical notion of dimension first came up in the context of Euclidean geometry, it is actually a topological concept applicable to quite a large class of topological spaces. Some basic problems of dimension theory were solved only in the first half of the 20th century and many others remain open. Some mathematicians think that problems associated with the notion of dimension are of the utmost importance not only in topology but also in all of mathematics. There is the famous dictum of Poincaré: "I think that the most important of the theorems of *Analysis situs* [an earlier name for topology that goes back to Leibniz and was very popular in the 19th century] is the one that asserts that space has three dimensions [1]."

We noted earlier that the roots of dimension theory go back to the Hellenistic period. In modern mathematics the concept of dimension was initially treated intuitively and nonuniformly, namely, there were distinct definitions of dimension in geometry and in linear algebra, as well as distinct definitions of the dimension of a simplex and of the dimension of a linear subspace. Nevertheless dimension was, and continues to be, one of the most important elements of the description of space and of geometric objects. The need for a nonintuitive, rigorous definition of dimension was highlighted by remarkable discoveries made at the turn of the 20th century. Thus in 1877 Georg Cantor showed that a segment and a square have the same cardinalities, i.e., that there exists a one-to-one mapping of a segment onto a square. This implied that dimension is not a set-theoretic concept. Again, a curve can be thought of as the result of a single stroke of a pencil. Hence Camille Jordan's definition of a curve as a subset of the plane that is a continuous image of an interval. Intuition dictated the notion that such an object is 1-dimensional. But in 1890 Giuseppe Peano constructed a curve, i.e., a continuous image of an interval, that filled a square. These results showed that the dimension of a Euclidean space is not preserved either by one-to-one or by continuous mappings. There remained the open problem of whether a topological mapping, i.e., a one-to-one bicontinuous mapping, can map a plane onto 3-space. We should add that at that time the question of dimension in topology was primarily a question of its relation to Euclidean geometry. The properties of plane figures are so different from the properties of solids that if the dimension of a Euclidean space were not a topological invariant then topology would tell us very little about its geometric properties. The problem of dimension was made more acute by the discovery of curious figures such as the Sierpiński carpet and the Menger cube (see [2], [4]), whose dimensions cannot be guessed at a glance. It was obvious that the notion of dimension had to be reinvestigated "from the ground up."

It is probably safe to say that a breakthrough was Luitzen E.J. Brouwer's proof, provided between 1911 and 1912, of the invariance of domain (see [2]). This implied that if $n \neq m$ then $\mathbb{R}^n$ and $\mathbb{R}^m$ are not homeomorphic. The significance of this result is obvious: since the intuitive notion of the dimension of a Euclidean space, given by the number of coordinates, turned out to be a topological invariant, it was reasonable to talk of dimension in topology. But the nature of the new invariant remained unclear because Brouwer did not use in his proof any topological characteristics of Euclidean dimension.

During the next two decades, i.e., until the mid 1930s, mathematicians such as Henri Lebesgue, Karl Menger, Pavel Uryson, and Eduard Čech developed the foundations of a topological theory of dimension by defining dimension in a rigorous way and by proving many theorems that determined its properties; the theory applies to all classical cases (see [5]). Unfortunately, the theory is not unique in that there are three different definitions of the dimension of a topological space, namely the covering dimension (dim), the small inductive dimension (ind), and the large inductive dimension (Ind). These three notions coincide for metric spaces with a countable basis but diverge for more general spaces.

Fortunately, the complications that arise in the general theory do not worry the physicists because the most general spaces used in physics are differentiable manifolds. In view of Whitney's immersion theorem such a manifold can always be represented as a subset of $\mathbb{R}^k$ for a large enough k. More precisely, if the manifold has dimension n, then it is a (regularly immersed) subset of $\mathbb{R}^{2n+1}$ (see [2]). In the sequel we will limit ourselves to $\mathbb{R}^n$ and its subsets. This means that we can use any one of the three definitions of dimension. These are rather complicated; the simplest and the most convenient for us is the definition of small inductive dimension. We present an abbreviated version of it.

1. The dimension of the empty set is -1 (ind$\varnothing = -1$).
2. If every point of a space X has an arbitrarily small neighborhood whose boundary has dimension $\leq n - 1$, then X has dimension $\leq n$ (ind $X \leq n$).
3. If ind $X \leq n$ but it is not true that ind $X \leq n - 1$, then X has dimension n (ind $X = n$).
4. If ind $X \leq n$ is false for all n, then the dimension of X is infinite (ind $X = \infty$).

If we look carefully at the above definition of dimension, then we see that it includes all the intuitive perceptions that we associate with this notion. If we remove from a line a point together with a small neighborhood of this point, then the boundary of that neighborhood consists of two points (if finitely many lines intersect at the point in question, then the corresponding boundary consists of finitely many points) and so has dimension 0. Similarly, if we remove from a plane (or, more generally, from a surface) a point together with a small neighborhood of this point, then the boundary of that neighborhood consists of a closed line and so has dimension 1. This observation can be carried over by induction to any number of dimensions.

The definition just given satisfies almost all of the requirements we are likely to associate with the notion of dimension. Its advantages are twofold. For one thing, the inductive definition of dimension of a set agrees with its "intuitive dimension." For example, the inductive dimension of "n-dimensional" figures such as a cube, a sphere, a ball, and a simplex is n (this is difficult to prove!). For another, the definition of the dimension ind enables us to assign dimension numbers to an extensive class of topological spaces that baffle our intuition. Examples of such spaces are the Cantor set (dimension 0), the Sierpiński carpet (ind = 1), and the Menger cube (ind = 1) (see [3]).

We used the words "almost all" in connection with the definition of ind because, upon closer inspection, it turns out to have certain flaws (we won't discuss them here) which led to the formulation of the two other definitions of dimension (chronologically, dim appeared in Lebesgue's papers before the appearance of ind). The latter took the place of ind in more general topological spaces.

To avoid possible misunderstandings, we wish to emphasize that, in spite of its name, the *fractal Hausdorff dimension*, now very fashionable in statistical physics, has nothing in common with the notion of *topological dimension* discussed in the present paper, for it is *not* a topological invariant.

In the sequel, whenever discussing Euclidean spaces and their subsets, we will write dim X in place of ind X.

It is clear that the definition of dimension does not single out any dimension (with the possible exception of the dimension of the empty set and the infinite dimension). This being so, it is natural to expect that the number of constructs will grow with the dimension of the space. A space of large dimension can be expected to have many relations and connections among its subsets, and the latter are likely to be marked by great complexity. In brief, we can expect that the greater the dimension, the greater the constructional possibilities. But one must also expect increased difficulties to be faced by the mathematician deprived of the assistance of intuition. On the whole, our expectations—and fears—are correct. This is confirmed by a host of theorems in geometry, in linear algebra, in the theory of manifolds, in the theory of differential equations on manifolds, and so on.

But is it *always* the case that increased dimension implies increased complexity and an increased supply of structures? The most obvious counterexample is that of vector spaces. And there are other, rather weighty counterexamples to this assertion as well as problems in which dimension plays a very subtle role. The rest of this paper is an account of these counterexamples. Their detailed analysis will enable us to answer in a mathematically reasonable manner the question in the title of this paper: *Does mathematics distinguish certain dimensions of spaces*?

3. CLASSIFICATION OF REGULAR POLYHEDRA. We begin with a problem which a mathematician is likely to dismiss as a mere curiosity, but which serves as the most elementary counterexample to the assertion that a space of large dimension is richer in structural possibilities than a space of small dimension. Specifically, we consider the classification of regular polyhedra and their higher- (and lower-) dimensional analogues in Euclidean spaces.

A complete definition of a polyhedron is too complicated to be given here. For our purposes it suffices to define a polyhedron as the intersection of a finite number of halfspaces. A regular polyhedron is a polyhedron whose faces are congruent polygons and one with the additional property that the number of faces at each vertex is the same. Informally, we could say that a regular polyhedron is a polyhedron which looks the same from every side. The ancients knew that there are exactly five regular polyhedra (Platonic solids), namely, the tetrahedron, the hexahedron (the cube), the octahedron, the dodecahedron, and the icosahedron (see [6], [7]).

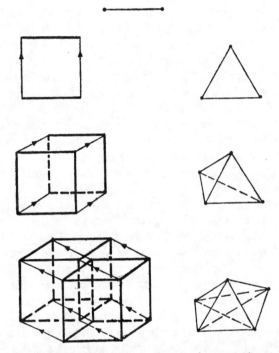

Figure 1. Lower- and higher-dimensional analogues of a cube (on the left) and a tetrahedron (on the right). From the figure we can read off the procedure for forming higher-dimensional cubes from lower-dimensional ones: a square (a 2-dimensional cube) is obtained from two appropriately located segments (1-dimensional cubes) by appropriately joining their vertices (their endpoints), a cube is obtained from two appropriately located squares by appropriately joining vertices, a hypercube is obtained from two cubes, etc. The analogues of a tetrahedron—simplexes—are the smallest convex sets containing respectively three noncollinear points (a triangle), four noncoplanar points (a tetrahedron), five points not in the same space (a 4-dimensional simplex), etc.

The role of regular polyhedra in the plane is played by regular polygons. The Greek mathematicians knew that their number is infinite. It is easy to define analogs of regular polyhedra in 4-dimensional space—the role of faces is played by ordinary 3-dimensional regular polyhedra. It turns out that there are just six such "regular cells" [6], [7]. What is surprising is that a Euclidean space of dimension 5 or higher contains just three regular cells! One of them is the n-dimensional analogue of a cube and is called a hypercube.

In order to describe the two remaining regular cells we introduce the notion of a p-dimensional face of an n-dimensional (not necessarily regular) polyhedron, $p \leq n - 1$, which is a "face of a face." For example, for $n = 3$ we have: $p = 0$ vertices, $p = 1$ edges, $p = 2$ ordinary faces of a polyhedron. It is possible to show that the number of p-dimensional faces of a hypercube is $2^{n-p}\binom{n}{p}$. We also introduce the notion of a simplex, which will be needed later as well. We take a Euclidean space of sufficiently large dimension and in it $n + 1$ points that do not lie on an $(n - 1)$-dimensional hyperplane (if $n = 3$ the four points are not coplanar). We now join the points of each pair by means of straight lines and obtain in this way an n-dimensional polyhedron whose vertices are the initial $n + 1$ points; the segments joining the vertices determine the p-dimensional faces of the polyhedron, $1 \leq p \leq n - 1$. This polyhedron is called an n-dimensional simplex. For $n = 0$ the simplex is a point, for $n = 1$ a segment, for $n = 2$ a triangle, for $n = 3$ a tetrahedron, etc. Now the second regular cell for dimension $n \geq 5$ is the n-dimensional regular simplex; it has $\binom{n+1}{p+1}$ p-dimensional faces. The third and last regular cell is the dual of the hypercube; its duality is reflected in the fact that the number of its p-dimensional faces is equal to the number of $(n - p)$-dimensional faces of the hypercube.

It would seem that the number of regular cells should increase with increasing dimension. But this is not the case. The number in question has the fixed value 3 for all $n \geq 5$. In this sense the dimensions 2, 3, and 4—and especially 2—are distinguished.

4. CLASSIFICATION OF MANIFOLDS. Our primary reason for considering the classification of manifolds is not to supply another counterexample to the assertion stated earlier. Rather, it is to illustrate the singular situations that turn up in spaces of different dimensions. The simplicity and considerable generality of the notion of a manifold has made it a leading concept of modern mathematics. One of the key advantages of manifolds is that it is possible to give their global description using methods characteristic of Euclidean spaces. Specifically, the technique created for flat spaces can be used for "curved" objects.

We will be mostly concerned with the notion of a topological manifold. An n-dimensional manifold is a Hausdorff space every point of which has a neighborhood U homeomorphic to all of $\mathbb{R}^n$ or to one of its open subsets. The mapping φ that takes the neighborhood U to an open subset of $\mathbb{R}^n$ is called a coordinate system, the pair (U, φ) is called a chart, and the value $\varphi(p) = (x^1, \ldots, x^n)$ of φ at a point $p \in U$ is referred to as the coordinates of p with respect to this chart. Let (U, φ) and (U, ψ) be two charts with the same domain but with different coordinate systems. Then the composition $\varphi \circ \psi^{-1}$ maps a certain open set in $\mathbb{R}^n$ onto another open set in $\mathbb{R}^n$. The definition of a topological manifold requires this composition to be a homeomorphism in $\mathbb{R}^n$. If, in addition, we require $\varphi \circ \psi^{-1}$ and its inverse to be infinitely differentiable, then we obtain the notion of a smooth manifold [2].

Physicists work most of the time with smooth manifolds, familiar from courses in analysis or in tensor calculus; a prototype of such a manifold is any smooth non-self-intersecting surface in Euclidean space. Here the term "smooth surface"

Figure 2. The sets in this figure are not manifolds. This is due to the presence of branch points or branch segments or to local changes of dimension.

is somewhat misleading, for "smoothness," i.e., the existence of a tangent plane, is not necessary for a surface to be a smooth manifold (a similar remark applies to higher-dimensional manifolds). A close look at the definition shows that a surface as rich in edges as the surface of a cube is also a smooth manifold! Indeed, this surface is homeomorphic to a sphere, and every topological space homeomorphic to a smooth manifold can be given the structure of a smooth manifold. However, it is convenient to stick to certain accepted images, and, when talking about differentiable manifolds, to have in mind objects such as an hyperboloid, a sphere, a torus, etc.

Let us go back to topological manifolds. A natural and very important problem is that of their classification. This means giving a sequence of nonhomeomorphic manifolds of the same dimension such that every manifold of this dimension is homeomorphic to one of the manifolds in the sequence. In the case of dimension 1 things are simple: every manifold is homeomorphic to a circle or to a straight line, so that the required sequence has just two elements. 2-dimensional manifolds, or surfaces, were classified at the beginning of the 20th century (in the case of dimensions 2 and higher one usually classifies only compact manifolds without a boundary, and these are the only manifolds we consider in this paper); here the sequence of representative manifolds is infinite. (More precisely, the sequence splits into two branches. The elements of one branch are connected sums of finitely many toruses, i.e., product spaces $S^1 \times S^1$, and the elements of the other branch are connected sums of finitely many projective planes.) As for 3-dimensional manifolds, the situation is as yet unclear; we do not know if these are classifiable. In 1982 William Thurston gave a partial—and very incomplete—classification of these manifolds (see [36] and the popular article [8]). We do not know what further progress is possible here.

Matters are clear for topological manifolds of dimension 4 and higher: in 1958 A.A. Markov proved that their classification is impossible [9], [10]. We will try to explain why this is so. Every smooth compact manifold can be triangulated, i.e., cut into simplexes (recall that simplexes of dimension 1, 2, and 3 are segments, triangles, and tetrahedra respectively). This implies the possibility of associating with a manifold a table listing its simplexes, their faces, and the ways in which they are connected; this table is a kind of code for the manifold. It is clear that such a code is not a unique "proof of identity" of the manifold it represents, for it is possible to cut up a given manifold in different ways and thus to associate with it different codes. This means that we would need a universal algorithm for deciding whether or not two given codes represent the same manifold (of course, the same up to a homeomorphism). If such an algorithm existed, then we could use it to divide all the codes into equivalence classes, with codes in each class corresponding to the same manifold. Markov's theorem is a nonexistence theorem: for 4-dimensional (and for higher-dimensional) manifolds there is no algorithm that would permit us to decide if two given codes correspond to homeomorphic manifolds. Thus the question of classification of manifolds is doubly hopeless. Not only is there no method of classification, but even if somebody were inspired and produced a sequence of nonhomeomorphic manifolds that exhausted all the possibilities, this result would be useless because we have no effective (i.e., universal, and requiring only a finite number of steps) way of checking which element of the sequence of representative manifolds a given manifold is homeomorphic to. In other words, even if we had a classificatory sequence of manifolds,

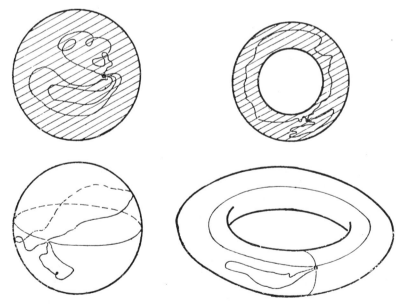

Figure 3. A disk and a sphere are simply connected, for their fundamental groups are trivial (a closed loop on either of them can be shrunk to a point). An annulus and a torus are not simply connected, but their fundamental groups are different. The fundamental group of an annulus is isomorphic to $\mathbb{Z}$ and the fundamental group of the torus is isomorphic to $\mathbb{Z} \oplus \mathbb{Z}$. In the case of the annulus the fundamental group is generated by closed paths that loop the hole and by their integral multiples. In the case of the torus we have meridians and parallels and their multiples.

the classification of a concrete manifold would require the carrying out of infinitely many comparison operations between this manifold and the successive elements of the sequence. We note that even if it turned out that 3-manifolds (this is a standard abbreviation) are likewise not classifiable, this impossibility would be different from that for manifolds of dimension four and up. This is so because 3-manifolds behave differently than their higher-dimensional "relatives."

Wherein lies the difference? A detailed answer to this question would require the presentation of a number of facts from algebraic topology. This we cannot do here. What we can do is describe a certain deep difference between 3-manifolds and 4-manifolds. The difference is that given a finitely generated group it is possible to construct a 4-manifold whose fundamental group is the given group, whereas for 3-manifolds this construction is in general impossible. More specifically, it has been shown that there are groups that cannot be the fundamental group of any 3-manifold [9]. One such is the direct sum $\mathbb{Z} \oplus \mathbb{Z} \oplus \mathbb{Z} \oplus \mathbb{Z}$ of four copies of $\mathbb{Z}$.

We see that topological manifolds of lowest dimension behave in a "singular" manner, in the sense that they can be classified. In this respect the dimension 4 plays the role of a limiting case.

It might seem that classification is a manifestation of excessive pedantry on the part of topologists whose love of order compels them to try to systematize the abundance of manifolds. But recently this classification has become of intense interest to physicists. The reason is that the topology of manifolds determines the possible interactions of the elementary components of matter. We give some relevant facts.

Until recently it was generally assumed that the elementary particles are pointlike. If so, then their history in spacetime is represented by worldlines. In classical accounts the interactions of particles are represented by intersecting or ramified worldlines. In quantum accounts they are represented graphically in a similar way by means of Feynman diagrams. A single line is a manifold, but an object in the form of an X or a Y is not, for a branch point, or a point of intersection, constitutes a singularity. This means that 1-dimensional topology, and in particular 1-dimensional manifolds, are useless for the description of interactions. The situation changes radically if the elementary objects are stringlike. Closed strings or loops, regarded as more interesting than open strings (segments), determine in spacetime worldtubes shaped like the surface of an asymmetric cylinder whose axis is a timelike line. We illustrate an interaction in which two strings collide and become one by means of a diagram that resembles a pair of trousers: two cylinders merging into one. The common feature of these and of more complex interactions is that the worldtubes of free and interacting strings are "proper" 2-manifolds without singular points. If a worldtube has the topology of a cylinder then the string is free. Every interaction results in a change in the topology. Topologically inequivalent worldtubes correspond to different interactions. This means that the topological classification of manifolds characterizes the possible Feynman diagrams of interacting strings (see [34]).

This reasoning carries over to higher-dimensional elementary objects. 2-dimensional membranes describe in spacetime 3-dimensional "worldsolids," and the difference between free and interacting membranes is reflected in the different topologies of their "solids." It is generally thought that one should consider strings and ignore membranes and higher-dimensional objects. The arguments in favor of such a position are purely technical and are based on certain formal properties of mathematical strings; their physical sense is unclear. We recall that it is impossible to classify 4-manifolds, and that the problem of classification of 3-manifolds offers difficulties that have not been surmounted thus far. This implies the impossibility of classification of Feynman diagrams corresponding to dimension 4 and our present inability to classify the Feynman diagrams corresponding to dimension 3. Is it possible that these facts offer the first serious argument in favor of strings and against membranes and other objects? The thought that there may be so intimate a connection between topology and physics is truly fascinating.

5. THE POINCARÉ CONJECTURE. This conjecture is an important problem in the topology of manifolds and is linked with the classification problem. The great Henri Poincaré created algebraic topology. In particular, he provided in the 1890s the foundations of homotopy and homology theories. To explain the Poincaré conjecture we must say something about the first of these theories.

We assume that the reader is familiar with the definition of the *fundamental group*, or *first homotopy group* $\pi_1(X)$ of a space X. The concept of a fundamental group is a typical example of the procedures used in algebraic topology: given a topological space, we associate with it various groups and reduce the study of the space to the study of these groups. What makes the fundamental group important is that it is a topological invariant, i.e., the fundamental groups of homeomorphic spaces are isomorphic.

In general, the problem of finding the fundamental group of a space is very difficult. Below we give the fundamental groups of some simple spaces.

1. $\mathbb{R}^n$, i.e., n-dimensional Euclidean space, is simply connected. Define an n-dimensional sphere S^n as all the points in $\mathbb{R}^{n+1}$ such that

$$(x^1)^2 + (x^2)^2 + \cdots + (x^{n+1})^2 = 1.$$

For $n \geq 2$, S^n is also simply connected. This means that the fundamental groups of S^n for $n \geq 2$ and of $\mathbb{R}^n$ for all n are trivial.

It would seem that a space is simply connected only if it has no "holes," for then every closed loop can be contracted to a point. This can be translated into the statement that the fundamental group, which counts inequivalent closed loops, counts "holes" as well. This is not quite true. The sphere S^2 minus a point is simply connected (it is homeomorphic to the plane) and so is $\mathbb{R}^3$ minus a point (or even minus a ball). In both cases the fundamental group fails to "notice" the hole.

2. The fundamental group of the circle S^1 is isomorphic to the additive group $\mathbb{Z}$ of the integers. Indeed, if $n \neq m$, traversing the circle n times yields a loop inequivalent to that obtained by traversing it m times. Other spaces with the same fundamental group are the plane minus a point, and, more generally, any figure in the form of an annulus or a cylinder. Yet another space with the same fundamental group is $\mathbb{R}^3$ minus a straight line.
3. On a torus $S^1 \times S^1$ there are three inequivalent types of closed loops: contractible closed loops, closed loops in a plane perpendicular to the axis of the torus, and closed loops in the plane of the axis of the torus. It follows

that the fundamental group of the torus is the direct sum $\mathbb{Z} \oplus \mathbb{Z}$. Another space with the same fundamental group is $\mathbb{R}^3$ minus two linked circles.

Back to our main topic. Around 1895 Poincaré first advanced a conjecture that he formulated rigorously in 1904. The conjecture was to the effect that *every simply connected compact 3-manifold without a boundary is homeomorphic to S^3*.

The conjecture sounds deceptively simple. Mathematicians have pondered it for close to a century without being able to prove it or to disprove it. In the process they generalized it to an arbitrary number of dimensions. This is not a trivial generalization, for the conjecture is false for manifolds that are just simply connected (for example: $S^2 \times S^2$ is a compact and simply connected manifold without boundary which is *not* homeomorphic to S^4). This led to the following version of the generalized Poincaré conjecture for topological manifolds: *a compact n-manifold without boundary that has the homotopy type of S^n is homeomorphic to S^n*.

We digress to explain what is meant by the "homotopic sameness" of spaces. (This concept is somewhat similar to that of the homotopic sameness of paths.) The definitions that follow are taken from [11]. I denotes $[0, 1]$.

Definition. Two continuous maps $\varphi_0, \varphi_1 : X \to Y$ are *homotopic* if and only if there exists a continuous map $\varphi : X \times I \to Y$ such that, for $x \in X$,

$$\varphi(x, 0) = \varphi_0(x),$$
$$\varphi(x, 1) = \varphi_1(x).$$

If two maps φ_0 and φ_1 are homotopic, we shall denote this by $\varphi_0 \simeq \varphi_1$. This is an equivalence relation. The equivalence classes are called *homotopy classes* of maps.

To better visualize the geometric content of the definition, let us write $\varphi_t(x) = \varphi(x, t)$ for any $(x, t) \in X \times I$. Then, for any $t \in I$,

$$\varphi_t : X \to Y$$

is a continuous map. Think of the parameter t as representing time. Then, at time $t = 0$, we have the map φ_0, and, as t varies, the map φ_t varies *continuously* so that at time $t = 1$ we have the map φ_1. For this reason a homotopy is often spoken of as a continuous deformation of a map [11, p. 64].

We can now define the "homotopic sameness" of spaces.

Definition. Two spaces X and Y are of the *same homotopy type* if there exist continuous maps (called *homotopy equivalences*) $f: X \to Y$, $g: Y \to X$ such that $gf \simeq$ identity: $X \to X$ and $fg \simeq$ identity: $Y \to Y$ [11, p. 84].

Before continuing, we recall several definitions of terms that appear in the sequel. (A useful reference is [27].) We assume familiarity with the notion of a topological manifold and of standard terms associated with this notion.

   α. A smooth manifold, also called a C^∞-manifold, is a manifold with C^∞ transition functions.

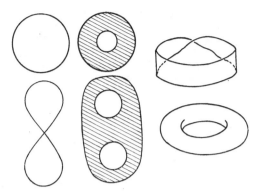

Figure 4. An annulus, a Möbius band, and a circle are homotopically equivalent. So too are a figure eight and an "annulus" with two holes. A figure eight and a circle are not homotopically equivalent. A sphere and a plane have the same fundamental groups but are not homotopically equivalent. A sphere minus a point and a plane are homotopically equivalent (and even homeomorphic). A torus minus a point, a plane minus two points, and a figure eight are homotopically equivalent.

β. A homeomorphism between open subsets of Euclidean space is PL (piecewise linear) if there is a triangulation of each open set into rectilinear simplexes such that the homeomorphism is linear on each simplex.

A manifold is PL if it admits a PL structure. A PL structure on a manifold is specified by an atlas such that the transition functions are all PL-homeomorphisms (such homeomorphisms between manifolds are defined in the usual way by using charts).

γ. If M and N are PL manifolds and $f: M \to N$ and f^{-1} are PL-homeomorphisms, then we say that M and N are PL-isomorphic.

δ. Let M and N be smooth manifolds. $f: M \to N$ is called a C^∞-homeomorphism if, for each $x \in M$, there are charts (U, φ) and (V, ψ) in the atlases of M and N, respectively, such that $x \in U$, $f(U) \subseteq V$, and $\psi \circ f \circ \varphi^{-1}$ is a C^∞-map between open subsets of $\mathbb{R}^n$. f is called a diffeomorphism from M to N if both f and f^{-1} are C^∞-homeomorphisms.

By now we are prepared to resume our account.

The generalized conjecture for topological manifolds was first proved for large values of n [12]. Between 1960 and 1962 Stephen Smale proved it for smooth manifolds of dimension $n \geq 5$ [13] while John Stallings and Christopher Zeeman proved it for PL manifolds of dimension $n \geq 5$. The only open cases left were $n = 3$ and 4. (For $n = 2$ the generalized conjecture follows immediately from the theorem on the classification of compact surfaces without boundary. The only such simply connected surface is S^2.) In 1964 Zeeman proved the following variant of the generalized conjecture: A compact PL-manifold M^n, $n \geq 6$, without boundary that has the homotopy type of S^n is PL-isomorphic to S^n.

These results were obtained by using powerful new methods for the study of topological spaces discovered in the 1950s. But their use is limited to manifolds of large dimension; there is "not enough room" in low-dimensional manifolds for these methods to work. That is why, surprisingly, the topology of low-dimensional manifolds is more difficult to study than the topology of high-dimensional ones; the techniques used to study the latter are not applicable to the former.

We note that, depending on the structures that can be introduced on manifolds, there are three formulations of the generalized Poincaré conjecture. The one stated earliest applies to topological manifolds. In the version for differentiable manifolds all the mappings are smooth and the homeomorphisms are replaced by diffeomorphisms. In the version for PL manifolds the mappings, homotopies, and isomorphisms are piecewise linear.

Here is what was known about the generalized Poincaré conjecture before 1982:

a. The topological case: the conjecture is known to be true for all dimensions other than the distinguished dimensions 3 and 4.
b. The PL case: the conjecture is known to be true for all dimensions other than the distinguished dimensions 3 and 4.
c. The differentiable case: the conjecture is known to be true for $n = 5$ and 6. In general, it is false for $n \geq 7$. This is a consequence of John Milnor's discovery [14] of the existence of "nondiffeomorphic" differentiable structures on spheres of dimension $n \geq 7$ (see Section 6).

In 1981 M. H. Freedman proved the topological variant of the conjecture for the case $n = 4$ [15]. In 1986 Colin Rourke and Eduardo Rego announced a proof of the classical Poincaré conjecture for $n = 3$. However, it turned out that their proof contained a technical error that could not be eliminated.

6. DIFFEOMORPHIC AND NONDIFFEOMORPHIC DIFFERENTIABLE STRUCTURES.

We recall the definition of a differentiable structure on a smooth manifold M: By an atlas on M we mean a set of charts (U, φ) such that the U's cover M. Two atlases are said to be *compatible* if their union is an atlas. Alternatively, two atlases are compatible if, whenever a chart (U, φ) in one atlas and a chart (V, ψ) in the other atlas satisfy $U \cap V \neq \emptyset$, then the composition

$$\varphi \circ \psi^{-1} : \psi(U \cap V) \to \varphi(U \cap V)$$

is a diffeomorphism. Compatibility of atlases is an equivalence relation, and an equivalence class of atlases is called a *differentiable structure* on M. Briefly, we say that two different atlases define the same differentiable structure if they are compatible in this sense.

It is a complete triviality to note that any differentiable manifold admits many distinct differentiable structures. In fact, let h be a completely arbitrary homeomorphism from the manifold M to itself. If the given differentiable structure is described by the atlas $\{(U, \varphi)\}$, then we can use h to construct a new atlas $\{(U, h \circ \varphi)\}$ and hence a new differentiable structure. This new differentiable structure will coincide with the original one only if h is actually a diffeomorphism.

Two differentiable structures that are related in this trivial way are said to be "diffeomorphic." That is, two differentiable structures on a topological manifold M are diffeomorphic if there is a homeomorphism from M to M that becomes a diffeomorphism when these structures are utilized.

As an example, if our manifold M is the real line $\mathbb{R}$ with its usual differentiable structure, then the homeomorphism

$$h(x) = x^3$$

does not have a differentiable inverse, and hence gives rise to a new differentiable structure. (This h is a diffeomorphism except at the origin, but one can equally well construct examples that are *nowhere* differentiable.)

Until 1956 it was generally assumed that this was the only way to generate distinct differentiable structures. In other words, it was expected that two differentiable manifolds that are homeomorphic would necessarily be diffeomorphic. Small wonder that John Milnor's discovery [14] that S^7 admits as many as 28 nondiffeomorphic differentiable structures was viewed as sensational. Soon mathematicians found out that, in this respect, S^7 is not an oddity, for higher-dimensional spheres have similar properties [16]. It was shown that it is possible to construct nondiffeomorphic differentiable structures on other manifolds as well. This was a great triumph of the methods of algebraic topology. A property viewed for a long time as natural turned out to be exceptional as soon as one looked at manifolds of sufficiently high dimension. Small wonder that people asked the obvious question: Is it possible to construct on $\mathbb{R}^n$ a structure nondiffeomorphic to the natural one? It required great effort to prove that for $n \neq 4$ the answer is negative. Mathematicians breathed more freely—at least here there were no surprises [17]. But what about $\mathbb{R}^4$? People were prepared for technical difficulties but assumed that more powerful methods would show that in this respect $\mathbb{R}^4$ resembles other $\mathbb{R}^n$. Not so! Freedman's papers on the Poincaré conjecture for dimension 4 made it possible for S. K. Donaldson to prove that there are nondiffeomorphic structures on $\mathbb{R}^4$ [15], [18]; one such is known as the *exotic* $\mathbb{R}^4$. Moreover, Donaldson and R. Gompf found other nondiffeomorphic structures [18], [19]. But many mathematicians were surprised when Gompf showed [20] that there is a countably infinite collection of nondiffeomorphic differentiable structures on $\mathbb{R}^4$. Then C. H. Taubes [28] proved that there are *uncountably many* such structures. Departures from the norm are one thing, but such extremism on the part of $\mathbb{R}^4$ . . . .

In this connection it is worth noting a fact that in the future may become more than just a curiosity. For most of the 20th century mathematics may be said to have parted company with physics, and situations in which physical problems inspire fundamental mathematical research are few and far between. One well-known exception is the rise of the theory of distributions, which was the mathematicians' response to the introduction by Dirac of the δ function into quantum mechanics. It turns out that the methods used by Donaldson and Gompf are another such exception. They used results and methods from various areas of mathematics, such as algebraic topology, differential geometry, and algebraic geometry, but they also used results and methods from the theory of gauge fields, a theory hitherto of interest to physicists alone. The application of methods of this theory was an essential element of their construction, and was viewed as a sensational departure by mathematicians unaccustomed to seeing purely mathematical problems solved by methods borrowed from theoretical physics. Given the present state of mathematical knowledge, Gompf and Donaldson could not have obtained their results if they were restricted to the "traditional" methods of algebraic topology. The theory of gauge fields bids fair to become the darling of topologists [18], [21], [22], [23].

What is the "meaning" of nondiffeomorphic differentiable structures? The following is a partial answer to this question.

Let $\mathscr{R}^4$ denote $\mathbb{R}^4$ with the "exotic" differentiable structure found by Donaldson. In this differentiable manifold one can find a compact set that cannot be surrounded by any *smoothly* embedded 3-sphere! It is easy to find *continuously* embedded 3-spheres: for example, choose any $\mathbb{R}^4$ metric and look at the 3-sphere $S(r)$ centered at the origin with radius r (i.e., the set of all points x in $\mathscr{R}^4$ with $|x| = r$). However, for r big enough $S(r)$ will be very jagged! This is very different from how our familiar $\mathbb{R}^4$ works, and indeed suffices to show that $\mathscr{R}^4$ has a differentiable structure nondiffeomorphic to that of the usual $\mathbb{R}^4$ [26].

A natural and complementary question to the one just asked is whether it is always possible to put a compatible differentiable structure on a topological manifold; in other words, whether it is always possible to define the notion of a *differentiable* function on a topological manifold in such a way that it is necessarily *continuous*. In dimensions 1, 2, and 3 the answer is always yes; in fact, each manifold of dimension 1, 2, and 3 admits just one differentiable structure (of course, up to a diffeomorphism) compatible with the topological structure. In dimensions 4 and higher the answer is no; for example, there are infinitely many compact 4-manifolds that admit no differentiable structure [26]. Similarly, it is known that any differentiable manifold can be given a compatible PL structure (i.e., it can be triangulated), and for dimensions $n \leq 6$, every PL n-manifold admits a unique (up to a diffeomorphism) compatible differentiable structure.

7. AN ATTEMPT AT A SUMMARY. It can be argued that the preceding examples have been tendentiously selected, and that one can find just as many—in fact more—theorems that behave in a "proper" manner in all dimensions. Moreover, for an arbitrary fixed n it is possible to find pathological situations precisely in n-dimensional space. Homotopy theory supplies many relevant examples.

The area of homotopy theory we have in mind is the one concerned with multidimensional knots and links. This theory arose as a natural generalization of 1-dimensional knots and links. An n-dimensional knot is homeomorphic to a sphere S^n, whereas an n-dimensional link is a disjoint sum of a finite number of knots. High-dimensional links can behave very differently. Consider, for example, a link made up of two spheres of dimension 50. In a space of dimension 102 or higher such links are splittable. Spaces of dimension 101, 100, 99, and 98 admit a nontrivial link, but spaces of dimension 97 and 96 don't! Nontrivial links can again be constructed in spaces whose dimension lies between 96 and 52 [24].

Can we nevertheless claim that mathematics singles out dimensions 3 and 4? Our cautious reply, dictated by the imprecision of terms such as "singles out" and "distinguishes" and by the inescapable element of subjectivism present in all interpretations, is that it does.

Unlike the theories of knots and links, the preceding problems belong to the most fundamental problems of topology, and, more generally, of mathematics. It is no accident that the topology of low-dimensional spaces (including the distinguished dimensions) has become an important branch of topology that relies on specific research methods. A surprising variety of difficulties, associated with low-dimensional versions of a great many theorems, forced mathematicians to create appropriate new methods and approaches. Specialization went even further: there came into being a topology of 3-manifolds and a distinct 4-dimensional topology. We emphasize that our discourse is limited to topology, for dimension is a topological concept and not a set-theoretic one.

We give other examples in which the dimensions 3 and 4 are singled out, examples less weighty from a mathematician's standpoint. We know from group theory that the group of rotations $SO(n)$ of $\mathbb{R}^n$ is simple for all n other than 4, when it is the direct product $SO(3) \times SO(3)$. Next an example from the theory of vector spaces of special importance for physicists: only in $\mathbb{R}^3$ can we define the cross product of vectors, and only there is the curl of a vector a vector. If we lived in a Euclidean (or almost Euclidean) space of another dimension, then the moment of a force and the moment of momentum would be antisymmetric tensors rather than vectors.

The question of how 3-dimensional space is distinguished by physical phenomena is completely outside the scope of our article. Nevertheless it is tempting to mention a few such phenomena. This problem was first investigated by Paul Ehrenfest in a famous article published in 1917 [25]. It turns out that the 3-dimensionality of physical space plays a key role for the most important phenomena of nature. If $n \neq 3$, then there are no stable atoms, so that everything connected with chemistry and biology ceases to be; elementary particles can coalesce into larger objects only under the action of gravitational and nuclear forces. But they cannot form planetary systems for the orbits of planets are unstable. Thus such a world would be completely different from the one we know. Only in $\mathbb{R}^3$ can wave phenomena be used for reliable transmission of information. Somewhat later Hermann Weyl showed that only in 4-dimensional spacetime are Maxwell's equations conformally invariant. There are more examples of this kind.

Given the present state of knowledge, there is no connection between problems that incline mathematicians to single out the dimensions 3 and 4 and the fact that physical spacetime—in its entirety or in its macroscopically observable part—is 4-dimensional. All attempts to relate these two facts are outside mathematics and physics, and belong to the realm of free philosophical reflection. The mathematician's assertion that "3- and 4-dimensional spaces are distinguished" is an abbreviated description of the fact that there are many independent and important problems whose nature or solution is atypical in such spaces. In no case can we say that such spaces are distinguished or singular "by their nature." Such a statement goes beyond the confines of mathematics. If anything, the connection with physics is even more baffling. Many physicists think that the deepest foundations of mathematics derive from the physical world. At least some mathematicians doubt this. Even if this were true (how is one to justify such a claim?), we have no idea of how the dimension of the universe we live in is reflected in the mathematical properties we discussed earlier.

On the other hand, the assumption that the coincidence of the dimension of the physical universe and of the dimension of spaces in which many important and independent topological problems have a singular character is entirely accidental and has no deeper foundation is hardly satisfactory. One thing is certain: the question of why the dimension of physical spacetime is what it is, is a genuine scientific question.

ACKNOWLEDGMENT. This is an abbreviated version, approved by the authors, of a Polish paper titled "Czy matematyka wyróżnia jakiś wymiar przstrzeni?" ("Does mathematics distinguish certain dimensions of spaces?"). The authors are Zdzisław Pogoda and Leszek M. Sokołowski. The paper appeared in *Postępy fizyki* (*Progress of physics*) 40 (1989), 407–433.

The translator wishes to thank John Milnor, Terry Gannon, and David Spring for their help.

REFERENCES

1. H. Poincaré, Pourquoi l'espace a trois dimensions?, *Revue de Metaphysique et de Morale* 20 (1912).
2. J. Stillwell, *Classical topology and combinatorial group theory*, second ed., Springer-Verlag, New York, 1993.
3. K. Menger, *Dimensionstheorie*, Teubner, 1928.
4. V. G. Boltyanskiĭ and V. A. Efremovich, *Intuitive combinatorial topology*, Springer, 2001, tr. A. Shenitzer.
5. R. Engelking, *Dimension theory*, North-Holland, Amsterdam, 1978.
6. H. S. M. Coxeter, *Introduction to geometry*, Wiley, New York, 1969.
7. D. Hilbert and S. Cohn-Vossen, *Geometry and the imagination*, Chelsea, New York, 1952, tr. P. Nemenyi.
8. S. Smale, On the structure of manifolds, *Amer. J. Math.* 84 (1962) 387–399.
9. A. T. Fomenko, *Differential geometry and topology*. Supplementary chapters, Moscow State University, Moscow, 1983. (Russian)
10. A. A. Markov, *Proceedings of the International Congress of Mathematicians*, 1958, 300–306.
11. W. S. Massey, *Algebraic topology: an introduction*, Harcourt, Brace & World, New York, 1967.
12. R. Duda, On the Poincaré conjecture, *Wiadomości Matematyczne* XVIII (1974) 19–39. (Polish)
13. S. Smale, Generalized Poincaré's conjecture in dimensions greater than four, *Ann. Math.* 74 (1961) 391–406.
14. J. Milnor, On manifolds homeomorphic to the 7-sphere, *Ann. of Math.* (2), 64 (1956) 399–405.
15. M. H. Freedman, The topology of four-dimensional manifolds, *J. Differential Geom.* 17 (1982) 357–454.
16. M. A. Kervaire and J. W. Milnor, Groups of homotopy spheres I, *Ann. of Math.* (2) 77 (1963) 504–537.
17. R. Kirby and L. Siebenmann, *Foundational essays on topological manifolds, smoothings and triangulations*, Ann. of Math. Studies 88, Princeton University Press, Princeton, 1977.
18. S. K. Donaldson, An application of gauge theory to four-dimensional topology, *J. Differential Geom.* 18 (1983) 279–315.
19. R. Gompf, Three exotic $\mathscr{R}^4$'s and other anomalies, *J. Differential Geom.* 18 (1983) 317–328.
20. _____, An infinite set of exotic $\mathscr{R}^4$'s, *J. Differential Geom.* 21 (1985) 283–300.
21. R. J. Stern, Instantons and the topology of 4-manifolds, *Math. Intelligencer* 5 (1983) 39–44.
22. _____, Gauge theories as a tool for low-dimensional topologists, 497–507, in: *Perspectives in mathematics*, Birkhäuser, Basel, 1984.
23. H. B. Lawson Jr., *The theory of gauge fields in four dimensions*, CBMS, AMS, Providence, 58 (1985).
24. D. Rolfsen, *Knots and links*, Publish or Perish, Berkeley, 1976.
25. P. Ehrenfest, In what way does it become manifest in fundamental laws of physics that space has three dimensions? *Proc. Amsterdam Acad.* 20 (1917).
26. D. S. Freed, K. K. Uhlenbeck, *Instantons and four-manifolds*, 2nd ed., Springer-Verlag, New York, 1991.
27. *Encyclopedic dictionary of mathematics*, 2nd edition, MIT Press, Cambridge MA, 1987.
28. C. H. Taubes, Gauge Theory on asymptotically periodic 4-manifolds, *J. Differential Geom.* 25 (1987) 363–430.
34. J. H. Schwarz, Superstrings, *Phys. Today* 40 (November 1987) 33–40.

Note: A glossary of physical terms used in this article may be found in pp. 225–228 of *Superstrings. A theory of everything?* Ed. P. C. W. Davies and J. Brown, Cambridge University Press, Cambridge, 1988.

Glimpses of Algebraic Geometry[1]

I. G. Bashmakova and E. I. Slavutin

PLANE ALGEBRAIC CURVES. Consider the equation
$$F(x,y) = 0, \qquad (1)$$
where $F(x, y)$ is a polynomial with rational coefficients that is irreducible over the field $\mathbb{Q}$ of rational numbers. The set of points of the real plane $\mathbb{R}^2$ whose coordinates satisfy the equation (1) is called a *plane (rational) algebraic curve*. If F is linear then we speak of a *rational line*. The points with rational coordinates are called *rational points*.

By the *degree* of the curve Γ defined by equation (1) we mean the degree n of the polynomial $F(x, y)$. The number of points of intersection of Γ and an arbitrary line $Ax + By + C = 0$ is exactly n. When counting the number of points of intersection we must consider multiplicities, complex points, and points at infinity. We give a few illustrative examples.

a. The curve $x^2 + y^2 = 1$ and the straight line $x + y = 10$ intersect in two complex points;
b. The curve $y^3 = 1 - x^3$ and the straight line $y = 1$ have the triple point of intersection $P(0, 1)$;

  (*Remark.* For a discussion of singular and multiple points see [1] and [5]. (Trans.))

c. The curve $y^2 = 4x^2 + x + 2$ and the straight line $y = 2x$ have two points of intersection, namely the point $M(-2, -4)$ and a point at infinity.

In order to define points at infinity we must introduce homogeneous coordinates, that is, essentially, we must go from the real plane $\mathbb{R}^2$ to the projective plane $\mathbb{P}^2$. A point of the projective plane is given by an ordered triple of real numbers (u, v, w) not all of which are 0. Proportional triples define the same point. The numbers in a triple (u, v, w) are called *homogeneous coordinates in* $\mathbb{P}^2$.

We now determine a (partial) correspondence between the points on $\mathbb{R}^2$ and on $\mathbb{P}^2$. Let (u, v, w) be a point on $\mathbb{P}^2$. If $w \neq 0$, then $(u/w, v/w, 1)$ determines the same point on $\mathbb{P}^2$. We associate with it the point on $\mathbb{R}^2$ with coordinates $x = u/w$, $y = v/w$. If $w = 0$, then the point $(u, v, 0)$ has no "partner" on $\mathbb{R}^2$. We call such points *points at infinity*. All points at infinity lie on the line at infinity $w = 0$.

In order to change equation (1) to an equation in homogeneous coordinates we put $x = u/w$, $y = v/w$. After obvious simplifications we obtain a homogeneous equation of the form
$$\Phi(u, v, w) = 0. \qquad (2)$$
Now points at infinity have the same status as ordinary points.

---

[1]This is a translation (by Abe Shenitzer, who also titled the piece) of part of the introduction to the monograph by Bashmakova and Slavutin titled *A History of Diophantine Analysis from Diophantus to Fermat*, published in 1984.

In terms of homogeneous coordinates, our curve $y^2 = 4x^2 + x + 2$ has the equation $v^2 = 4u^2 + uw + w^2$. Putting $w = 0$, we obtain its two rational points at infinity $M_1(1, 2, 0)$ and $M_2(1, -2, 0)$. The line $v = 2u$ passes through the point M_1. This is its second point of intersection with our curve.

The classification of curves by degree is of great significance. It was introduced by Descartes (who put in the same class curves of degree $2n$ and $2n - 1$) and made more precise by Newton.

The fundamental theorem related to degree is due to Bezout. It states that *the number of points of intersection of a curve of order m and a curve of degree n is mn*. Of course, here we must take into consideration multiplicities, complex points, and points at infinity.

Notwithstanding its importance, the classification of curves by degree alone is rather crude for purposes of diophantine analysis. Two curves of the same degree can have very very different sets of rational points. Thus the curve Γ with equation $x^2 + y^2 = 1$ has infinitely many rational points (with coordinates $x = 2k/(k^2 + 1)$, $y = (k^2 - 1)/(k^2 + 1)$, k rational), whereas the curve $x^2 + y^2 = 3$ has none.

The notion of greatest importance for diophantine analysis is that of birational equivalence of curves.

Definition 1. Two curves $f(x, y)$ and $g(u, v)$ are said to be *birationally equivalent* if the coordinates of each of them are expressible in terms of the coordinates of the other as rational functions with rational coefficients:

$$x = \varphi(u, v), \quad u = \varphi_1(x, y),$$
$$y = \psi(u, v), \quad v = \psi_1(x, y).$$

It is clear that the respective sets of rational points of two birationally equivalent curves coincide with the possible exception of a finite number of points. Birationally equivalent curves can have different degrees, that is, the degree of a curve is not a birational invariant. For example, the quartic curve

$$y^2 = x^4 - x^3 + 2x - 2 = (x - 1)(x^3 + 2)$$

can be transformed by means of the substitution

$$x = (1 + u)/u, \quad y = v/u^2$$

into the cubic

$$v^2 = 3u^3 + 3u^2 + 3u + 1,$$

with u and v rationally expressible in terms of x and y:

$$u = 1/(x - 1), \quad v = y/(x - 1)^2.$$

We will see that a quadratic curve with at least one rational point is birationally equivalent to a rational straight line.

It was Henri Poincaré who first called attention to the fundamental significance of birational transformations in the study of the arithmetic of algebraic curves. In the introduction to his famous paper "On the arithmetical properties of algebraic curves" he wrote: "I asked myself if it is not possible to connect many problems of analysis on a systematic basis by introducing a new classification of homogeneous polynomials of higher degree, analogous in a sense to the classification of quadratic forms.

This classification would have to be built on the foundation of the group of birational transformations admitted by the algebraic curve."[2]

One of the basic invariants of the group of birational transformations is the genus of a curve. To define it, we introduce first the notion of a simple double point on a curve.

The *singular points* on a curve Γ given by (1) are the points whose coordinates satisfy the equations

$$f_x(x, y) = 0, \qquad f_y(x, y) = 0.$$

An algebraic curve has only finitely many such points. A singular point $P(x_0, y_0)$ is called a *double point* if at least one of the second partial derivatives f_{xx}, f_{xy}, and f_{yy} does not vanish at P. Finally, a *simple double point* is a double point at which the curve has two noncoincident tangents (see Figure 1). When defining the genus of a curve we will assume that its only singular points are simple double points. This is not a serious restriction, for it can be shown that an algebraic curve is birationally equivalent to one with only simple double points.

We can now define the genus of a curve.

Definition 2. By the *genus* of a plane algebraic curve Γ of degree n we mean the number

$$g = \frac{(n-1)(n-2)}{2} - d, \tag{3}$$

where d is the number of simple double points on the curve.

It is clear that g is an integer. It can be shown that $g \geq 0$. If the degree is 1 or 2, then $g = 0$. Such curves are called *rational*. The reason for this is that if a curve Γ of genus 0 with equation

$$F(x, y) = 0$$

has a rational point $P(x_0, y_0)$, then the coordinates x and y can be expressed in the form $x = \varphi(t)$, $y = \psi(t)$, where φ and ψ are rational functions with rational coefficients and $F(\varphi(t), \psi(t)) \equiv 0$. Moreover, $t = \chi(x, y)$, where χ is also a rational function with rational coefficients.

One also says that curves of genus 0 can be *uniformized by means of rational functions*.

If $n = 1$, that is, in the case of straight lines, it is clear that any two rational straight lines $Ax + By + C = 0$ and $A_1 x + B_1 y + C_1 = 0$ are birationally equivalent, that is, there is just one class of birationally equivalent straight lines.

If $n = 2$, that is, if the curve is a conic section, and if there is a rational point $P(x_0, y_0)$ on it, then the curve is *birationally equivalent to a straight line*. To see this it suffices to take an arbitrary rational straight line D and to establish a one-to-one correspondence between the points M on the conic and the points M' on D so that the points P, M, M' are collinear. Since every conic with a rational point is equivalent to a rational line, all such conics are (birationally) equivalent to each other, and so form a single class that includes all rational lines. This implies that if a conic has a rational point then it has infinitely many rational points.

There are a great many equivalence classes of conics without rational points.

Poincaré proved the following theorem: "Every curve of genus 0 and degree $n > 2$ is birationally equivalent to a curve of degree $n - 2$." (Hilbert and Hurwitz

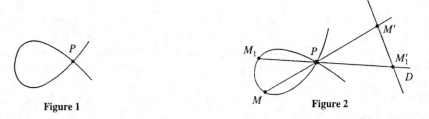

Figure 1 Figure 2

proved a similar result 10 years earlier; see [3].) Hence a rational curve of genus 0 is always equivalent to a straight line or to a conic.

If a cubic curve has genus 0, then

$$\frac{(3-1)(3-2)}{2} - d = 0,$$

that is, $d = 1$. But then the curve must have a simple double point which is clearly rational. A straight line passing through the double point P intersects the curve Γ in just one other point. We will show that in this case the cubic Γ is birationally equivalent to a rational line. To this end we take a rational line D and establish a one-to-one correspondence between the points M on Γ and the points M' on D so that the three points M, M', and the double point P are collinear (see Figure 2).

This shows that a cubic curve Γ with a simple double point is birationally equivalent to a rational line. As such, it can be uniformized by means of rational functions. For example, $P(0,0)$ is a double point on the curve $y^2 = x^3 - 2x^2$. If we pass through P the lines $y = kx$, then we obtain $k^2x^2 = x^3 - 2x^2$, whence $x = k^2 + 2$, $y = k(k^2 + 2)$.

Now we consider curves of genus 1.

It can be shown that curves of genus 1 cannot be uniformized by means of rational functions but can be represented by means of elliptic functions of one variable. Hence the name *elliptic* is attached to such curves.

If an equation

$$f(x, y) = 0 \tag{4}$$

determines a curve Γ of genus 1 with a rational point $P(x_0, y_0)$, then it is possible to reduce it by means of birational transformations to the form

$$y^2 = x^3 + ax + b. \tag{5}$$

This is the so-called *Weierstrass normal form*. In this case it is possible to express x and y in terms of the Weierstrass functions

$$x = \wp(t), \qquad y = \wp'(t).$$

Thus the coordinates of the rational points on a cubic curve cannot, in general, be expressed as rational functions of a single parameter. However, if we know one or two rational points on such a curve, then we can find yet another rational point on it. To do this one makes use of two methods, known, respectively, as the *method of tangents* and the *method of secants*.

1. If P is a rational point on a cubic Γ, then by drawing at P the tangent to Γ we obtain a rational straight line (the slope of this line is rational) that intersects Γ in a third rational point. This is the *method of tangents*.
2. If P_1 and P_2 are rational points on Γ, then the rational line $P_1 P_2$ intersects Γ in a third rational point P_3. This is the *method of secants*.

The fundamental theorem about curves of genus 1 was proved by Poincaré. It asserts that: **Every rational curve of genus 1 with a rational point is birationally equivalent to a cubic curve.**

Thus cubics are a model for the study of the arithmetic of curves of genus 1.

Let $\mathscr{M}$ be the set of rational points on an elliptic curve. Using the tangent and secant methods it is possible to impose on it the structure of an abelian group. In essence, this was done by Jacobi in 1835 in [**4**]. A deeper study of this group was carried out by Poincaré [**2**], who surmised that this group has a finite number of generators. He called this number the *rank* of the cubic curve. It was later shown that the rank of a curve is an invariant of the group of birational transformations. Poincaré posed the question of the possible values of the rank of a cubic curve. This question remains open. The English mathematician L. J. Mordell proved the deep result that the rank of an elliptic curve is always finite.

Poincaré showed that the group of an elliptic curve can contain elements of finite order (that is, that it is a group with torsion). In essence, this was already known to Fermat and Euler.

We conclude by considering the geometric sense of the group operations associated with the method of secants and the method of tangents. We first reduce a cubic curve Γ with rational points to the form (5). Let A and B be rational points on Γ and let C' be the point in which the straight line AB intersects Γ. Then we call the point C, symmetric to C' with respect to the x-axis, the sum of the points A and B:

$$A \oplus B = C.$$

Thus if C' has coordinates (x, y), then C will have coordinates $(x, -y)$. The transition from C' to C is of vital importance. It is only then that the operation of addition acquires the group properties, namely associativity, the existence of a zero element, and the existence of an additive inverse for each of its elements. The commutativity property of our operation is obvious.

To add A to itself, that is, to obtain $2A$, we use the method of tangents. We define the point $2A = D$ to be the point symmetric to the point D' in which the tangent at A intersects Γ.

It remains to find the point that plays the role of the zero element. No finite point will do. When we go to homogeneous coordinates by putting $x = u/w$, $y = v/w$, (5) becomes

(6) $$v^2 w = u^3 + auw^2 + bw^3.$$

If w and u are 0, then v is arbitrary and we can put $v = 1$. We denote by $\mathscr{O}$ the point at infinity on Γ with coordinates $(0, 1, 0)$. It is clear that the point $\mathscr{O}'$, symmetric to the point $\mathscr{O}$ with respect to the axis of abscissas, coincides with $\mathscr{O}$.

We show that $\mathscr{O}$ plays the role of zero. To this end we show that all vertical lines $u = mw$ intersect at $\mathscr{O}$. Indeed, if w and u are 0, then we can put $v = 1$. Now let $A(x_0, y_0)$ be a rational point on Γ. Then, according to what has just been shown, the straight line through A and $\mathscr{O}$ is vertical, that is, its equation is $x = x_0$. This straight line intersects Γ in the three points A, $\mathscr{O}$, and $A'(x_0, -y_0)$, the latter symmetric to A with respect to the x-axis. According to our definition, the sum of the points A and $\mathscr{O}$ is the point symmetric to A', that is, A itself. Thus $A \oplus \mathscr{O} = A$.

Finally, the inverse of A is $A'(x_0, -y_0)$. Indeed, the straight line joining these two points is vertical, and so intersects Γ at $\mathscr{O}$, that is, $A \oplus A' = \mathscr{O}$.

Points of finite order are characterized by the fact that $nA = A$ for some n, that is, there is a return to the initial point.

REFERENCES

1. F. Kirwan, *Complex algebraic curves*, London Math. Soc. Student Texts 23, Cambridge (1992). (Included by the translator.)
2. H. Poincaré, Sur les propriétés arithmétiques des courbes algébriques, *J. Math.*, 5e série, 1901, **7**, 161–233.
3. D. Hilbert, A. Hurwitz, Über die diophantische Gleichungen vom Geschlecht Null, *Acta Math.*, 1890, **14**, 217–224.
4. C. G. J. Jacobi, De usu theoriae integralium ellipticorum et integralium abelia-norum in analysi Diophantea, *Crelle's Journal für die reine und angew. Mathematik*, 1835, **13**, 353–355. Gesammelte Werke, Bd. 2, S.53–55.
5. J. H. Silverman, J. Tate, *Rational points on elliptic curves*, Springer (1992). (Included by the translator.)

Four Significant Axiomatic Systems and Some of the Issues Associated with Them

Stefan Mykytiuk and Abe Shenitzer

A. GREEK AXIOMATICS AND EUCLID'S GEOMETRY. One of the greatest intellectual achievements of the Greeks was the axiomatic method, a method for the systematic discovery of presumably absolute truths based on the application of logic to postulates and axioms. Postulates, to the Greeks, were "request(s) that something be allowed." More specifically, they were elementary, presumably obvious, truths relating to a particular discourse. (For example, the first of Euclid's postulates is: "A straight line can be drawn from any point to any point.") Axioms, to the Greeks, were elementary, presumably obvious, truths of a general nature. (For example, the first of Euclid's axioms is: "Things which are equal to the same thing are also equal to one another.") Euclid preceded his postulates and axioms with "Initial explanations and definitions" that suggest meanings and images the reader should attach to the terms of the discourse. They make it clear that one is dealing with abstractions from various physical objects. Euclid's geometry is the first known *extensive* example of what we now call an axiomatic structure.

Given some of the uses they made of Euclidean geometry, it is safe to say that the Greeks regarded it as a blueprint for the metric relations in the real world. Its uniqueness was unquestioned.

B. HYPERBOLIC GEOMETRY AND SOME EFFECTS OF ITS DISCOVERY. Euclid relied on five postulates. The fifth of these, a kind of fly in the ointment, is the famous Euclidean parallel postulate:

> If a straight line falling on two straight lines makes the interior angles on the same side together less than two right angles, the two straight lines, if produced indefinitely, meet on that side on which the angles are together less than two right angles.

"Now whatever else this postulate may be, self-evident it is not, and this was early perceived." [2] The commentator Proclus (5th century AD) objected to it by citing the asymptotic behavior of certain lines (= curves) and asking: "May not the same thing be possible in the case of straight lines . . . ?" Attempts to deduce the parallel postulate from the remaining postulates and the axioms were undertaken by various mathematicians for over 2000 years. All of them ended in failure.

A variant of Euclid's parallel postulate is the assertion: If l is a straight line and P a point not on l, then the number of straight lines through P that do not intersect l is just one. Its negation, obtained by replacing "is just one" by "is greater than one," is the so-called hyperbolic parallel postulate.

Around 1800 a few mathematicians began to experiment with the system of postulates and axioms obtained from Euclid's system by replacing the Euclidean parallel postulate by the hyperbolic parallel postulate. The latter is not an abstraction from sense impressions but a *logical* alternative to the Euclidean parallel postulate. It was Gauss, Lobachevskiĭ and Bolyai who explored the new "not-Euclidean" geometry, based on this modified foundation, in greatest depth. Lobachevskiĭ's investigations were more varied and extensive than those of either Gauss or Bolyai. Lobachevskiĭ and Gauss* carried out inconclusive physical experiments to determine which of the two geometries fitted physical space best, and formulated views of geometry radically different from the traditional ones inherited from the Greeks. Gauss' view is made clear by the following well-known quotations from his letters:

> I come more and more to the view that the necessity of our geometry cannot be proved.... Perhaps we shall come to another insight in another life into the nature of space, which is unattainable for us now. But until then one must not rank Geometry with Arithmetic which is truly a priori, but with Mechanics.... (From a letter to Olbers in 1817.)
>
> It is my deepest conviction that the positions of the science of space and of the pure science of magnitude vis-à-vis our knowledge a priori differ greatly; our knowledge of the former has none of the complete conviction of necessity (and thus also of absolute truth) associated with the latter; we must admit in all humility that while number is the product of our mind alone, space has also a reality outside our mind whose laws we cannot completely prescribe a priori. (From a letter to Bessel in 1830.)

Lobachevskiĭ's sophisticated and far less well-known view is made clear by the following quotation and comment:

> "In theory, nothing prevents us from assuming that the angle sum of a rectilinear triangle is less than two right angles.... The assumption that the angle sum of a triangle is less than two right angles is admissible only in Analytics, for measurement in nature does not reveal the slightest deflection of this sum from a half circle."
>
> This means...that Lobachevskiĭ views the generalized geometry as a mental, imaginary construction which makes sense only as an analytic generalization. What justifies it is not its possible use for purposes of measurement but its usefulness for all mathematics.... [For him its] acceptability...derives from his view of a mathematical theory as a method.] [5]

The first published accounts of hyperbolic geometry—by Lobachevskiĭ in 1829 and by Bolyai in 1832—had no immediate effect on the work of other mathematicians. Some of the mathematicians who were aware of the new geometric system were inclined to regard it as an aberration rather than as, in some sense, a valid alternative to Euclidean geometry. This began to change around 1860, as a result of the publication of the correspondence between Gauss and Schumacher.

One of Gauss' letters referred to his abiding interest in, and contributions to, hyperbolic geometry and to Lobachevskiĭ's masterly development of that geometry. Coming from Gauss, this letter generated a wave of interest in hyperbolic geometry. This interest was a key factor that ushered in a series of momentous discoveries and ideological changes not only in geometry but in all of mathematics. All this occurred in the short period between 1868 and 1872.

---

*In his *Gauss, a Biographical Study* (Springer, 1981), W. K. Bühler doubts the claim that Gauss tried to determine the angular defect of a triangle determined by three mountain peaks (see p. 100).

In 1868 Beltrami, who had familiarized himself with the work of Lobachevskiĭ, used the methods of differential geometry to establish the then surprising result that the intrinsic geometry of the pseudosphere, a surface with constant negative curvature, is *locally* hyperbolic. While doing this he also introduced an incomplete model of the hyperbolic plane in the interior of the unit disk. The gaps in Beltrami's model were filled in 1871 by Felix Klein, who arrived at *his* disk model along a projective route. *The Beltrami-Klein disk model showed that hyperbolic geometry is as consistent as Euclidean geometry. As a result, the status of Euclidean geometry as a unique system of absolute geometric truths was destroyed once and for all.*

The multiplicity of systems called geometries—Euclidean, projective and hyperbolic geometries, the geometries of surfaces in space, the geometries introduced by Riemann—gave rise to the question of what is a geometry. In his Erlangen Program of 1872 Klein gave a comprehensive (but not all-embracing) answer to this question by defining a geometry as the totality of invariants of the subsets of a set with respect to a group of permutations of that set.

"Without knowing of Klein's work, Henri Poincaré expressed similar ideas in 1880. He too was interested in [hyperbolic] geometry, and was aware of its usefulness in connection with the theory of differential equations. He picked up Beltrami's idea that on [a surface of constant curvature] it is possible to move figures without deforming them, and added that these motions form a group. Partly because of its visionary imprecision, his paper had a tremendous impact; it made the role of groups in geometry known far and wide." [6] Some twenty years later, the demonstrated importance of groups in Galois theory, in geometry, and in analysis paved the way for group theory as a distinct area of mathematics.

Hyperbolic motions play a vital role in the theory of automorphic functions, initially developed in the 1880s by Klein and Poincaré (see Chapter 1 in [7] and the paper [8] (which also discusses recent work involving so-called hyperbolic manifolds)). They also play a key role in 4-dimensional Minkowskian geometry, the mathematical setting of the special theory of relativity first presented by Minkowski in lectures in 1907. Specifically, the group of motions of H^3 (= hyperbolic 3-space) is isomorphic to the group of homogeneous motions (= homogeneous Lorentz transformations) of Minkowskian 4-space (see Chapter 7 in [9]).

As a postscript to this account of the effects of the discovery of hyperbolic geometry we might add that it had a liberating effect not only on mathematics but also on mathematicians. As H. Weyl put it, "the individual mathematician feels free to define his notions and to set up his axioms as he pleases."

C. PEANO'S AXIOMS AND "THE GREATEST INTELLECTUAL DISCOVERY OF THE 20TH CENTURY."
In the late 19th century, mathematicians managed to axiomatize arithmetic, and therefore, in a sense, all of mathematics. The first such axiomatization was achieved by Dedekind in 1888. Peano, working independently, published his clearer axiomatization of arithmetic a year later. This was the last triumphant step in a kind of "backward development"—from the complex to the simple. Specifically, in the 1830s Hamilton gave a rigorous definition of the complex numbers in terms of the reals, and in the 1870s Dedekind defined the reals as cuts[1] in the system of rationals, the field of quotients of the integers. Modern textbooks usually reverse the historical process and go by rigorous steps

> **Descriptions of Some Technical Terms**
>
> [1]**Cuts**. Split the rational numbers into two nonempty classes A and B such that every element of A is less than every element of B and such that B has no least element. Every such pair (A, B) is called a (Dedekind) *cut*.
>
> There are natural definitions of addition and multiplication of cuts that make them into an isomorphic replica of the real number system.
>
> [2]**Cardinals**. The *cardinal number* $|A|$ of a set A is, in some sense, a measure of its size. In fact, in the case of a finite set A, $|A|$ is just the number of its elements.
>
> If there is a 1-1 correspondence between sets A and B, we write $|A| = |B|$. If A is finite or $|A| = |N|$ (where N denotes the natural numbers), then we say that A is *countable*. Otherwise A is *uncountable*.
>
> [3] and [5] **Well-ordered sets and ordinals**. For some ordered infinite sets, the natural numbers suffice to describe the positions of the elements. For example, in the usual ordering $1, 2, 3, \ldots$ of the natural numbers, each number is both an element of the ordered set and a description of its position in the ordering.
>
> Now consider the ordering $1, 3, 5, \ldots, 2, 4, 6, \ldots$ of the natural numbers. Here we run out of natural numbers after describing the positions of the odd numbers (1 is in the first position, 3 in the second, and so on). Cantor proposed the symbol ω for the position of 2, $\omega + 1$ for the position of 4, and so on.
>
> Other orderings of the natural numbers led Cantor to introduce still more order symbols. For the ordering $3, 6, 9, \ldots, 1, 4, 7, \ldots, 2, 5, 8 \ldots$ (that is, first all numbers of the form $3k$, then of the form $3j + 1$, and finally of the form $3i + 2$) he used $\omega \cdot 2$ for the position of 2, $\omega \cdot 2 + 1$ for the position of 5, and so on.
>
> Each of the orderings of the natural numbers which Cantor considered has the property that every nonempty subset has a least element. (Note that the integers with their usual ordering do not have this property.) He called ordered sets with this property *well-ordered*, and the new symbols he introduced to describe position in such orderings of the natural numbers, *countable ordinals*.
>
> [4]**The continuum** is the set of all real numbers.
>
> [6]**The axiom of choice** states that given a family of nonempty disjoint sets, a set can be constructed containing exactly one element from each set in the family.

from the realm of the discrete to the realm of the continuous, from the natural numbers to the real and complex numbers.

A remarkable insight into the nature of the system of Peano's axioms, and therefore of mathematics, was achieved by Kurt Gödel in 1931. To describe it, we begin with certain preliminaries about systems of axioms.

We want a system of axioms to be *consistent*, that is, free of contradictions. If an axiom is implied by the other axioms then we can dispense with it, so it is natural to require each axiom to be *independent* of the others. Another property of a system of axioms is its *completeness*. This means that we have enough axioms to decide the truth or falseness of each statement of the system.

While we would like to know that we are working with a consistent system of axioms, we don't always strive for completeness; for example, the usual group axioms are not complete. On the other hand, it would be nice to know that Peano's axioms, the usual axiomatic basis of arithmetic, form a complete axiom set. This brings us to what is arguably the greatest intellectual discovery of the 20th century, namely the Gödel Incompleteness Theorems. They were discovered in

1931 by the 25-year-old Kurt Gödel who proved that

> For any consistent and finitely axiomatizable formal system F which contains the natural number system [with + and ·] there are undecidable propositions in F. One such undecidable proposition is the consistency of F.

F. de Sua described these remarkable insights in the following witty manner:

> Suppose we loosely define a *religion* as any discipline whose foundations rest on an element of faith, irrespective of any element of reason which may be present. Quantum mechanics for example would be religion under this definition. But mathematics would hold the unique position of being the only branch of theology possessing a rigorous demonstration of the fact that it should be so classified. [3]

The one island of presumed certainty of human thought was *proved* uncertain.

D. THE ZERMELO-FRAENKEL AXIOMATIZATION OF SET THEORY AND PAUL COHEN'S INDEPENDENCE RESULTS. What we refer to as naive (= pre-axiomatic) set theory was greatly advanced by Georg Cantor between 1872 and 1897. Unlike the post-Zeno Greeks, Cantor accepted the actual infinite without hesitation. He made sets the ultimate components of all things mathematical and provided a calculus of sets of arbitrary "size." By the end of the century his results enjoyed wide acceptance. Then came the difficulties in the form of paradoxes (Burali-Forti, Russell, and others) and such seemingly intractable problems of set theory as the problem of the continuum hypothesis (is the cardinality[2] of the set of countable ordinals[3] equal to the cardinality of the continuum?[4]) and the problem of well-ordering[5] the continuum. These two problems troubled most mathematicians more than the paradoxes, which they viewed as somewhat esoteric difficulties.

The question of well-ordering the continuum was solved in 1904 by Zermelo, who showed that if one accepts the axiom of choice,[6] then *all* sets can be well-ordered. But the axiom of choice had "side effects"—it led to various paradoxical subdivisions of figures (for example, a ball can be subdivided into five pieces (one of which is a single point) which can be reassembled into two balls each congruent to the original ball). The problem of the continuum hypothesis remained intractable.

Many of the logical difficulties associated with Cantor's set theory were overcome as a result of Zermelo's axiomatization, introduced by him in 1908 and later refined by A. Fraenkel, T. Skolem, and Zermelo himself. While its consistency is unprovable (because it effectively includes Peano's axioms), it is accepted by most mathematicians as a foundation for all mathematics more basic than Peano's axioms. The axiom of choice is now generally accepted. The continuum hypothesis remains open. In 1938, Gödel showed that these two are consistent both with each other and with the other axioms of set theory. In 1963 Paul Cohen did the same for their independence. This means, among other things, that mathematicians are free to adopt different mathematics! [4]

E. A SUMMARY. It is useful to juxtapose key past and present views.

Until the discovery of hyperbolic geometry it was thought that postulates and axioms are abstractions from experience and, together with their logical consequences, are at least approximately true of certain objects in the real world. Consistency of the postulates and axioms was taken for granted. These were "gut feelings" as well as "official" views.

The discovery of hyperbolic geometry initiated revolutionary changes in these views of a factual as well as a philosophical nature. There are now many axiomatic systems, including a whole hierarchy of set theories. The "official" view of postulates (or axioms—we now use the terms interchangeably) is that they are assumptions about some undefined primitive terms, hence results based on them are relative *logical* truths devoid of any outer physical meaning. The consistency of mathematics, whether we base it on Peano's axioms or on the Zermelo-Fraenkel axioms, is *in principle* unprovable. Just as the discovery of the independence of the parallel postulate split geometry in two, so too, more than a century later, the discovery of the independence of the axiom of choice and the continuum hypothesis from one another and from the remaining axioms of set theory split mathematics. So much for facts and "official" views. Now we come to feelings.

It is safe to say that almost every mathematician is a least a "residual Platonist," and this makes him more or less the intellectual brother of the ancient Greek mathematicians and an "emotional" opponent of formalism. Dieudonné described one variant of this syndrome in the following words:

> On foundations we believe in the reality of mathematics, but of course, when philosophers attack us with their paradoxes we rush to hide behind formalism and say "mathematics is just a combination of meaningless symbols...." Finally we are left in peace to go back to our mathematics and do it as we have always done, with the feeling each mathematician has that he is working with something real. This sensation is probably an illusion, but it is very convenient. That is Bourbaki's attitude toward foundations. (Quoted in [**1**].)

(The Platonism of other prominent mathematicians is more robust than that of Bourbaki.)

REFERENCES

1. M. J. Greenberg, *Euclidean and non-Euclidean geometries*, Freeman, 1974 (second edition) (Chapter 8).
2. J. L. Coolidge, *A history of geometricial methods*, Dover, 1963 (Chapter IV).
3. H. Eves, *Great moments in mathematics*, the MAA, Dolciani Mathematical Expositions, volumes 5 and 7 (Lectures 7, 26, 27, 35 and 38).
4. N. Ya. Vilenkin, *In search of infinity*. (Chapter 4), Birkhäuser, 1995, tr. A. Shenitzer.
5. V. Ya. Perminov, *The philosophical and methodological thought of N. I. Lobachevskiĭ*. (Russian, 1993. An English translation of this paper will appear in "Philosophia Mathematica.")
6. *Geschichte der Algebra*, ed. E. Scholz, BI-Wissenschaftsverlag, 1990 (Section 11.4 (pp. 307–309)).
7. J. Lehner, *Discontinuous groups and automorphic functions*, AMS, 1964 (Chapter 1).
8. J. Milnor, *Hyperbolic geometry: the first 150 years*, BAMS, v.6, #1, Jan. 1982.
9. N. V. Efimov, *Higher geometry*, Mir, 1980 (Chapter 7).
10. R. Rucker, *Infinity and the mind*, Birkhäuser, 1982.

Foundations of Mathematics in the Twentieth Century

Victor W. Marek and Jan Mycielski

1. INTRODUCTION AND EARLY DEVELOPMENTS. Logic and foundations are a domain of mathematics concerned with basic mathematical structures (in terms of which one can define all other mathematical structures), with the correctness and significance of mathematical reasoning, and with the effectiveness of mathematical computations. In the twentieth century, these areas have crystallized into three large chapters of mathematics: *set theory*, *mathematical logic* (*including model theory*), and *computability theory*, which are intertwined in significant ways. In this paper we describe the evolution and present state of each of them. In modern times the study of logic and foundations has attracted eminent mathematicians and philosophers such as Cantor, Cohen, Frege, Gödel, Hilbert, Kleene, Martin, Russell, Solovay, Shelah, Skolem, Tarski, Turing, Zermelo, and others, and has given rise to a large body of knowledge. Although our paper is only a brief sketch of this development, we discuss essential results such as Gödel's theorem on the completeness of first-order logic and his theorems on the incompleteness of most mathematical theories, some independence theorems in set theory, the role of axioms of existence of large cardinal numbers, Turing's work on computability, and some recent developments.

There are still many interesting unsolved problems in logic and foundations. For example, logic *does not* explain what "good" mathematics is. We know that mathematics has a very precise structure: axioms, definitions, theorems, proofs. Thus we know what is correct mathematics but not why the works of certain mathematicians delight us while others strike us as downright boring. Nor do foundations tell us *how* mathematicians construct proofs of their conjectures. Since we have no good theoretical model of the process for constructing proofs, we are far from having truly effective procedures for automatically proving theorems, although spectacular successes have been achieved in this area. We mention other unsolved problems at the end of this paper.

We now give a brief sketch of the history of logic and foundations prior to 1900. The ancient Greeks asked: *what are correct arguments?* and: *what are the real numbers?* As partial answers they created the theory of syllogisms and a theory of commensurable and incommensurable magnitudes. These questions resurfaced in the 18th century, due to the development of analysis and to the lack of sufficiently clear concepts of sets, functions, continuity, convergence, etc. In a series of papers (1878–1897) Georg Cantor created set theory. In 1879 Gottlob Frege described a formal system of logic that explained precisely the logical structure of all mathematical proofs. In 1858 Richard Dedekind gave a definitive answer to the question: *what are the real numbers?* by defining them in terms of sets of rational numbers. He proved the axiom of continuity of the real line, an axiom accepted hitherto (beginning with the Greeks) without proof.

As for the question of characterizing correct arguments or proofs, Aristotle created the theory of syllogisms, which codifies certain forms of proof. In terms of modern logic, these were rules pertaining to unary relations. A relevant example is the syllogism: *if Socrates is a Greek and if all Greeks are mortal then Socrates is mortal*. Written in modern logical notation this syllogism takes the form

$$[Greek(Socrates) \wedge \forall x(Greek(x) \Rightarrow mortal(x))] \Rightarrow mortal(Socrates).$$

In this formula $Greek(\cdot)$ and $mortal(\cdot)$ are symbols for unary relations. We used the universal quantifier $\forall x \ldots$, which means *for all x we have* …. Similarly, one introduces the existential quantifier $\exists x \ldots$ which means *there exists x such that* …. For two thousand years the theory of syllogisms was the core of logic. In the 17th century G. W. Leibniz hoped to create a "characteristica universalis," a language that would make it possible to express all mathematical sentences, and a "calculus ratiocinator," which would reduce all reasoning to computation. In the middle of the 19th century G. Boole, guided by the algebra of numbers, introduced an algebra of sentences. We owe further evolution of these ideas to A. De Morgan, C. S. Pierce, G. Frege, and other mathematicians and philosophers. In particular, it became clear that the logic used in mathematics goes beyond syllogisms. Indeed, mathematics involves relations between two or more objects. For example, "less than" relates two numbers and is thus a *binary* and not a *unary* relation.

Cantor made the most important steps toward the determination of fundamental mathematical structures. He showed that (almost) all objects used by mathematicians can be thought of as sets. Moreover, it turned out that such an interpretation removed all ambiguities previously encountered in mathematics. For example, Zermelo and von Neumann used the concept of set to define the concept of a natural number. In turn, natural numbers could be used to define integers and rational numbers. B. Bolzano, R. Dedekind, and Cantor showed independently how to use sets of rational numbers to construct the real numbers and to prove their fundamental properties. Then functions were defined as sets of ordered pairs, as we explain in Section 2. This new definition was simpler and more general than the earlier approach, which treated a function f as an algorithm for computing the value $f(x)$ for a given element x in the domain of f. In the 1880s and 1890s Cantor used this general concept of function to prove many theorems of set theory. He extended the concept of a cardinal number to infinite sets, analyzed notions such as linear order, and introduced the concept of well-ordering and the related concept of an ordinal number. He also used transfinite induction in definitions and proofs. This was a new method that went beyond induction over the natural numbers, already used in Euclid's *Elements*.

There was no theory of computability at the end of the 19th century. The original concept of functions as algorithms was developed only later, beginning with the work of Turing in 1935. However, issues of effectiveness have concerned mathematicians and philosophers since ancient times. There were many examples of algorithms, such as Euclid's algorithm for finding greatest common divisors, Cardano's algorithm for solving cubic equations. Newton's algorithm for finding zeros of differentiable functions, and Gauss's algorithm involving the arithmetic-geometric mean. Some computing devices existed in antiquity. In many cultures there were types of abaci, some of which are still in use today. Mechanical computing devices for addition and multiplication were developed for use in censuses. Science, artillery, and finance required many accurate computations. The demand for computing devices for astronomy, engineering, taxation bureaus, banks, etc., grew steadily. In the first half of the 19th century C. Babbage proposed the construction of a machine similar to a modern stored-program computer, but he was not able to complete this project. Up to the middle of the 19th century, many great analysts did not accept existential statements unsupported by algorithms. For example, to show that the initial value problem $y' = -y$, $y(1.3) = 2.4$, has a solution, one had to know how to construct it, i.e., one had to give a general prescription such that given any real number x one could compute the value $y(x)$ of the solution function y. Eventually, in the 20th century, the mathematical concept of a computable function was created by A. Church, K. Gödel, E. Post, and in a more definitive way by A. Turing.

An interesting prelude to the development of logic and foundations in the 20th century occurred in 1900. At the International Congress of Mathematicians in Paris, David Hilbert presented a lecture in which he stated 23 unsolved problems. Three of them, Nos. 1, 2, and 10, deal with logic and foundations:

> Problem 1 belongs to set theory and was originally posed by Cantor. It is called the *continuum hypothesis* and asks: *are there infinite sets of real numbers whose cardinality differs from the cardinality of the set of integers as well as from the cardinality of the set of all reals?*
> Problem 2 belongs to logic and asks: *are the axioms of analysis consistent?* In his lecture, Hilbert asked about the consistency of arithmetic. But by arithmetic he meant what is now called analysis: the arithmetic of the integers augmented by set variables and a full comprehension scheme that we discuss in Section 2.
> Problem 10 belongs to the theory of computability and asks: *is there an algorithm for deciding the existence of integral solutions of multivariable polynomial equations with integral coefficients?*

Thus the three areas of foundations of mathematics appeared in Hilbert's lecture. In the 20th century these three problems were solved in very surprising ways: K. Gödel and P. J. Cohen showed that the continuum hypothesis is independent of the generally accepted axioms of set theory (Table I). Gödel showed that Problem 2 is unsolvable. Moreover, no consistent theory containing enough combinatorics, whose axioms are explicitly given, can prove its own consistency. And M. Davis, Yu. Mativasevich, H. Putnam, and J. Robinson showed that Problem 10 has a negative answer: no such algorithm exists.

Before continuing our presentation of logic and foundations we add some remarks about the relationship of this paper to the accounts of other historians and philosophers.

Some authors divide the study of logic and foundations into three orientations: intuitionism, logicism, and formalism. There are reasons to think that this division is misleading; first, because these three concepts are not of the same kind, and second, because logicism and formalism are terminological oxymorons. Specifically, intuitionism is not an attempt to explain what mathematics is, but rather a proposal for an alternative mathematics. On the other hand, logicism and formalism are such attempts, but the difference between them is rather insignificant. Since these terms turn up so frequently in the literature, it makes sense to discuss them in greater detail.

Intuitionism was invented in 1908 by L. E. J. Brouwer as a reaction to the freedom of imagination proposed by Cantor, Hilbert, and Poincaré (*in mathematics, to exist is to be free of contradiction*), and to the program formalizing (that is, defining in mathematical terms) the rules of logic and the axioms of mathematics. Brouwer thought that freedom of imagination and formalization are incompatible: he also thought that many objects of classical mathematics are not sufficiently constructive. Developing these ideas, he rejected the existence of least upper bounds of nonempty bounded sets of reals (that is, the continuity of the real line) and many other nonconstructive statements. He proposed an alternative mathematics, called intuitionism, which was intended to be more meaningful than classical mathematics and to have a more meaningful logic (for example, the general theorem of classical logic *p or not p* was rejected; *p or not p* would be proved only by proving *p* or proving *not p*), and this logic would undergo perpetual development.

This program attracted some outstanding mathematicians, such as H. Poincaré and H. Weyl (although in their practice neither accepted all the strictures of intuitionism) and E. Bishop (who accepted them fully). However, intuitionism proved to be rather unsuccessful. It had two unpleasant technical features: its weak logic made it hard to prove theorems, so that many theorems of classical mathematics were rejected outright; and it has a plethora of concepts that are equivalent in classical mathematics but not under the weak logic of intuitionism. But Brouwer raised an issue that, if valid, would make up for these technical difficulties. He claimed that many concepts and objects of classical mathematics are meaningless. For example, he accepted the integers but rejected well-orderings of the real line.

Even today many mathematicians who are not familiar with logic and foundations would say that this sounds reasonable. But in 1904 Hilbert observed that the infinite sets of pure mathematics are actually imagined, in the sense that they are like containers whose intended content has not yet been constructed (imagined). The same view of sets was also held by Poincaré. This observation removed the existential (ontological) problem of infinite sets. Moreover, in 1923 the logic of Frege was extended by Hilbert to one in which quantifiers cease to be primitive concepts; they become abbreviations for certain quantifier-free expressions. In this way the apparent reference to infinite universes in pure mathematics, suggested by quantifiers, becomes a metaphor for a finite constructive process actually occurring in the human imagination (see Section 3). Thus the critique of classical mathematics raised by Brouwer and his followers collapsed. The authors of this article know of no clear motivation for intuitionism. The integers and their algebra (if we include some very large integers such as $10^{10^{10}}$) are no more constructive and finitary than any other mathematical objects and their "algebras." All we can say today in defense of Brouwer's idea is that some concepts of mathematics are useful in science and others are not; the latter must be regarded as pure art. But both kinds of objects are equally solid constructions of the human imagination and they become more permanent when expressed in writing.

Next we turn to logicism and formalism. Unlike intuitionism, they are attempts to explain what mathematics is. The first is attributed to G. Frege, B. Russell, and A. N. Whitehead, and its spirit is close to the philosophers of the Vienna circle. The second is attributed to D. Hilbert and his collaborators and to Th. Skolem, and is close to the Polish school of logic. The dividing line between logicism and formalism is not very significant because it depends merely on a terminological difference: whether set theory, or some of its variants (such as the theory of types), is included in logic. The term *logic* was used for a long time in the narrower sense (often called *first-order logic*), which does not include set theory; the growth of set theory and of model theory has motivated the modern terminology. An important fact that favors the more inclusive sense of logic (of Frege et al.) is the naturalness of the axioms of set theory; see Table I. They are so intuitive that they belong to the mental logic of almost all people who work in pure mathematics or use mathematics in science. Another reason why the term logicism is not used any more is that its program—to derive mathematics from logic in the wider sense—has been fully accomplished. This fundamental achievement is due mainly to Cantor and Dedekind. We say more about it in the next section.

To complicate matters, the term formalism has several meanings. First, it is associated with the problem posed by Hilbert in 1900 of proving the consistency of stronger (more expressive) theories in weaker theories, because this requires formalization of the stronger theory in the weaker one. Although in 1931 Gödel proved that such proofs do not exist, the idea of Hilbert gave rise to the problem of classification of mathematical theories according to their strength: we say that S is *stronger* than T if T is interpretable in S (e.g., analysis in set theory) but not vice versa. Gödel's results imply that S is stronger than T if the consistency of T can be proved in S. Work on the ensuing classification, which is sometimes called the *Hilbert program*, is still in progress. Second, formalism is used as a name for the philosophy of Hilbert that we discussed previously in connection with intuitionism. Brouwer called Hilbert a *formalist*, meaning that Hilbert professed that pure mathematics is just a formal game of symbols (Stalin levelled the same accusation against western art, which he called bourgeois art). This was unfair, since Hilbert insisted that mathematics is, first and foremost, a structure of thought-objects in our minds rather than of symbols on paper, and, if the word *game* can be applied to it, then the adjectives *interesting*, *beautiful*, and *often applicable* would have to be added as well. Thus, contrary to Brouwer's misnomer *formalism* for Hilbert's philosophy, the role of mathematics as an interface between science and art was essential to Hilbert.

2. SET THEORY. In the first decade of the 20th century, mathematicians clarified the concept of set. A naive approach to this concept, the *full comprehension scheme* of G. Frege and B. Russell, claims that "every class one can think of is a set," in symbols $\exists x \, \forall y [y \in x \Leftrightarrow \varphi(y)]$, where φ is any property expressed in the language of set theory. Russell himself found that this led to a contradiction when $\varphi(y)$ is the property $y \notin y$. Indeed, suppose that R is the collection $\{y : y \text{ is a set and } y \notin y\}$. It follows immediately from this definition that if R is a set then

$$R \in R \Leftrightarrow R \notin R.$$

This is a contradiction, hence R cannot be a set. It became clear that the concept of set had to be refined so as to avoid such contradictions. This is not to imply that the mathematical community, Cantor in particular, ever accepted the full comprehension scheme. Cantor knew that this scheme was too liberal and he had other examples to prove it. Collections defined by arbitrary formulas φ of the language of set theory are called *classes*. Cantor found that if the class of ordinal numbers were a set, then there would be an ordinal number not in that class—a contradiction; and if the class of all sets were a set, then its power set would have a cardinal number greater than itself—again a contradiction.

Something had to be done to make the concept of set clear and useful. This problem was solved by Russell and Whitehead, who constructed a weak set theory called the *theory of types*, and in a more satisfying way by E. Zermelo. In 1908 Zermelo proposed a system of axioms describing sets and methods for constructing them. These axioms do not imply that Russell's R, or Cantor's other classes, are sets. Zermelo's system, with certain improvements introduced by Th. Skolem and A. Fraenkel, became the generally accepted axiomatization of set theory. Almost all of mathematics can be developed within this theory, called Zermelo-Fraenkel set theory, or ZFC; see Table I.

These axioms describe a hierarchy of sets, among which are the natural numbers $0, 1, 2, \ldots$. First, taking any formula φ that is always false in axiom A4, we get a set with no elements. By axiom A1 this set is unique; we call it the *empty*

TABLE I. The Axioms of Set Theory ZFC

A1	**Extensionality**	(Sets containing the same elements are equal):
		$\forall z[z \in x \leftrightarrow z \in y] \to x = y$
A2	**Union**	(The union u of all elements z of a set x is a set; $u = \bigcup(x)$):
		$\exists u \, \forall y[y \in u \leftrightarrow (\exists z \in x)[y \in z]]$
A3	**Powerset**	(The collection p of all subsets of a set x is a set; $p = \mathcal{P}(x)$):
		$\exists p \, \forall y[y \in p \leftrightarrow (\forall z \in y)[z \in x]]$
A4	**Replacement**	(The image r of a set d by a mapping φ is a set):
		$\forall x \, \exists z \, \forall y[\varphi(x, y) \to y = z] \to \forall d \, \exists r \, \forall y[y \in r \leftrightarrow (\exists x \in d)\varphi(x, y)]$
	(The axiom A4 is a scheme: for each formula φ we get a separate axiom. The formula φ may contain free variables other than x and y but not the variables d, r, or z.)	
A5	**Infinity**	(There exists an infinite set s):
		$\exists s[\exists x[x \in s] \land (\forall x \in s)(\exists y \in s)\forall z[z \in y \leftrightarrow (z \in x \lor z = x)]]$
A6	**Regularity**	(Every non-empty set has an $\in$-minimal member):
		$\exists y[y \in x] \to (\exists y \in x)(\forall z \in y)[z \notin x]$
A7	**Choice**	(Every set x of non-empty, pairwise-disjoint sets y possesses a selector s):
		$\{(\forall y \in x)\exists z[z \in y] \land \forall yzt[y \in x \land z \in x \land t \in y \land t \in z \to y = z]\}$ $\to \exists s(\forall y \in x)\exists t \, \forall u[(u \in y \land u \in s) \leftrightarrow u = t)]$

set and denote it by $\emptyset$ or 0. By axiom A3, we have a set $\mathscr{P}(\emptyset) = \{\emptyset\}$ (also denoted by 1), which has the single element $\emptyset$. Then we have the set $\mathscr{P}(\{\emptyset\}) = \{\emptyset, \{\emptyset\}\}$ (also denoted by 2). And given any two sets u and v we can build the set $\{u, v\}$ by using axiom A4 with $d = \{\emptyset, \{\emptyset\}\}$ and

$$\varphi(x, y) := [(x = \emptyset \to y = u) \wedge (x = \{\emptyset\} \to y = v)].$$

We denote $\bigcup(\{u, v\})$ by $u \cup v$ and say that $x \subseteq y$ if and only if $x \cup y = y$. The ordered pair (x, y) is $\{\{x\}, \{x, y\}\}$ and the Cartesian product $a \times b$ is $\{(x, y): x \in a \wedge y \in b\}$.

Another important definition, due to von Neumann, is the class *Ord* of ordinal numbers. These are sets α (such as the sets 0, 1, and 2) with the property:

$$\forall x, y [x \in y \in \alpha \to x \subseteq y \subseteq \alpha].$$

The class *Ord* of ordinal numbers is not a set. One proves that *Ord* is the smallest class containing $\emptyset$, closed under the operation $x + 1 = x \cup \{x\}$ and under arbitrary unions of sets of its elements. An ordinal λ is called a *limit* if $\bigcup(\lambda) = \lambda$. We denote the least infinite ordinal number by ω or N, so $\omega = N = \{0, 1, 2, \ldots\}$.

The members of an ordinal are well-ordered by the $\in$ relation, and in general we define a *well-ordering* to be any ordering isomorphic to an ordinal, or to *Ord* itself. The axiom of choice A7 implies that any set can be well-ordered (and the converse also holds); Zermelo introduced A7 for precisely this purpose.

Every set x can be assigned a unique ordinal number called its *rank* and defined as follows $rank(\emptyset) = 0$ and $rank(x) = \bigcup(\{rank(y) + 1: y \in x\})$. The set of all sets of rank at most α exists and is denoted by V_α. We can define V_α (where α is any ordinal) inductively by:

$$V_0 = \emptyset, \quad V_{\alpha+1} = \mathscr{P}(V_\alpha), \quad V_\lambda = \bigcup_{\xi < \lambda} V_\xi \text{ for limit } \lambda.$$

The class of all sets is denoted by V, so $V = \bigcup_\alpha V_\alpha$; see Figure 1.

It is a surprising and important fact (established by Cantor, Dedekind, von Neumann, and others) that all mathematical objects can be interpreted in a natural way as sets in V. A simple but important step of this interpretation was the construction (by Wiener and Kuratowski) of ordered pairs (x, y) as $\{\{x\}, \{x, y\}\}$. Later, D. Scott constructed cardinal numbers without using the axiom of choice, by defining the cardinal number of x to be the set of all sets y of least possible rank such that y can be bijectively mapped onto x.

The axiom of choice, A7, troubled some mathematicians. This axiom asserts that for any set x of nonempty and pairwise disjoint sets there is a set with just one element in common with each member of x. This generalizes an obvious property of finite collections of sets, but it leads to sets that are not explicitly definable. Mathematicians now use the axiom of choice in many of its equivalent forms, such as Zorn's lemma or the well-ordering principle, but of course accepting the axiom implies accepting its consequences, such as the *paradoxical decomposition of a ball* found by S. Banach and A. Tarski in 1924. Improving on an earlier theorem of F. Hausdorff, they showed that a ball can be divided into five disjoint sets that can be moved by isometries to form two balls of the same size as the initial ball. Of course, this contradicts physical experience. Thus one can say that the mathematical concept of a set is far more liberal than the concept of a physical body. In

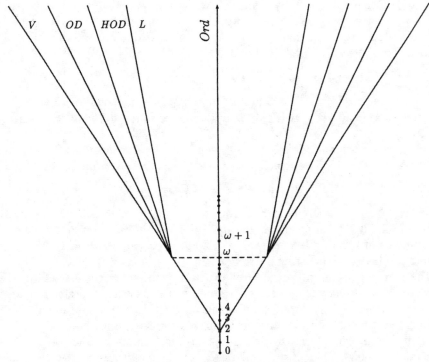

$Ord \subseteq L \subseteq HOD \subseteq OD \subseteq V$;
ZFC is true in V, in HOD, and in L;
ZFC + GCH + V = L is true in L;
It is consistent with ZFC that not all real numbers are in OD;
$Ord: 0 = \emptyset; \quad \xi + 1 = \xi \cup \{\xi\}; \quad \lambda = \bigcup_{\xi < \lambda} \xi, \quad \lambda$ limit;
$V: V_0 = \emptyset; \quad V_{\xi+1} = \mathcal{P}(V_\xi), \quad V_\lambda = \bigcup_{\xi < \lambda} V_\xi, \quad \lambda$ limit;

Figure 1. Ordinal-definable, hereditarily ordinal-definable and constructible sets in the universe of Set Theory.

general, the consequences of the axiom of choice (for uncountable families of sets) are less relevant to applications of mathematics than are theorems whose proofs do not involve this axiom.

There is a weaker axiom of choice (called the *axiom of dependent choices* or DC) that does not have such odd consequences. DC says that if $A_0, A_1, \ldots$ is a sequence of nonempty sets and $R_i \subseteq A_i \times A_{i+1}$ are relations such that

$$(\forall x \in A_i)(\exists y \in A_{i+1})[(x, y) \in R_i],$$

then there exists a sequence $a_0, a_1, a_2, \ldots$ such that

$$\forall i[(a_i, a_{i+1}) \in R_i].$$

The classification of all sets by rank is paralleled by other classifications, not necessarily of all sets, by *ordinal definability* or *constructibility*. All of these classifications can be defined in the theory ZFC minus the axiom of choice, which is called ZF.

A set x is *ordinal-definable* if, for some ordinal α and some formula $\varphi(y)$, x is the set of elements y of V_α that satisfy the formula $\varphi(y)$ in V_α. The class of ordinal-definable sets is denoted by OD. HOD denotes the class of *hereditarily ordinal-definable sets*, i.e., sets such that they, their elements, the elements of their elements, and so on, are in OD. Gödel showed, without using the axiom of choice, that HOD satisfies all the axioms of ZFC. Thus if ZF is consistent then so is ZFC. The assumption $V = OD$ implies $V = HOD$ and the axiom of choice.

As mentioned in Section 1, the continuum hypothesis says that any infinite set of real numbers has either the cardinal number $\aleph_0$ of the set of integers or else the cardinal number $\mathfrak{c}$ of the set of all real numbers. ZFC implies that this hypothesis is equivalent to $\mathfrak{c} = \aleph_1$, where $\aleph_1$ is the cardinality of the shortest well-ordered uncountable set. Attempts to decide whether $\mathfrak{c} = \aleph_1$ have led to many interesting results. In the late 1930s Gödel showed that the continuum hypothesis is consistent with the axioms of ZFC, and in 1964 Paul J. Cohen showed that the axiom of choice does not follow from ZF and that the continuum hypothesis does not follow from ZFC.

Gödel's technique amounts to reversing the objections to the axiom of choice. Instead of asking how to define choice sets (selectors), he admits only the sets *constructible* by means of the following transfinite process. Working in ZF, we inductively define "constructible levels" L_α so that sets at a particular level are definable in the structure consisting of the objects at previous levels. Thus $L_0 = \emptyset$, $L_{\alpha+1} = Def(L_\alpha)$, and $L_\lambda = \bigcup_{\alpha < \lambda} L_\alpha$ for limit λ. Here $Def(\mathscr{A})$ is the operation that adds to $\mathscr{A}$ new elements, representing all subsets of $\mathscr{A}$ definable in terms of elements of $\mathscr{A}$.

This definition of L_α can be formalized in ZF because definitions are formulas, and formulas can be encoded by natural numbers. In fact, we can even construct a well-ordering of the universe $L = \bigcup_\alpha L_\alpha$ constructible sets. This is done by induction. It is enough to define the ordering of a given level L_α. First we order "k-tuples" of objects from the previous levels. Then we order the given level by comparing pairs consisting of a code of a formula and a sequence of parameters (from the previous levels). This well-ordering of L shows that the axiom of choice holds in L.

To obtain the continuum hypothesis in L, Gödel proved a deeper theorem: all constructible real numbers occur in levels L_α with indices α that are countable in L. This readily implies the continuum hypothesis in L. What remains to be shown is that all the axioms of ZFC hold in L. We have already shown that the axiom of choice holds in L. Showing the validity of the other axioms of ZF in L is easy. Hence the axioms of ZFC remains consistent after adding the continuum hypothesis. Even the *generalized* continuum hypothesis—$2^{\aleph_\alpha} = \aleph_{\alpha+1}$ for every ordinal α—holds in the universe L of constructible sets. Gödel also showed that all constructible sets (and only those) are constructible in L. Thus all the consequences of ZFC, and of the sentence "$V = L$" asserting that all sets are constructible, are true in L.

Although it is consistent to assume that $V = L$, it is more natural to assume that $V \neq L$, in which case the class L can be enlarged by adding sets from $V \setminus L$ and using them in an inductive definition of a larger universe. For example, inclusion of all the real numbers yields an interesting universe denoted by $L[\mathbf{R}]$; see Figure 2.

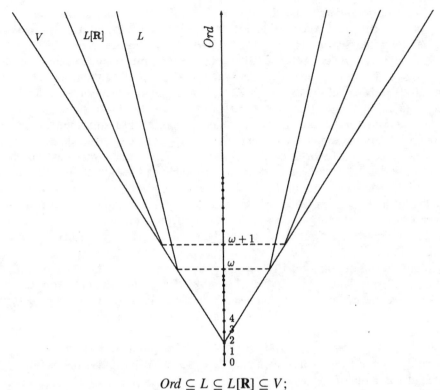

$Ord \subseteq L \subseteq L[\mathbf{R}] \subseteq V$;
ZFC is true in V and in L, ZF is true in $L[\mathbf{R}]$;
Under suitable large cardinal axioms ZF + $\forall X[AD(X)]$ + DC is true in $L[\mathbf{R}]$;

Figure 2. Sets that are constructible and constructible from reals in the universe of Set Theory.

Cohen's argument for the independence of the axiom of choice and of the continuum hypothesis is more complicated than Gödel's. It can be explained as the construction of a *Boolean model*—a class of functions whose values belong to the Boolean algebra of open-closed subsets of a topological space. For suitable definitions of the membership and equality relations between such functions, the construction assigns the Boolean value **1** to all the axioms of ZFC. For a suitable choice of topological space the value of the continuum hypothesis in this model is **0**. But it can be shown that whatever is provable from formulas of value **1** in this model likewise has value **1**. Hence the continuum hypothesis is an independent sentence—it is not provable from ZFC. In a similar way, Cohen proved the independence of A7 (the axiom of choice) from ZF. His method, called *forcing*, was later used to obtain other independence proofs, for many well-known problems in set theory, real analysis, and algebra.

While these results show the independence of various sentences in ZFC, the truth or falsehood of most of them can be established in the two natural subuniverses L and $L[\mathbf{R}]$. The proofs of these theorems in $L[\mathbf{R}]$, however, require some extensions of ZFC. Specifically, they require cardinal numbers so large that their existence cannot be proved in ZFC.

The role of different universes, i.e., of different models of ZF, becomes particularly clear if we look at the *axiom of determinacy* introduced by J. Mycielski and H. Steinhaus in 1962. Let X be a set of infinite sequences of 0s and 1s, and consider the following game between two players I and II. Player I chooses 0 or 1, then player II chooses 0 or 1, then I again chooses 0 or 1, etc. When making their choices, both players know X and they know the previous choices. Player I wins if the sequence of 0's and 1's belongs to X, while player II wins if the sequence does not belong to X. The axiom of determinacy for X, $AD(X)$, states that one of the players has a winning strategy. Using the axiom of choice, one can show that there are sets X for which $AD(X)$ is false, i.e., for which neither of the players has a winning strategy. But if a model of ZF satisfies $AD(X)$ for all X, then this model is closer to physical reality than any model of ZFC. For example, the Banach-Tarski paradoxical decomposition of a ball is impossible. In 1989, D. A. Martin, J. R. Steel, and H. Woodin showed that if one assumes ZFC plus appropriate large cardinal axioms, then $L[\mathbf{R}]$ satisfies $AD(X)$ for all X, as well as the axiom DC of dependent choices. Whether $AD(X)$ holds for all $X \in OD$ remains an open problem; see the problems at the end of this paper.

The universes HOD, L, and $L[\mathbf{R}]$ are very different. For example, in HOD and L there are definable non-Lebesgue measurable sets of reals, while (under the assumption of the existence of a suitable large cardinal) in $L[\mathbf{R}]$ all sets of reals are Lebesgue measurable and $|\mathbf{R} \cap HOD| = \aleph_0$.

For these reasons, and because of their intrinsic interest, large cardinals have become an important topic. What are they? Here we introduce one type, called *measurable cardinals*. We call a cardinal number κ *measurable* if there is a measure, with values 0 and 1, defined on the family of all subsets of a set X of cardinality κ, such that the measure of every one-element set is 0, the measure of the whole set X is 1, and the measure of the union of fewer than κ sets of measure 0 is still 0. $\aleph_0$ is the smallest measurable cardinal. In ZFC it is not possible to prove the existence of measurable cardinals greater than $\aleph_0$. In fact, D. Scott showed in 1961 that $V = L$ implies that they do not exist. Thus we see that $V \neq L$ is a more natural assumption than $V = L$ since it does not rule out the natural idea of measurable cardinals.

But why are measurable cardinals called "large?" The reason is a 1930 theorem of Ulam that any measurable cardinal κ can "model the universe," i.e., the sets of rank less than κ satisfy the axioms of ZFC. This, like the theorem of Scott just mentioned, implies that existence of measurable cardinals cannot be proved in ZFC. However, their existence (and that of other large cardinals) has interesting and deep consequences in real analysis, in the combinatorics of infinite sets, and even in the combinatorics of finite sets.

Practicing mathematicians, whether algebraists, geometers, or analysts, almost never use axioms beyond ZFC. This is surprising, since set theory was invented by Cantor more than 100 years ago and ZFC expresses only Cantor's assumptions. It is even more surprising, since mathematics is constantly enriched by definitions and theorems suggested by the physical sciences. Nonetheless, famous problems such as the four-color conjecture, Fermat's last theorem, Bieberbach's conjecture, and many others, were solved in ZFC; in fact, in weak subtheories of ZFC. It may be of some interest that large cardinal existence axioms (e.g., those used by Martin et al.) are not inspired by physical experience, but rather by concepts arising in the human imagination for which there are no ready physical interpretations. On the other hand, the axiom of determinacy was motivated by H. Steinhaus' wish to build a system of mathematics closer to physical experience than ZFC.

Independence of a sentence in set theory tells us nothing about its future, since it is possible that new developments in mathematics will decide this sentence. Independence tells us only that the sentence is false in some model of ZFC and true in another. If we limit ourselves to the universe of constructible sets, then we obtain theorems about sets of real numbers that are very different from those true in the universe of sets with a measurable cardinal.

Today, set theory plays a role similar to that played by Euclidean geometry for over 15 centuries (up to the time of the construction of mathematical analysis by Newton and Leibniz). Namely, it is a universal axiomatic theory for modern mathematics. It bears repeating that its axiomatization is not complete, and that new fundamental principles about sets are sometimes discovered or invented. Large cardinal axioms and the axiom of determinacy for the class $L[\mathbf{R}]$ are relevant examples. The evolution of mathematics decides which axioms become generally accepted.

Some axiomatic theories weaker than ZFC are still very strong in the sense that large chapters of mathematics can be developed in them. The simplest is PA (*Peano arithmetic*; see Table II). PA is so powerful that it allows us to prove all known theorems of classical number theory. However, there are theorems of finite combinatorics (which can be expressed somewhat artificially as number-theoretic statements) that are provable in ZFC but not in PA. We meet such theorems in the next two sections.

The theory PA has a natural extension called Z_2 or *second order arithmetic*; see Table III, Z_2 is obtained from PA by adding a second sort of variables $X, Y, \ldots$ called *set variables*, a binary symbol $\in$, an axiom of extensionality, the comprehension scheme, and the set form of the induction axiom PA6.

TABLE II. The Axioms of Peano's Arithmetic, PA
(Primitive symbols $1, +, \cdot$)

PA1:	$x + 1 \neq 1$;
PA2:	$x \neq y \rightarrow x + 1 \neq y + 1$;
PA3:	$x + (y + 1) = (x + y) + 1$;
PA4:	$x \cdot 1 = x$;
PA5:	$x \cdot (y + 1) = (x \cdot y) + x$;
PA6:	$\varphi(1) \wedge \forall x[\varphi(x) \rightarrow \varphi(x + 1)] \rightarrow \forall x \varphi(x)$, for every formula $\varphi(x)$, which may contain free variables other than x.

TABLE III. The Axioms of Second Order Arithmetic, Z_2
(Primitive symbols $1, +, \cdot$ and $\in$)

$Z_2 1$–$Z_2 5$:	Formulas PA1–PA5 of table II;
$Z_2 6$:	$\forall x[x \in X \leftrightarrow x \in Y] \rightarrow [X = Y]$;
$Z_2 7$:	$\exists X \, \forall y[y \in X \leftrightarrow \psi(y)]$ where ψ is any formula of the extended language without the free variable X;
$Z_2 8$:	$1 \in X \wedge \forall x[x \in X \rightarrow x + 1 \in X] \rightarrow \forall x[x \in X]$.

The theory Z_2 is a natural framework for the development of mathematical analysis, but it is much weaker than ZFC.

3. LOGIC AND MODEL THEORY. We now turn to mathematical logic and model theory, and to Hilbert's second problem. While Frege discovered the formalism of logic as early as 1879, introduction of a good system of notation took considerable time. In their *Principia Mathematica*, the first volume of which appeared in 1910, B. Russell and A. N. Whitehead popularized the modern notation for the Fregean syntax of logic. However, their system of logic also included a part of set theory. The first truly modern textbook of mathematical logic, written by D. Hilbert and W. Ackermann, was published in 1928. Then, in 1929, Gödel showed in his doctoral dissertation that Frege's rules of proof suffice to prove all logically valid sentences. We call this result the *completeness theorem for first-order logic*. It motivated the modern terminology according to which logic is the system of axioms and rules of Frege.

In 1924 Hilbert extended Frege's system by adding an operator ϵ for creating functions. If $\varphi(\bar{x}, y)$ is any formula (here $\bar{x}$ denotes any finite sequence of free variables), then Hilbert's operator gives a function symbol $\epsilon y\varphi$ of the variables $\bar{x}$; see axiom L6 of Table IV. This formalizes the mathematical practice of naming new objects and functions with desired properties: in this case, "a y such that $\varphi(\bar{x},y)$ holds." If there is no such y, then the value $(\epsilon y\varphi)(\bar{x})$ does not matter and L6 holds automatically. Similarly, L7 allows any formula $\varphi(\bar{x})$ to be abbreviated by a single relation symbol $R_\varphi(\bar{x})$. Hilbert's extension has three advantages: (a) it simplifies the rules of logic, (b) choice functions are part of logic, so the axiom of choice follows from the remaining axioms of ZFC, and, most importantly, (c) it allows us to solve the ontological mystery of quantification over infinite universes.

First order logic generalizes the languages of ZFC, whose primitives are the relation symbols $\in$ and $=$, and of PA, whose primitives are the constant 1, the

TABLE IV. The Axioms and Rules of First Order Logic
(Essentially after G. Frege and D. Hilbert)

Primitive symbols: $\neg$ negation; $\rightarrow$ implication; $=$ equality symbol; sequences of variables $\bar{x} = (x_1, \ldots, x_n)$ and $\bar{y} = (y_1, \ldots, y_n)$, $(n \geq 0)$, ϵ an operator such that if $\varphi(\bar{x}, y)$ is any formula then $\epsilon y\varphi$ is a function symbol of n arguments (or a constant symbol if $n = 0$).

L1: $(\varphi \rightarrow \psi) \rightarrow [(\psi \rightarrow \chi) \rightarrow (\varphi \rightarrow \chi)]$;
L2: $\varphi \rightarrow (\neg \varphi \rightarrow \psi)$;
L3: $(\neg \varphi \rightarrow \varphi) \rightarrow \varphi$;
L4: If φ and $\varphi \rightarrow \psi$ are proved then ψ is proved;
L5: If $\varphi(x)$ is proved, then $\varphi(f(\bar{y}))$ is proved, where $f(\bar{y})$ is any term;
L6: $\varphi(\bar{x}, y) \rightarrow \varphi(\bar{x}, (\epsilon y\varphi)(\bar{x})))$;
L7: $R_\varphi(\bar{x}) \leftrightarrow \varphi(\bar{x})$;
L8: $x = x$;
L9: $x = y \rightarrow y = x$;
L10: $x = y \rightarrow (y = z \rightarrow x = z)$;
L11: $[x_1 = y_1 \wedge \ldots \wedge x_n = y_n] \rightarrow [R(\bar{x}) \leftrightarrow R(\bar{y})]$, where R is any relation symbol of n variables;
L12: $[x_1 = y_1 \wedge \ldots \wedge x_n = y_n] \rightarrow [f(\bar{x}) = f(\bar{y})]$, where f is any term of n variables.

function symbols $+$ and $\cdot$, and the relation symbol $=$, to an arbitrary system of relation symbols R_i ($i \in I$) and functions symbols f_j ($j \in J$). Our generalized language allows us to express properties of structures (called *interpretations* of the language), of the form:

$$M = \langle A, \tilde{R}_i, \tilde{f}_j \rangle_{i \in I, j \in J},$$

where A is a nonempty set and the $\tilde{R}_i$ and $\tilde{f}_j$ are relations and functions on A. Given any interpretation M of the language and any sentence σ in the language (for example, an axiom of ZFC or PA) we can talk about the *truth of σ in M*. M is called a *model* of a theory T if all the axioms of T are true in M.

In Table IV we list all the rules and axioms of logic. Notice that L1–L7, L11, and L12 are rules since φ, ψ, and χ are arbitrary formulas and R and f are arbitrary relation or function symbols. Similarly, A4 in Table I and PA6 in Table II were also rules; the other statements in Table I and II are single statements (axioms). The system presented in Table I and Table IV formalizes almost all of mathematics.

It is natural to add various abbreviations to Hilbert's system. The following are particularly important, and some of them have been used already.

$$(\varphi \vee \psi) := (\neg \varphi \to \psi);$$
$$(\varphi \wedge \psi) := \neg(\varphi \to \neg \psi)$$
$$(\varphi \leftrightarrow \psi) := (\varphi \to \psi) \wedge (\psi \to \varphi);$$
$$(x \neq y) := \neg(x = y);$$
$$(x \notin y) := \neg(x \in y);$$
$$\{y : y \in z \wedge \varphi(\bar{x}, y)\} = \epsilon t \, \forall y [y \in t \leftrightarrow y \in z \wedge \varphi(\bar{x}, y)](\bar{x}) \text{ (in the case of ZFC)};$$
$$\exists y \varphi(\bar{x}, y) := \varphi(\bar{x}, (\epsilon y \varphi)(\bar{x}));$$
$$\forall y \varphi(\bar{x}, y) := \varphi(\bar{x}, (\epsilon y(\neg \varphi))(\bar{x})).$$

Since this formalism avoids any reference to infinite universes, it defuses Brouwer's criticism of classical mathematics. Indeed, variables are merely places for substituting constant terms. And the quantifiers $\exists$ and $\forall$ are defined by the last two formulas. This elimination of quantifiers in favor of Hilbert's ϵ terms can be used to prove the completeness of first order logic, by first showing that any consistent set T of sentences has a model; this justifies Hilbert's slogan that *existence is freedom from contradiction*.

In 1929, Gödel proved the following fundamental theorem. If A is a set of sentences (axioms) then a sentence σ is derivable by rules L1–L12 from A if and only if for every model M of A, σ is true in M. This result is called the *completeness theorem for first order logic*. This result justifies the current significance of the word *logic*.

In 1950 Alfred Tarski initiated a systematic study of first-order theories and their models. Model theory has applications in algebra, algebraic geometry, functional analysis, and other areas of mathematics. One such application is *nonstandard analysis*. While 17th and 18th-century mathematicians used infinitesimals without defining them in a clear way, and while Cauchy introduced the $\epsilon - \delta$ definition of continuity that allowed mathematicians to avoid their use, in the 1960s Abraham Robinson used model theory to introduce infinitesimals by clear and precise definitions and he made good use of them. However, nonstandard analysis is conservative in the sense that if a theorem is stated in the standard concepts and can be proved using nonstandard concepts, then it can also be proved without them.

In mathematical practice we almost always use many-sorted logic, that is, a language with various sorts of variables restricted to appropriate kinds of objects. Second order arithmetic, defined at the end of Section 2, is a good example. Likewise, when we analyze a ring, then, as a rule, we discuss not only the set of its elements but also the set of its ideals, which is a family of its subsets. Thus again we use two sorts of variables. However, one can reduce many-sorted logics to first-order logic with additional unary relation symbols.

Hilbert's second problem was to prove the consistency of the theory Z_2 of Section 2. How could this consistency be established? Hilbert recognized that as soon as a theory T is axiomatized and its rules of proof are spelled out, the consistency of T becomes a combinatorial problem. Since PA can express all problems of finite combinatorics, Hilbert proposed proving the consistency of Z_2 in PA. PA appears to be consistent, because its only problematic part is the induction scheme PA6, and PA6 is corroborated by the following experiment: if one knocks down the first in a row of dominoes, and if every falling piece knocks down the next, then, no matter how many pieces there are, they all fall down.

Generalizing, we can ask for a proof of the consistency of ZFC, i.e., of most of mathematics, in PA. Hilbert thought that PA is complete and so he believed that such proofs can be found. But in 1931 Gödel proved that PA is *not* complete and, moreover, that one cannot prove the consistency of PA *within* PA. Hence it is impossible to prove the consistency of Z_2 or of ZFC in PA, and the solution sought by Hilbert does not exist.

The theorems that wrecked Hilbert's program are called Gödel's *first* and *second incompleteness theorems*. He proved them by encoding symbols, formulas, and sequences of formulas by numbers. When this is done appropriately we can say, *in the language of* PA, what it means for a number to be the code of a correct proof. Then the language of PA can express a sentence Con(PA) that says that PA is consistent; Con(PA) says "There is no proof of the formula $1 \neq 1$ from the axioms of PA."

Gödel showed that if PA is consistent, then it contains no proof of Con(PA)! More generally, he showed that a consistent theory T containing PA, in which the consistency of T can be expressed by a sentence Con(T), cannot prove Con(T). Of course, we can prove Con(PA) in ZFC. We can also show in ZFC that the negation $\neg$ Con(PA) has no proof in PA (because ZFC proves that PA has a model). Hence PA is not a complete theory, and neither is any consistent extension of PA by a set of axioms whose set of codes is definable in the language of PA. In particular, none of these theories proves its own consistency. In this strong sense, Gödel showed that Hilbert's second problem of proving consistency cannot be solved.

This brings us to an important ordering of theories, in which $S < T$ means that Con(S) is provable in T:

$$\text{RCF} < \text{PA} < \cdots < A_i < \cdots < \text{ZFC} < \cdots < \text{ZFC} + \text{LC}_i < \cdots$$

RFC (called the theory of real closed fields) is the theory of the field of real numbers $\langle \mathbf{R}, +, \cdot, < \rangle$, which happens to be a complete theory. PA was defined in

Table II; A_i are various axiomatic systems of analysis (Z_2 among them); ZFC was defined in Table I; LC_i are various axioms of existence of large cardinals (some were discussed in Section 2).

Various interesting combinatorial theorems of ZFC that can be stated in the language of PA have been found to be independent of PA. This is the case for some strong forms of Ramsey's theorem on the coloring of graphs. In fact, H. Friedman, L. Harrington, R. Laver, J. Paris, and R. Solovay have proved that some such sentences are intimately related to the existence of large cardinals.

4. COMPUTABILITY THEORY. Now we turn to problems of computability, and to Hilbert's 10th problem. While proving the incompleteness theorem discussed in Section 3, Gödel considered a class of functions now called computable or recursive (though without claiming that they captured the intuitive notion of computability). A function $f: N \to N$ is *computable* if there exists a formula $\varphi(x, y)$ of the language of PA such that $\varphi(n, m)$ is a theorem of PA if $f(n) = m$, and $\neg \varphi(n, m)$ is a theorem of PA if $f(n) \neq m$. Here m and n denote any terms of the form $1 + \cdots + 1$ and $f(n)$ is also represented in this way. This definition of computability is robust in the sense that replacing PA by a stronger theory, e.g., by ZFC, gives the same class of computable functions. Around 1935, Post, Church, and Turing proposed alternative definitions of computable functions and later all these definitions were proved to be equivalent.

Turing's definition constitutes a theoretical model of the modern computer. In 1935 he introduced the concept of a machine, now called a *Turing machine*, and proved the existence of a *universal machine*: a machine that, given an appropriate code, is capable of simulating the work of any other Turing machine. Turing's definition of computability does not depend on the notion of proof and leads to many interesting concepts, theorems, and still unsolved problems.

A *Turing machine* M is a function $M: A \times S \to A \times S \times C$, where A is a finite set called an *alphabet*. We assume without loss of generality that A consists of only three letters, 0, 1, and the blank symbol $\square$. S is a finite set called the *set of states* or *memory* of M, and C consists of three symbols: R, L, Q called *right*, *left*, and *quit*, which we call *commands*. The letter $\square$ and a certain state s_0 in S play special roles. Such a machine M acts upon an infinite tape

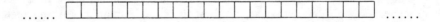

whose squares are filled with single letters of A. At time 0 we have a doubly infinite string of letters $\ldots a_{-1}^0 a_0^0 a_1^0 \ldots$. The machine always starts at the location 0 and in the state s_0; thus it first looks at a_0^0 and evaluates $M(a_0^0, s_0) = (a_0^1, s_1, c)$. It erases a_0^0, and puts a_0^1 in its place. Then it moves left or right or quits depending on the value of the command c. If it quits, it declares that its job is finished. If it moves right, it looks at the current content of the cell with the index $+1$, and if it moves left, it looks at the current content of the cell with the index -1. Then, if this content is a, it evaluates $M(a, s_1)$, and acts again in the same way. M works in this way until (if ever) the third component c becomes Q, in which case it halts. Notice that only a single cell on the tape is affected at each time step. The machine creates the sequence of consecutive tapes $\ldots a_{-1}^n a_0^n a_1^n \ldots$ where $n = 0, 1, \ldots$ is the number of steps.

The initial tape $\ldots a_{-1}^0 a_0^0 a_1^0 \ldots$ is called the *input*, and the final tape (if M reaches Q) is called the *output*. From now on we assume that the input is of the form $\ldots \square \square w \square \square \ldots$ where w is a finite string of letters from the alphabet A, which we denote briefly by w. We also assume that at time 0 the first letter of w is at the location 0.

Any machine M is determined by a table of function values proportional in size to its set S of states, hence M can be coded by a finite sequence of 0s and 1s of length at most $K|S|\log|S|$, where K is an absolute constant. We denote the code of the machine M by $code(M)$.

A *universal* Turing machine M_0 is one that, given any input of the form $code(M)\square w$, works like M on the imput w. Thus M_0 halts for that input $code(M)\square w$ if and only if M halts for the input w, and it yields the same output. An important theorem of Turing says that universal Turing machines exist.

The concept of a Turing machine has philosophical significance because of the following suggestive analogies:

$$\frac{\text{Turing machine}}{\text{its tape}} = \frac{\text{organism}}{\text{its environment}}$$

$$\frac{\text{universal Turing machine}}{\text{the code part of its input}} = \frac{\text{human being}}{\text{its knowledge}}.$$

These analogies are supported by the fact that for every real-world computational procedure (algorithm) there exists a Turing machine that can perform that procedure. A universal Turing machine M_0 is a theoretical model for a programmable computer.

Turing machines permit us to define the concept of computable function as follows. A function $f: D \to N$, where $D \subseteq N$, is called *computable* if there is a Turing machine M which converts any input $n \in D$ to output $f(n)$, and does not halt for input $n \notin D$. Turing proved that if $D = N$ then this definition is equivalent to Gödel's definition stated at the beginning of this section: the main difficulty was to show that there is a Turing machine that enumerates all the codes (Gödel numbers) for theorems of PA and only such numbers.

The idea of a machine enumerating objects is formalized by the concept of a *recursively enumerable set*, which may be defined as the domain of a computable function. The set of consequences of a finite set of axioms is a typical example of a recursively enumerable set. More generally, the consequences of a recursively enumerable set of axioms (such as PA) are recursively enumerable. The latter set is one for which the membership problem is *undecidable*: although there is a Turing machine that lists all theorems of PA, there is no machine that lists all nontheorems, and hence no machine that correctly distinguishes theorems from nontheorems.

The negative solution of Hilbert's 10th problem comes from a new characterization of recursive enumerability, achieved by the efforts of many mathematicians in the years 1950–1970. M. Davis, Y. Matiyasevich, H. Putnam, and J. Robinson showed that for every recursively enumerable set $X \subseteq N$ there is a polynomial $p(x, x_1, \ldots, x_n)$ with integer coefficients such that $x \in X$ if and only if there exist integers $y_1, \ldots, y_n$ such that $p(x, y_1, \ldots, y_n) = 0$. It follows (by taking p to represent a set X with undecidable membership problem) that there is no algorithm for deciding, for each x, whether the equation $p(x, x_1, \ldots, x_n) = 0$ has an integer solution. Thus even elementary number theory is a profound subject, so profound that there is no uniform method for solving its most natural problems.

Another important concept of computability theory, due independently to Solomonoff, Kolmogorov, and Chaitin, is that of an *incompressible sequence*. This formalizes the idea of a sequence that cannot be generated by a program shorter than itself. Let $|w|$ denote the length of the sequence w. Then a finite sequence v of 0s and 1s is *incompressible* relative to a universal Turing machine M_0 if, for every code w such that M_0 with the input $w \square 0$ yields the output v, we have $|w| \geq |v|$. Since there are 2^n sequences of length n and $2^n - 1$ sequences of length $< n$, there exists an incompressible sequence of any length n.

Chaitin showed that for every theory T with a recursively enumerable set of axioms, and every universal machine M_0, there are only finitely many sequences that T can prove to be incompressible relative to M_0. This is another concrete form of the Gödel incompleteness theorem: only *finite many* sentences of a certain type can be proved, although infinitely many of them are true. Thus proving such sentences always remains an art, rather than a science.

Here is Chaitin's interesting proof. Let M_T be a Turing machine that enumerates the theorems of T. For each natural number c we upgrade M_T to a Turing machine M_T^c that, given the input 0, searches for a sequence s with $|s| \geq c$ that T proves to be incompressible relative to M_0, and if it finds such a sequence it prints s and stops. We can define M_T^c so that the code of M_T^c relative to M_0 is no longer than

$$|code(M_T)| + K_1 \log_2 c + K_2,$$

where K_1, K_2 are constants independent of c. Now choose c so that

$$|code(M_T)| + K_1 \log_2 c + K_2 < c.$$

Then M_0 applied to the input $code(M_T^c) \square 0$ does not halt; otherwise its output s would be compressible, contrary to the definition of M_T^c. Therefore, T proves no theorem of the form "s incompressible" for $|s| \geq c$.

We now turn to some other important concepts of computability theory, formalizing the *time and space complexity* of problems. An algorithmic problem in mathematics typically consists of an infinite set L of questions, each of which may be encoded by a finite sequence of 0s and 1s. A solution may be embodied in a Turing machine M that takes each such question w as input and eventually gives the answer (say, giving output 1 for "yes" and 0 for "no"). In this case the set L is said to be *decidable*. Similarly, a theory T is called decidable if the set of its theorems is decidable.

If we are interested not only in the existence of a solution, but also in its *feasibility*, then we impose a bound on the time or space used in the computation. We say that L is *decidable in polynomial time* ($L \in$ PTIME) if there is a Turing machine M and a polynomial $p(x)$ such that M decides whether $w \in L$ in no more than $p(|w|)$ steps. An example is the problem of checking multiplication of integers: given an equation $c = ab$ of length n, we can multiply the numbers a and b, and can then compare the result with c, in around n^2 steps. Hence checking multiplication is a PTIME problem.

The class PSPACE (*polynomial space*) is defined similarly, except that now we require M to visit no more than $p(|w|)$ cells of the tape; the number of steps does not matter.

Finally, we say that $L \in$ NPTIME (*nondeterministic polynomial time*) if there is a Turing machine M and a polynomial $p(x)$ such that $w \in L$ if and only if there is a v such that, given the input $w \square v$, M stops in at most $p(|w|)$ steps.

The classes PTIME and PSPACE are natural candidates for classes of *feasibly decidable* problems, with respect to time and space, respectively. The significance of NPTIME is that M is allowed to incorporate a "guess" v in its computation. This is the nondeterministic ingredient, and we illustrate its naturalness by two examples.

4.1 The problem of deciding whether an n-digit integer c is composite is not known to be PTIME. However, a Turing machine M can solve this problem in around n^2 steps with the help of a correct "guess" $c = ab$, i.e., by giving M the guess $a \square b$. Then M needs to check only the multiplication, so the problem of recognizing composite numbers is in NPTIME.

4.2 Gödel's incompleteness theorem implies that no consistent, recursively enumerable theory containing PA is decidable. We can of course decide, given a sentence σ and integer n, whether σ has a proof with fewer than n symbols. In fact, this is an NPTIME problem, because if a correct proof is guessed, one can check it in polynomial time. But again we do not know whether this problem is in PTIME. Gödel posed this question in a letter to von Neumann; most mathematicians believe that the answer is negative.

It is easily seen that

$$\text{PTIME} \subseteq \text{NPTIME} \subseteq \text{PSPACE}.$$

An important theorem of S. Cook says that if problem 4.2 is in PTIME, then PTIME = NPTIME. There are many other such problems (called NP-*complete*), which are important in combinatorics, computer science, and applications of mathematics.

5. CONCLUDING REMARKS. What is mathematics? Mathematical practice and logic and foundations indicate that mathematics is the *art*, rather than the *science*, of constructing deductive theories. For example, while the Goldbach conjecture (every even number greater than 2 is a sum of two primes) is a well established empirical fact, mathematicians do not include it as an axiom because it is not a basis for an interesting theory. Also, it may turn out to be provable in PA or ZFC. Thus mathematics is the art of deduction rather than a list of facts. On the other hand, the study of nature often suggests new mathematical conjectures and sometimes even their proofs, and from time to time it suggests new fundamental axioms that yield interesting theories. In particular, logic is study of ways in which nature has prepared us to describe reality, a study of natural intelligence. Even the concept of infinite sets is suggested by some apparently unending processes, and by the physical space-time continuum. Thus mathematics is not based on human imagination alone. Mankind in contact with nature generates the art of mathematics.

The main achievements of logic and foundations in the 20th century are:

(A) A precise language for mathematics has been constructed, so that we know what mathematical sentences are and what is a proof of a sentence based on a set of axioms and definitions. Gödel's completeness theorem for first-order logic shows that this concept of proof matches the set-theoretic semantics developed later by Tarski. When proving theorems, mathematicians have always appealed to logical intuition. In the 20th century this intuition has been fully and precisely described; see Table IV.

(B) An axiomatic set theory ZFC, whose primitive notions are those of logic plus the membership relation, has been constructed. Modern mathematics is based on this theory. Its axioms are so simple that they can be presented on a single page; see Table I. On the other hand, as shown by Gödel in 1931, and later by Cohen and others, many questions in finite combinatorics, in the arithmetic of natural numbers, in analysis, and in set theory cannot be decided in ZFC. The work of extending ZFC by new axioms with interesting consequences continues.

(C) A theory of computable functions and algorithms has been constructed. We have a clear distinction between the general classes of functions and their subclasses of computable functions, defined by means of Turing machines. More concrete classes concerned with the time and space complexity of computation, such as PTIME, NPTIME, and PSPACE, are also being investigated.

Thus the fundamental questions posed in the 19th century have been answered. Those answers are surprising and beautiful. One can ask whether mathematics and computer science in the 21st century will be founded on the same concepts as in the 20th century. We conjecture that set theory will remain the most useful and inspiring universal theory on which all of mathematics can be based.

Finally, let us formulate three open problems in logic and foundations that seem to us of special importance.

1. *To develop an effective automatic method for constructing proofs of mathematical conjectures, when these conjectures have simple proofs!* Interesting methods of this kind already exist but, thus far, automated theorem proving procedures are not dynamic: they do not use large lists of axioms, definitions, theorems, and lemmas that mathematicians could provide to the computer. Also, the existing methods are not yet powerful enough to construct most proofs regarded as simple by mathematicians, and conversely, the proofs constructed by these methods do not seem simple.

2. *Are there natural large cardinal existence axioms LC such that ZFC + LC implies that all OD sets X of infinite sequences of 0s and 1s satisfy the axiom of determinacy AD(X)?* This question is similar to the continuum hypothesis, in the sense that it is independent of ZFC plus all large cardinal axioms proposed thus far.

3. *Is it true that* PTIME ≠ NPTIME, *or at least, that* PTIME ≠ PSPACE? An affirmative answer to the first of these questions would tell us that the problem of constructing proofs of mathematical conjectures in given ax-

iomatic theories (and many other combinatorial problems) cannot be fully mechanized in a certain sense.

In May 2000 the Clay Mathematics Institute announced a prize of one million dollars for a proof or disproof of the conjecture PTIME ≠ NPTIME. See http://www.claymath.org/prize_problems/index.htm for details.

ACKNOWLEDGEMENTS This paper grew out of a shorter version published in the popular Polish magazine *Delta* in the issues 9 and 11 of 1999, and 1 and 3 of 2000. We are indebted to Abe Shenitzer, who translated the *Delta* version into English.

SUGGESTED FURTHER READING

1. General Information
   - J. Barwise, ed., *Handbook of Mathematical Logic*, North Holland, Amsterdam and New York, 1977.
     A comprehensive presentation of all three areas of logic and foundations at the beginning of the fourth quarter of the 20th century.
   - Y. L. Ershov, S. S. Goncharov, A. Nerode, and J. B. Remmel, eds., *Handbook of Recursive Mathematics*, Elsevier, Amsterdam and New York, 1998, Vol. I and II.
     An extensive and comprehensive presentation of effective mathematics.
   - D. Gabbay and F. Guenther, eds., *Handbook of Philosophical Logic*, Reidel, Dordrecht and Boston, 1985, vol. I–IV.
     A very extensive presentation of aspects of logic and foundations as studied by philosophers (often inconsistent with the views expressed in our paper).
   - J. van Heijenoort, ed., *From Frege to Gödel, A Source Book in Mathematical Logic, 1879–1931*, Harvard University Press, Cambridge, MA and London, 1967.
     Translations and commentary on most of the important papers in logic up to 1931.
   - J. van Leeuwen, ed., *Handbook of Theoretical Computer Science*, Elsevier, Amsterdam and New York, and MIT Press, Cambridge, MA, 1990, vols. I and II.
     A very extensive presentation of syntactic and semantic issues of modern theoretical computer science.

2. Set Theory
   - P. J. Cohen, *Set Theory*, W. A. Benjamin, New York and Amsterdam, 1966.
     An excellent introduction to axiomatic set theory and independence proofs.
   - A. A. Fraenkel, Y. Bar-Hillel, and A. Levy, *Foundations of Set Theory*, North Holland, Amsterdam, 1976.
     An interesting treatment of the history of axiomatic set theory.
   - T. Jech, *Set Theory*, Academic Press, New York, 1977.
     A comprehensive course on classical set theory.
   - K. Kunen, *Set Theory, An introduction to independence proofs*. North Holland, Amsterdam and New York, 1980.
     A very useful course on set theory and on Cohen's method of forcing.

- K. Kuratowski and A. Mostowski, *Set Theory: with an introduction to descriptive set theory*, 3rd ed., North Holland, Amsterdam and New York, 1976.
  An advanced course on axiomatic set theory.
3. Model Theory
    - C. C. Chang and H. J. Keisler, *Model Theory*, North Holland, Amsterdam and New York, 1990.
      A very beautiful exposition of model theory.
    - W. Hodges, *Model Theory*, Cambridge University Press, Cambridge and New York, 1993.
      A comprehensive monograph on model theory with a large bibliography.
3. Computability Theory
    - P. Odifreddi, *Classical Recursion Theory*, vols. I and II, North Holland, Amsterdam and New York, 1989–1999.
      An extensive presentation of classical recursion theory.
    - H. Rogers, Jr., *Theory of Recursive Functions and Effective Computability*, MIT Press, Cambridge, MA, 1987.
      A very beautiful presentation of computable function theory up to 1967.
    - R. I. Soare, *Recursively Enumerable Sets and Degrees: a Study of Computable Functions and of Computably Generated Sets*, Springer-Verlag, Berlin and New York, 1987.
      A comprehensive study of degrees of computability.

The Development of Rigor in Mathematical Probability (1900–1950)

Joseph L. Doob

1 INTRODUCTION. This paper is a brief informal outline of the history of the introduction of rigour into mathematical probability in the first half of this century. Specific results are mentioned only in so far as they are important in the history of the logical development of mathematical probability.

The development of science is not a simple progression from one advance to the next. Judged by hindsight, the development is slow, proceeds in a zigzag course, with many wrong turns and blind alleys, and frequently moves in directions condemned by leading scientists. In the 1930's Banach spaces were sneered at as absurdly abstract, later it was the turn of locally convex spaces, and now it is the turn of nonstandard analysis. Mathematicians are no more eager than other humans to embrace new ideas, and full acceptance of mathematical probability was not realized until the second half of the century. In particular, many statisticians and probabilists resented the mathematization of probability by measure theory, and some still place mathematical probability outside analysis. The following quotations (in translation) are relevant.

Planck: *A new scientific truth does not triumph by convincing its opponents and making them see the light, but rather because its opponents eventually die, and a new generation grows up with it.*

Poincaré: *Formerly, when one invented a new function, it was to further some practical purpose; today one invents them in order to make incorrect the reasoning of our fathers, and nothing more will ever be accomplished by these inventions.*

Hermite: (in a letter to Stieltjes) *I recoil with dismay and horror at this lamentable plague of functions which do not have derivatives.*

Probability theory began, and remained for a long time, an idealization and analysis of certain real life phenomena outside mathematics, but gradually, in the first half of this century, mathematical probability became a normal part of mathematics. The mathematization of probability required new ideas, and in particular required a new approach to the idea of acceptability of a function. In view of the above quotations it is not surprising that acceptance of this mathematization was slow and faced resistance. In fact even now some probabilists fear that mathematization has removed the intrinsic charm from their subject. And they are right in the sense that the charm of the old, vague probability-mathematics, based on nonmathematical definitions, has split into two quite different charms: those of

---

Reprinted with kind permission of Birkhäuser Verlag AG, from *Development of Mathematics 1900–1950*, edited by Jean-Paul Pier, Basel, 1994, ISBN 0-8176-2821-5

real world probability and of mathematical precision. But it must be stressed that many of the most essential results of mathematical probability have been suggested by the nonmathematical context of real world probability, which has never even had a universally acceptable definition. In fact the relation between real world probability and mathematical probability has been simultaneously the bane of and inspiration for the development of mathematical probability.

2 WHAT IS THE REAL WORLD (NONMATHEMATICAL) PROBLEM?

What is usually called (real world) probability arises in many contexts. Besides the obvious contexts of gambling games, of insurance, of statistical physics, there are such simple contexts as the following. Suppose an individual rides his bicycle to work. The rider would be surprised if, when the bicycle is parked, the valve on the front tire appeared in the upper half of the tire circle 10 successive days, just as surprised as if 10 successive tosses of a coin all gave heads. However, it is clear that (tire context) if the ride is very short, or (coin context) if the coin starts close to the coin landing place and the initial rotational velocity of the coin is low, the surprise would decrease and the probability context would become suspect. The moral is that the specific context must be examined closely before any probabilistic statement is made. If philosophy is relevant, an arguable question, it must be augmented by an examination of the physical context.

3 THE LAW OF LARGE NUMBERS.

In a repetitive scheme of independent trials, such as coin tossing, what strikes one at once is what has been christened the *law of large numbers*. In the simple context of coin tossing it states that in some sense the number of heads in n tosses divided by n has limit $1/2$ as the number of tosses increases. The key words here are *in some sense*. If the law of large numbers is a mathematical theorem, that is, if there is a mathematical model for coin tossing, in which the law of large numbers is formulated as a mathematical theorem, either the theorem is true in one of the various mathematical limit concepts or it is not. On the other hand, if the law of large numbers is to be stated in a real world nonmathematical context, it is not at all clear that the limit concept can be formulated in a reasonable way. The most obvious difficulty is that in the real world only finitely many experiments can be performed in finite time. Anyone who tries to explain to students what happens when a coin is tossed mumbles words like *in the long run*, *tends*, *seems to cluster near*, and so on, in a desperate attempt to give form to a cloudy concept. Yet the fact is that anyone tossing a coin observes that for a modest number of coin tosses the number of heads in n tosses divided by n seems to be getting closer to $1/2$ as n increases. The simplest solution, adopted by a prominent Bayesian statistician, is the vacuous one: never discuss what happens when a coin is tossed. A more common equally satisfactory solution is to leave fuzzy the question of whether the context under discussion is or is not mathematics. Perhaps the fact that the assertion is called a *law* is an example of this fuzziness. The following statements have been made about this law (my emphasis):

Laplace: (1814) *This theorem, implied by common sense, was difficult to prove by analysis.*

THE DEVELOPMENT OF RIGOR IN MATHEMATICAL PROBABILITY (1900–1950)   249

Ville: (1939) *One sees no reason for this proposition to be true; but as it is impossible to prove experimentally that it is false, one can at least safely state it.*

Bauer: (translated from the context of dice to that of coins) *It is an experimentally established fact that the quotient ... exhibits a deviation from* $1/2$ *which approaches* 0 *for large n.*

These statements illustrate the enduring charm of discussions of real world probability. Mathematicians, unfortunately, have felt forced to think about the following question, or at least to write about it.

4 WHAT IS PROBABILITY?   Here are some attempts to answer this question and to discuss the teaching of the subject.

Poincaré: (1912) *One can scarcely give a satisfactory definition of probability.*

Mazurkiewicz: (1915) *The theory of probability is not an independent element of mathematical instruction; nevertheless it is very desirable that a mathematician knows its general principles. Its fundamental concepts are incompletely determined. They contain many unsolved difficulties.*

v. Mises: (1919) *In fact, one can scarcely characterize the present state other than that probability is not a mathematical discipline.* (He proceeded to make it into a mathematical discipline by basing mathematical probability on a sequence of observations («Beobachtungen») with properties that cannot be satisfied by a mathematically well defined sequence. In a lighter mood he is said to have defined probability as a number between 0 and 1 about which nothing else is known.)

E. Pearson: (1935) (Oral communication) *Probability is so linked with statistics that, although it is possible to teach the two separately, such a project would be just a tour de force.*

Uspensky: (1937) In a useful textbook, he gave the following once common textbook definition. *If, consistent with condition S, there are n mutually exclusive, and equally likely cases, and m are favorable to the event A, then the mathematical probability of A is defined as* m/n.

The foregoing should make obvious the advisability of separating mathematical probability theory from its real world applications. Note however that no one doubts the real world applicability of mathematical probability. Gambling, genetics, insurance and statistical physics are here to stay.

Only *mathematical* probability will be discussed below, except for the following remark on coin tossing. Newtonian mechanics provides a partial mathematical model for coin tossing. In coin tossing, a solid body falls under the influence of gravity. Its motion is determined in Newton's model by his laws, and any discussion of what the coin does cannot be complete unless these laws are applied. Only these laws, rather than philosophical remarks, can explain the quantitative influence and importance of the initial and final conditions of the coin motion in order to justify allusions to equal likelihood of heads and tails. Of course these laws can at best reduce the analysis to considerations of the initial and final conditions of the toss, but these conditions can show what the «equal likelihoods» depends on and thereby give it a plausible interpretation.

5 MATHEMATICAL PROBABILITY BEFORE THE ERA OF PRECISE DEFINITIONS.
There were many important advances in mathematical probability before 1900, but the subject was not yet mathematics. Although nonmathematical probabilistic contexts suggested problems in combinatorics, difference equations and differential equations, there was a minimum of attention paid to the mathematical basis of the contexts, a maximum of attention to the pure mathematics problems they suggested. This unequal treatment was inevitable, because measure theory, needed for mathematical modeling of real world probabilistic contexts, had not yet been invented.

It was always clear that, however classical mathematical probability was to be developed, the concept of additivity of probability as applied to incompatible real world events was fundamental. Additive functions of sets were of course familiar to mathematicians from concepts of volume, mass and so on, long before 1900. It was realized that contexts involving averages led to probability. It was frequently clear how to use the contexts to suggest problems in analysis, but it was not clear how to formulate an overall mathematical context, that is, how to define a mathematical structure in which the various contexts could be placed.

A weaker condition than additivity was less familiar but turned out to be essential later. The standard loose language will be used here. If $x_1, x_2, \ldots$ are numbers obtained by chance, and if A is a set of numbers, consider the probability that at least one of the members of this sequence lies in A, or, in more colorful language, consider the probability that an orbit of this motion through points of a line hits A. The usual calculation (ignoring here all notions of rigour), defines a function $A \to \phi(A)$ which in general is not additive. In fact ϕ satisfies the inequality

$$\phi(A) + \phi(B) - \phi(A \cup B) \geq \phi(A \cap B), \tag{5.1}$$

whereas additivity of ϕ would imply equality in (5.1). The point is that the left side of (5.1) is the probability that the sequence $x\bullet$ hits both A and B, a probability at least equal to, and in general greater than, $\phi(A \cap B)$, the probability that the sequence hits $A \cap B$. The inequality (5.1), the *strong subadditivity* inequality, is satisfied also by the electrostatic capacity of a body in $\mathbf{R}^3$, and this fact hints at the close connection between potential theory and probability, developed in great detail in the second half of the century with the help of Choquet's theory of mathematical capacity.

6 THE DEVELOPMENT OF MEASURE THEORY.
Recall that a *Borel field* ($= \sigma$ algebra) of subsets of a space is a collection of subsets which is closed under the operations of complementation and the formation of countable unions and intersections. The class of *Borel sets* of a metric space is the smallest set σ algebra containing the open sets of the space. A *measurable space* is a pair, $(S, \mathbb{S})$, where S is a space and $\mathbb{S}$ is a σ algebra of subsets of S. The sets of $\mathbb{S}$ are the *measurable sets* of the space. In the following, if S is metric, the coupled σ algebra making it into a measurable space will always be the σ algebra of its Borel sets. In particular $(\mathbf{R}^N, \mathbb{R}^N)$ denotes N dimensional Euclidean space coupled with its Borel sets. The

superscript will be omitted when $N = 1$. A measurable function from a measurable space $(S_1, \mathbb{S}_1)$ into a measurable space $(S_2, \mathbb{S}_2)$ is a function from S_1 into S_2 with the property that the inverse image of a set in $\mathbb{S}_2$ is a set in $\mathbb{S}_1$.

Measure theory started with Lebesgue's thesis (1902), which extended the definition of volume in $\mathbf{R}^N$ to the Borel sets. Radon (1913) made the further step to general measures of Borel sets of $\mathbf{R}^N$ (finite on compact sets). These measures are usually extended to slightly larger classes than the class of Borel sets, by *completion*. Finally Fréchet (1915), 13 years after Lebesgue's thesis, pointed out that all that the usual definitions and operations of measure theory require is a σ algebra of subsets of an abstract space on which a measure, that is, a positive countably additive set function, is defined. At each step of this progression not necessarily positive countably additive set functions—signed measures—were incorporated into the theory. As noted below, the Radon-Nikodym theorem (1930), which gives conditions necessary and sufficient that a countably additive function of sets can be expressed as an integral over the sets, turned out to be the final essential result needed to formulate the basic mathematical probability definitions. Thus it was 28 years before Lebesgue's theory was extended far enough to be adequate for the mathematical basis of probability. This extension was not developed in order to provide a basis for probability, however. Measure theory was developed as a part of classical analysis, and applications in analysis were immediate, for example to the (Lebesgue measure) almost everywhere derivability of a monotone function.

There has been criticism of the fact that mathematical probability is usually prescribed not only to be additive but even to be countably additive. The question whether real world probability is countably additive, if the question is to be meaningful, asks whether a mathematical model of real world probabilistic phenomena *necessarily* always involves *countably* additive set functions. In fact there may well be real world contexts for which the appropriate mathematical model is based on finitely but not countably additive set functions. But there have been very few applications of such set functions in either mathematical or nonmathematical contexts, and such set functions will not be discussed further here.

7 EARLY APPLICATIONS OF EXPLICIT MEASURE THEORY TO PROBABILITY. Some probabilistic slang will be needed, enduring relics of the historical background of probability theory. A probability space is a triple $(S, \mathbb{S}, \mathrm{P})$, where $(S, \mathbb{S})$ is a measurable space and P is a measure on $\mathbb{S}$ with $\mathrm{P}(S) = 1$. A measure with this normalization is a *probability measure*. A *random variable* is a measurable function from a probability space $(S, \mathbb{S}, \mathrm{P})$, into a measurable space $(S', \mathbb{S}')$. The space S', or, when one writes carefully, $(S', \mathbb{S}')$, is the *state space* of the random variable. Mutual independence of random variables is defined in the classical way. The *distribution* of a random variable x is the measure P_x on $\mathbb{S}'$ defined by setting

$$P_x(A') = P\{s \in S : x(s) \in A'\}.$$

The joint distribution of finitely many random variables defined on the same probability space is obtained by making x into a vector and specifying $\mathbb{S}'$ and S' correspondingly. A stochastic process is a family of random variables $\{x(t, \bullet), t \in \mathcal{I}\}$ from some probability space $(S, \mathbb{S}, \mathrm{P})$, into a state space $(S', \mathbb{S}')$. The set $\mathcal{I}$ is the *index set* of the process. Thus a stochastic process defines a function of two variables, $(t, s) \to x(t, s)$, from $\mathscr{I} \times S$ into the state space. The function $x(t, \bullet)$ from S into S' is the tth random variable of the process; the function $x(\bullet, s)$ from $\mathscr{I}$ into S' is the sth sample function, or sample path, or sample sequence if $\mathscr{I}$ is a sequence.

Borel (1909) pointed out that in the dyadic representation $x = x_1 x_2 \ldots$ of a number x between 0 and 1, in which each digit x_j is either 0 or 1, these digits are functions of x, and if the interval $[0, 1]$ is provided with Lebesgue measure, a probability measure on this interval, these functions miraculously become random variables which have exactly the distributions used in calculating coin tossing probabilities. That is, 2^{-n} is the probability assigned to the event that, in a tossing experiment, the first n tosses yield a specified sequence of heads and tails, and 2^{-n} is also the total length (= Lebesgue measure) of the finite set of intervals whose points x have dyadic representations with a specified sequence of 0's and 1's in the first n places. Thus a mathematical version of the law of large numbers in the coin tossing context is the existence in some sense of a limit of the sequence of function averages $\{(x_1 + \cdots + x_n)/n, \; n \geq 1\}$. Classical elementary probability calculations imply that this sequence of averages converges in measure to $1/2$, but a stronger mathematical version of the law of large numbers was the fact deduced by Borel—in an unmendably faulty proof—that this sequence of averages converges to $1/2$ (Lebesgue measure) for almost every value of x. A correct proof was given a year later by Faber, and much simpler proofs have been given since. [Fréchet remarked tactfully: «Borel's proof is excessively short. It omits several intermediate arguments and assumes certain results without proof.»] This theorem was an important step, an example of a new kind of convergence theorem in probability. Observe that (fortunately) pure mathematicians need not interpret this theorem in the real world of real people tossing real coins. Some of the quotations given above indicate that they not only need not but should not.

Daniell (1918) used a deep approach to measure theory in which integrals are defined before measures to get a (rather clumsy) approach to infinite sequences of random variables by way of measures in infinite dimensional Euclidean space.

The Brownian motion stochastic process in $\mathbf{R}^3$ is the mathematical model of Brownian motion, the motion of a microscopic particle in a fluid as the particle is hit by the molecules of the fluid. The process is normalized by supposing it starts at the origin of a cartesian coordinate system in $\mathbf{R}^3$, and a (normalized) Brownian motion process in $\mathbf{R}$ is the process of a coordinate function of a normalized process in $\mathbf{R}^3$, vanishing initially. A (normalized) Brownian motion process in $\mathbf{R}^N$ is a process defined by N mutually independent Brownian motion processes in $\mathbf{R}$. It was well known what the joint distributions of the random variables of a Brownian motion process should be, and it had been taken for granted that in a proper mathematical model the class of continuous paths would have probability 1. By 1900, Bachelier had even derived various important distributions related to the Brownian motion process in $\mathbf{R}$, such as that of the maximum change during a time interval, by finding corresponding distributions for a certain discrete random walk and then going to the limit as the walk steps tended to 0. More precisely, what Bachelier derived were distributions valid for a Brownian motion process if in fact there was such a thing as a Brownian motion process, and if it was approximable by his random walks. Observe that there was no question about the existence of Brownian motion; Brownian motion is observable under a microscope. But there was as yet no proof of the existence of a stochastic process, a mathematical construct, with the desired properties. Wiener (1923) constructed the desired Brownian motion process, now sometimes called the *Wiener process*, by applying the Daniell approach to measure theory to obtain a measure with the desired properties on a space S of continuous functions: if $x(t,\bullet)$ is the random variable defined by the value at time t of a function in S, the stochastic process of these random variables is a stochastic process with sample functions the members of S, and with the joint random variable distributions those prescribed for the Brownian motion process.

Bachelier's results remained unnoticed for years, and in fact were rediscovered several times. Wiener's work, like his fundamental work in potential theory, had little immediate influence because it was published in a journal which was not widely distributed. It was an aspect of his genius that he carried out his Brownian motion research then and later without knowledge of the slang and some of the useful elementary mathematical techniques of probability theory.

Steinhaus (1930) demonstrated that classical arguments to derive standard probability theorems could be placed in a rigorous context by taking Lebesgue measure on a linear interval of length 1 as the basic probability measure, interpreting random variables as Lebesgue measurable functions on this interval, and expectations of random variables as their integrals. No new proofs were required; all that was required was a proper translation of the classical terminology into his context. If this were all mathematization of probability by measure theory had to offer, the scorn of rigorous mathematics expressed by some nonmathematicians would be justified.

8 KOLMOGOROV'S 1933 MONOGRAPH. Kolmogorov (1933) constructed the following mathematical basis for probability theory.

(a) The context of mathematical probability is a probability space $(S, \mathbb{S}, P)$. The sets in $\mathbb{S}$ are the mathematical counterparts of real world events; the points of S are counterparts of elementary events, that is of individual (possible) real world observations.

(b) Random variables on $(S, \mathbb{S}, P)$, are the counterparts of functions of real world observations. Suppose $\{x(t,\bullet), t \in \mathscr{I}\}$ is a stochastic process on a probability space $(S, \mathbb{S}, P)$, with state space S'. A set of n of the process random variables has a probability distribution on S'^n. Such finite dimensional distributions are mutually compatible in the sense that if $1 \leq m < n$, the joint distribution of $x(t_1,\bullet), \ldots, x(t_m,\bullet)$ on S'^m is the m-dimensional distribution induced by the n-dimensional distribution of $x(t_1,\bullet), \ldots, x(t_n,\bullet)$ on S'^n.

(c) Conversely, Kolmogorov proved that given an arbitrary index set $\mathscr{I}$, and a suitably restricted measurable space $(S'', \mathbb{S}')$ (for example, the measurable space can be a complete separable metric space together with the σ algebra of its Borel sets) and a mutually compatible set of distributions on S'^n, for integers $n \geq 1$, indexed by the finite subsets of $\mathscr{I}$, there is a probability space and a stochastic process $\{x(t,\bullet), t \in \mathscr{I}\}$ defined on it, with state space S', with the assigned joint random variable distributions. To prove this result he constructed a probability measure on a σ algebra of subsets of the product space $S'^{\mathscr{I}}$, the space of all functions from $\mathscr{I}$ into S', and obtained the required random variables as the coordinate functions of $S'^{\mathscr{I}}$.

(e) The expectation of a numerically valued integrable random variable is its integral with respect to the given probability measure.

(f) The classical definition of the conditional probability of an event (measurable set) A, given an event B of strictly positive probability, is $P(A \cap B)/P(B)$. In this way, for fixed B, new probabilities are obtained, and expectations of random variables for given B are computed in terms of these new *conditional* probabilities. More generally, given an arbitrary collection of random variables, conditional probabilities and expectations relative to given values of those random variables are needed, functions of the values assigned to the conditioning random variables. If $(S, \mathbb{S}, P)$ is a probability space, and if a collection of random variables is given, let $\mathbb{F}$ be the smallest sub σ algebra of $\mathbb{S}$ relative to which all the given random variables are measurable. This σ algebra is the *σ algebra generated by conditions imposed on the given random variables*. A reasonable interpretation of a measurable real valued function of the given collection of random variables is a measurable function from $(S, \mathbb{F})$ into **R**. The Kolmogorov *conditional expectation* of a real valued integrable random variable x on $(S, \mathbb{S}, P)$, relative to a σ algebra $\mathbb{C}$ of measurable sets, is a random variable which is measurable relative to $\mathbb{C}$ and has the same integral as x on every set in $\mathbb{C}$. The existence of such a random variable, and its uniqueness up to P-null sets, is assured by the Radon-Nikodym theorem. The conditional expectation of x relative to a collection of random variables is defined as the conditional expectation of x relative to the σ algebra generated by conditions on the random variables. A conditional probability of a measurable set A is defined as the conditional expectation of the random variable which is 1 on A and 0 elsewhere.

Kolmogorov's 1933 exposition paints a discouraging picture of mathematical progress. In the first pages of his monograph he states explicitly that real valued random variables are measurable functions and expectations are their integrals. Even as late as 1933, however, he must have thought that mathematicians were not familiar with measure theory. In fact in the body of his monograph, when he comes to the definition of a real valued random variable, he does not simply refer back to the first pages of the monograph and say that a random variable is a measurable function. Instead he actually defines measurability of a real valued function, and similarly when he defines the expected value of a random variable he does not simply state that it is the integral of the random variable with respect to the given probability measure, but he actually defines the integral. Later in the monograph, when he needs Lebesgue's theorem allowing taking limits of convergent function sequences under the sign of integration, he does not simply refer to Lebesgue but gives a detailed proof of what he needs. As confirmation of Kolmogorov's caution in invoking measure theory, the author recalls his student experience in 1932 when there were professorial disapproving remarks on the extreme generality of a seminar lecture given by Saks on what is now called the Vitali-Hahn-Saks theorem, a theorem which has since become an important tool in probability theory. [He also recalls that he did not understand the point of Kolmogorov's measure on a function space until long after he had read the monograph.]

It was some time before Kolmogorov's basis was accepted by probabilists. The idea that a (mathematical) random variable is simply a function, with no romantic connotation, seemed rather humiliating to some probabilists. A prominent statistician in 1935 wondered whether two orthogonal real valued random variables with zero means (integrals) are necessarily independent, as they are under the added hypothesis that they have a bivariate Gaussian distribution. He was rather surprised by the example of the sine and cosine functions on the interval $[0, 2\pi]$, with probability measure defined as Lebesgue measure divided by 2π. These two functions, orthogonal and with zero means but not independent, are not the kind of random variables probabilists were used to. Some analysts may be gratified, some humiliated, to learn that in discussing Fourier series they can be accused of discussing probabilities and expectations.

9 EXPANSION BACKWARDS OF THE KOLMOGOROV BASIS. Kolmogorov's basis for mathematical probability can be expanded, and should be expanded in the view of some probabilists, who want to start with some not necessarily numerical mathematical version of the confidence of observers that certain events will occur, and to proceed postulationally to numerical evaluations of this confidence, and finally to additivity. Such an analysis may be enlightened in discussing the appropriateness of mathematical probability as a model for real world phenomena, but any approach to the subject which ends with a justification of the classical calculations and is mathematically usable, will end with Kolmogorov's basis, however phrased, because all the measure basis to probability does is to give a formal precise mathematical framework for the classical calculations and their present refinements. This framework had made it possible to apply mathematical probability in many other mathematical fields, for example to potential theory and partial differential equations. Although such applications were made in the past before the acceptance of measure theory as the basis of probability, the probabilistic context served only to suggest mathematics and was not an integral part of the mathematics. The meaning of solutions as probabilities and expectations could not be formulated and exploited.

10 UNCOUNTABLE INDEX SETS. If the index set $\mathscr{I}$ of a stochastic process $\{x(t,\bullet),\ t \in \mathscr{I}\}$ is an interval of the line, and if the state space of the random variables is $\mathbf{R}$, the class of continuous sample functions may not be measurable. This difficulty arises in the processes derived by the Kolmogorov construction of a measure on a function space, for example, whatever the choice of joint distributions of the process random variables. To understand the difficulty, observe that if the index set $\mathscr{I}$ of a stochastic process with state space $\mathbf{R}$ is an interval, and if $\mathscr{J}$ is a subset of $\mathscr{I}$, the function $s \to \sup_{t \in \mathscr{J}} x(t, s)$ is measurable if $\mathscr{J}$ is countable, but need not be measurable if $\mathscr{J}$ is uncountable. If boundedness and continuity of sample functions are to be discussed, some modification of the probability relations of the random variables of a stochastic process should be devised to make such suprema measurable functions. A clumsy approach was proposed by Doob (1937) but a more usable one was not devised until after 1950.

11 RELUCTANCE TO ACCEPT MEASURE THEORY BY PROBABILISTS. There was considerable resistance to the acceptance and exploitation of measure theory by probabilists, both in Kolmogorov's day and later. The following quotation is an example of the reluctance of some mathematicians to separate the mathematics from the context that inspired it.

Kac (1959) *How much fuss over measure theory is necessary for probability theory is a matter of taste. Personally I prefer as little fuss as possible because I firmly believe that probability theory is more closely related to analysis, physics and statistics than to measure theory as such.*

12 NEW RELATIONS BETWEEN FUNCTIONS MADE POSSIBLE BY THE MATHEMATIZATION OF PROBABILITY.

Probability theory suggested new relations between functions. For example consider the sequence $x_1, x_2, \ldots$ of real valued integrable random variables on a probability space $(S, \mathbb{S}, P)$ and suppose that the conditional expectation of x_n given $x_1, \ldots, x_{n-1}$ vanishes (P) almost everywhere, for $n > 1$, that is, the integral of x_n over any set determined by conditions on the preceding random variables vanishes. If these random variables are square integrable, this condition is equivalent to the condition, much stronger than mutual orthogonality, that x_n is orthogonal to every square integrable function of $x_1, \ldots, x_n$. Bernstein (1927) seems to have been the first to treat such sequences systematically. This condition on a sequence of functions means that in a reasonable sense the sequence of partial sums of the given sequence is the counterpart of a fair game. In fact, the partial sums $y_1, y_2, \ldots$ are characterized by the property that the expectation of y_n relative to $y_1, \ldots, y_{n-1}$ is equal to y_{n-1} almost everywhere on the probability space. Processes with this property, called *martingales*, first used explicitly by Ville (1939), have had many applications, for example to partial differential equations, to derivation, and to potential theory. Another important class of sequences of random variables is the class of sequences with the Markov property. These sequences are characterized by the fact that when $n \geq 1$ the conditional probabilities for x_n relative to $x_1, \ldots, x_{n-1}$ are equal almost everywhere to those for x_n relative to x_{n-1}. Roughly speaking, the influence of the present, given the past, depends only on the immediate past. The Markov property, introduced in a very special case by Markov in 1906, (named in his honor by others) has proved very fruitful, for example, leading in the second half of the century to a probabilistic potential theory, generalizing and including classical potential theory.

13 WHAT IS THE PLACE OF PROBABILITY THEORY IN MEASURE THEORY, AND MORE GENERALLY IN ANALYSIS?

It is considered by some mathematicians that if one deals with analytic properties of probabilities and expectations then the subject is part of analysis, but that if one deals with sample sequences and sample functions then the subject is probability but not analysis. These authors are in the interesting situation that in considering a function of two variables, $(t, s) \to x(t, s)$—as in considering stochastic processes—they call it analysis if the family of functions $x(t, \bullet)$ as t varies is studied, but call it probability and definitely not analysis if the family of functions $x(\bullet, s)$ as s varies is studied. More precisely, they regard discussions of distributions and associated questions as analysis, but not discussions in terms of sample functions. This point of view is expressed in the following quotation.

Protter *By developing his integral in 1944 with stochastic processes as integrands, Itô was able to study multidimensional diffusions with purely probabilistic techniques, an improvement over the analytic methods of Feller.*

The following remark on the convergence of a sum of orthogonal functions illustrates the difficulty in separating (mathematical) probability from the rest of analysis. The measure space is a probability space, but with trivial changes the discussion is valid for any finite measure space.

If $x_\bullet$ is an orthogonal sequence of functions, on a probability measure space, and if x_n^2 has integral σ_n^2, then (Riesz-Fischer) $\sum x_n$ converges in the mean if

$$\sum \sigma_n^2 < +\infty. \tag{13.1}$$

The orthogonal series converges almost everywhere if either (Menšov-Rademacher) (13.1) is strengthened to

$$\sum \sigma_n^2 \log^2 n < +\infty, \tag{13.2}$$

or (Lévy, 1937) the condition (13.1) is kept but the orthogonality condition is strengthened to the condition in Section 12.

The reader should judge which of these results is measure theoretic and which is probabilistic, whether there is any point in evicting mathematical probability from analysis, and if so whether measure theory should also be evicted.

The Evolution of Methods of Convex Optimization

V. M. Tikhomirov

It is surprising that the story of convex optimization began only some fifty years ago. What is even more surprising is that the great analysts and specialists in the calculus of variations did not consider constraints in the form of inequalities. Before the 1940s there were very few papers devoted to the theory of inequalities. Of these, the most significant are a paper by Fourier (1823) and a paper by Vallée-Poussin (1911) (see [1]). Since that time, tens of thousands of papers have been devoted to this subject! A great many of them deal with convex problems.

The earliest nontrivial problems involving constraints in the form of inequalities appeared in a paper by L. V. Kantorovich. This happened as follows. In the spring of 1939 Kantorovich, then a young 26-year-old professor at Leningrad University, was approached by engineers of a plywood factory. They wanted to make more efficient use of their machine tools but lacked the background to deal with the mathematical version of their problem. The problem they put before the future recipient of a Nobel prize was very simple; in fact it was virtually a high school problem. But the young scholar treated it very seriously. Some 35 years later he wrote: "As it turned out, this was not a casual problem. I noticed a great many problems of different contents and of much the same mathematical character." They all fitted the scheme

$$\langle c, x \rangle \to \sup, \qquad \sum_{i=1}^{n} c_i x_i \to \sup,$$
$$\Leftrightarrow \qquad \qquad \qquad \qquad \qquad (1)$$
$$Ax \leq b \qquad \sum_{i=1}^{n} a_{ji} x_i \leq b_j, 1 \leq j \leq m$$

and came to be known as linear programming problems.

In the same year, in 1939, Kantorovich wrote a monograph [2] devoted to methods of solution of problem (1). He tried to interest the Soviet authorities in his investigations because he thought that they could be of use for the development of the Soviet economy. But according to some ideological doctrine, an abstract subject like mathematics was of no conceivable use to so life-related a subject as economics. As a result, Kantorovich was rudely told to keep out of this range of issues.

The vigorous development of linear programming, and then of convex optimization in a wider sense, began in 1947 in the United States. The story of linear programming was told by G. B. Dantzig in one of the papers in [3] devoted to the origins and evolution of mathematical programming. According to Dantzig, the swift progress of linear programming was due primarily to such eminent scholars as von Neumann, Kantorovich, Leont'ev and Koopmans. To these names one should definitely add Dantzig's own name. One of his greatest contributions was the development of a remarkable algorithm for the solution of linear programming problems known as

THE SIMPLEX METHOD. The essence of Dantzig's method is very simple. A careful look at problem (1) shows that the feasible vectors in this problem, that is vectors x for which $Ax \leq b$, form a polyhedron in $\mathbf{R}^n$ (which may or may not be bounded; if bounded it is called a convex polytope). The polyhedron is the intersection of a finite number of halfspaces. A linear function on such a set (satisfying the additional assumption that it contains no straight lines) attains its maximum (if at all) at one of its vertices. This means that we need only look over the values of the linear function in (1) which is to be maximized on the set of vertices and choose the largest. However, in applied problems the number of vertices can reach astronomical proportions, hence the need for a systematic search. Such a search procedure was devised by Dantzig. What follows is a description of Dantzig's simplex method in the so-called *nondegenerate* case.

Suppose we know a vertex. (There are effective methods for finding vertices.) We shall assume that this vertex is *nondegenerate*. This means that at this vertex exactly n "linearly independent" inequalities become equalities. An explanation is in order.

Without loss of generality, we can assume that the first n inequalities become equalities, that is

$$\langle a^j, \bar{x} \rangle = b_j, 1 \leq j \leq n, \langle a^j, \bar{x} \rangle < b_j, j \geq n+1, \qquad (i)$$
$$a^j := (a_{j1}, \ldots, a_{jn}), j = 1, \ldots, m.$$

Nondegeneracy of the vertex means that the vectors $\{a^1, \ldots, a^n\}$ form a basis of $\mathbf{R}^n$, that is, the matrix $A_n := (a_{ij})_{1 \leq i, j \leq n}$ is nonsingular. Put $\bar{b} = (b_1, \ldots, b_n) \in \mathbf{R}^n$. Then

$$A_n \bar{x} = \bar{b} = (b_1, \ldots, b_n) \Leftrightarrow \bar{x} = A_n^{-1} \bar{b}. \qquad (ii)$$

We solve the equation

$$A_n^T \bar{\lambda} = c \qquad (iii)$$

(A_n^T denotes the transpose of A_n.) We have the alternative: (I) $\bar{\lambda} = (\lambda_1, \ldots, \lambda_n) \geq 0$ or (II) some of the components of $\bar{\lambda}$ are negative.

We shall show that in the first case $\bar{x}$ is a solution of (1). In fact, let x be a feasible vector ($Ax \leq b$ (iv)). Put $\lambda = (\lambda_1, \ldots, \lambda_n, 0 \ldots 0)$. Then

$$\langle c, x \rangle \stackrel{(iii)}{=} \langle A_n^T \bar{\lambda}, x \rangle \stackrel{Id}{=} \langle \bar{\lambda}, A_n x \rangle \stackrel{Id}{=} \langle \lambda, Ax \rangle \stackrel{(iv) \text{ and } \lambda \geq 0}{\leq}$$
$$\leq \langle \lambda, b \rangle \stackrel{Id}{=} \langle \bar{\lambda}, \bar{b} \rangle \stackrel{(ii)}{=} \langle \bar{\lambda}, A_n \bar{x} \rangle \stackrel{Id}{=} \langle A_n^T \bar{\lambda}, \bar{x} \rangle \stackrel{(iii)}{=} \langle c, \bar{x} \rangle,$$

which was to be shown. We consider next the alternative possibility.

Suppose that, say, $\lambda_1 < 0$. We find a nontrivial solution of the homogeneous system

$$\langle a^2, y \rangle = \ldots = \langle a^n, y \rangle = 0. \qquad (v)$$

In view of the nonsingular nature of A_n we find that $\langle a^1, y \rangle \neq 0$. With possibly a change of sign we have

$$\langle a^1, y \rangle = -\varepsilon < 0. \qquad (vi)$$

Then for small values of $t > 0$ we have

$$\langle a^j, \bar{x} + ty \rangle < b_j, j = 1, j \geq n+1, \langle a^j, \bar{x} + ty \rangle = b_j, 2 \leq j \leq n,$$

that is the vector $\bar{x} + ty$ is feasible. Also,

$$\langle c, \bar{x} + ty \rangle \stackrel{(v),(vi)}{=} \langle c, \bar{x} \rangle + t \langle c, A_n^{-1}(-\varepsilon, 0, \ldots, 0) \rangle \stackrel{Id}{=} \langle c, \bar{x} \rangle$$
$$+ t \langle (A_n^{-1})^T c, (-\varepsilon, 0, \ldots, 0) \rangle \stackrel{Id}{=} \langle c, \bar{x} \rangle - t\varepsilon \lambda_1.$$

This means that $\langle c, \bar{x} + ty \rangle > \langle c, \bar{x} \rangle$ for all $t > 0$.

If $\bar{x} + ty$ is feasible for all $t > 0$ then the supremum for the problem is $+\infty$. Otherwise, $\langle a^j, \bar{x} + t_0 y \rangle = b_j$ for some value t_0 of t and for some $j \geq n+1$. Then $\bar{x} + t_0 y$ will take the place of $\bar{x}$ and we can perform another iteration. And so on. This completes the description of one version of the nonsingular nondegenerate simplex method.

The simplex method has played an important role in the history of numerical methods of optimization. For a long time it was not known whether or not the problem in (1) is *polynomial*, that is, whether or not there are for this problem algorithms that require, in all cases, just a polynomial number of operations (in terms of the size of the input). In 1970 Klee and Minty constructed examples which showed that in some situations the simplex method requires an exponential number of steps. But for some reason problems of this kind don't turn up in applications! Many mathematicians (including Dantzig) have said that it is something of a miracle that the simplex method has worked essentially without a hitch for about 50 years in countless applied situations. Dantzig himself said: "The tremendous power of the simplex method is a constant surprise to me."

Some years after his invention of the simplex method Dantzig decided to write a comprehensive monograph (in addition to the book [1]) devoted to linear programming. It was to be a survey of all papers devoted to inequalities and their applications to extremal problems. In particular, the monograph [2] of Kantorovich attracted his attention. Dantzig thought highly of the monograph and saw to it that it was translated. In this way the scientific world learned about Kantorovich's—and Koopman's—contributions to the theory. In 1975 the two scientists were awarded the Nobel prize in economics for their contribution to the development of linear programming and for its application to economics. Next we discuss other approaches to linear and convex optimization. We begin with a description of

THE METHOD OF CENTRAL SECTIONS. I was a witness to the story I am about to tell. In 1962 I worked in Voronezh, a big town in central Russia. At that time, the Voronezh school of mathematics, headed by M. A. Krasnosel'skiĭ, flourished. The school pursued important scientific objectives and tried to enter various applied areas. In particular, in the early sixties, Krasnosel'skiĭ signed a contract with an applied mathematics firm headed by David B. Yudin. Yudin's name was known to specialists in optimization from the monograph [4] written by him and his then coworker E. G. Gol'shteĭn. Yudin posed the following problem: *Find an effective algorithm that minimizes a sum of exponentials (with positive weights) on a compact polyhedron.*

The problem attracted the attention of Anatoli Yu. Levin, one of the main heroes of our story, and a member of Krasnosel'skiĭ's group. Levin pondered this, as well as more general problems, for a long time. He discussed them with colleagues, including myself, and one day hit on a remarkable idea for an algorithm that turned out to be applicable to the following more general problem: *find the minimum of a convex (as well as a quasiconvex) function f on a finite-dimensional convex body A*, i.e.

$$f(x) \to \inf; \ x \in A.$$

This is, essentially, the general problem of convex optimization. What follows is a description of Levin's algorithm for a smooth convex function.

We denote A by A_1. We determine $x_1 = grA_1$, the center of gravity of A_1. Then we compute $f'(x_1)$. If this is the zero vector, then the problem is solved. Otherwise we eliminate the part of A_1 in the halfspace $\Pi'_1 := \{x | \langle f'(x_1), x - x_1 \rangle > 0\}$. (This step is justified as follows: For a convex f it is easy to show that $f(x) - f(x_1) \geq \langle f'(x_1), x - x_1 \rangle$. But then for $x \in A_1 \cap \Pi'_1$ we have $f(x) > f(x_1) > \min$.) We denote the remaining part of A_1 by A_2 and repeat the procedure. And so on.

If we take as ξ_m a point of $\{x_1, \ldots, x_m\}$ at which $f(\xi_m)$ is not less than any of the values $f(x_i)$, $1 \leq i \leq m$, then it can be shown that $f(\xi_m)$ tends to the minimum of f on A and the error in the value of f decreases at the rate of a geometric progression. Also, the volume of A_m decreases exponentially. This fact is a consequence of the following result in convex geometry due to Grünbaum: *Let A be a convex body in R^n, and let $\xi = grA$ be its center of gravity. Every hyperplane passing through ξ divides A into two parts such that the volume of each of these parts is no less than the fraction $1 - (1/e)$ of the volume of A.*

We note that in the one-dimensional case the method just described reduces to halving of intervals.

Levin delayed the writing up of his result for publication. (Some of the reasons for this delay will be given below.) In the meantime the American D. J. Newman independently hit on the idea of the method of central sections. The Levin and Newman papers [5, 6] appeared simultaneously in 1965.

Some time later, my student A. I. Kuzovkin and I supplemented Levin's algorithm and showed that it is possible to obtain an exponential rate when computing the value of f rather than its gradient [7].

After that the issue was dormant. The next important event occurred some 15 years after the discovery of the method of central sections.

THE METHOD OF CIRCUMSCRIBED ELLIPSOIDS OF NEMIROVSKIĬ-YUDIN-SHOR. Let me introduce to the reader one more person who played an important role in the subsequent evolution of convex optimization. This is Arkadi Nemirovskiĭ, a graduate of Moscow University and one of the last students of G. E. Shilov.

After the completion of graduate studies in 1974 Nemirovskiĭ began to work for Yudin. Yudin put before his young coworker the issue of the complexity of the solution of the problem of convex optimization. In November of 1974, (while walking in the forest) Nemirovskiĭ hit on yet another method of solution of the problem of convex optimization. It came to be known as the method of circumscribed ellipsoids. The method was presented in a paper. The key idea of this method was found independently (and somewhat later) by the well-known specialist on convex optimization, the Kiev mathematician Naum Z. Shor; see [**18**]. That is why the method of circumscribed ellipsoids is sometimes referred to as the *Nemirovskiĭ-Yudin-Shor method.*

The method combines two ideas. One is the method of sections discussed above and the other is the geometric fact that *half an ellipsoid can be put in an ellipsoid of smaller volume than the initial ellipsoid.*

We now describe in greater detail the Nemirovskiĭ-Yudin-Shor algorithm. We denote by E_0 an ellipsoid circumscribed about A. If the center c_0 of this ellipsoid lies outside A, then we pass through it a halfspace not containing points of A and eliminate the half of the ellipsoid that doesn't intersect A. If $c_0 \in A$, then we compute $f'(c_0)$, carry out a section à la Levin-Newman,, and again end up with half an ellipsoid which we denote by E'_0. Next we circumscribe about E'_0 an ellipsoid of smaller volume than the volume of E'_0, denote it by E_1, and repeat our procedure.

The rate of the solution just described is that of a geometric progression. Yudin and Nemirovskiĭ showed in their paper that the Levin-Newman method of central sections cannot be substantially improved in the class of convergent algorithms of minimization of convex functions. The method of circumscribed ellipsoids is somewhat inferior to the method of central sections in terms of the rate of convergence but has the advantage that it obviates the need for finding centers of gravity of polyhedrons.

All this was told by Nemirovskiĭ and Yudin at various seminars. Once they described this method at the seminar of E. G. Gol'shteĭn in the Central Economics-Mathematics Institute. Gol'shteĭn liked the Nemirovskiĭ-Yudin lecture very much but observed that similar ideas had been advanced earlier. "Where?" asked the lecturers. "In A. Yu. Levin's paper" was the answer. "Which Levin?" asked Yudin nervously. "The same Levin who solved your problem 15 years ago..." replied Gol'shteĭn. Accordingly, Nemirovskiĭ and Yudin included in their paper [**8**] a reference to A. Yu. Levin's paper [**5**].

THE PAPER OF L. G. KHACHIAN. A few years passed during which only one event occurred that bears on our story. L. Levin, Nemirovskiĭ's one time fellow-student, decided to emigrate to America and Nemirovskiĭ went to bid him farewell. He brought with him a copy of his paper [**8**] on the method of ellipsoids and gave it to Levin. At the time Levin was not in a mood for mathematics and put the paper at the bottom of his valise.

One early morning I happened to be passing the Computing Center of the Academy of Sciences in Moscow and noticed the racing figure of a former fellow-student of mine who worked at the time at the Computer Center. I exclaimed: "What happened?" Without slowing down he shouted incoherently "Khachian...New York Times...Press conference...I am late...."

This is what happened. A young associate at the Computing Center, Leonid G. Khachian, had published a note [**9**] titled "A polynomial algorithm in linear programming" in the Proceedings of the Soviet Academy of Sciences. No one reacted to this paper for a long time. But then a conference of specialists in the area of convex optimization took place in the U.S. One of the participants presented a survey of recent progress in the area and mentioned the paper of Khachian. While describing the latter, he remarked: "Khachian shows that the linear programming problem can be solved in polynomial time . . . ." He was about to continue his lecture but was interrupted by the Hungarian mathematician Peter Gács: "Would you mind repeating what you just said?" The lecturer obliged. Gács exclaimed: "But this is the solution of a famous problem, the one about the polynomial nature of the problem of linear programming!" It turned out that Khachian's result could indeed be interpreted as a justification of the claim of the polynomial nature of the problem of linear programming. The algorithm turned out to be an application of the method of ellipsoids to the problem of linear programming. The press got interested in Khachian's paper and this led to *The New York Times* press conference in the Computing Center of the Soviet Academy of Sciences . . . .

Khachian's paper contained a reference to Shor's paper. This reminded L. Levin of the copy that Nemirovskiĭ had given him at the time of his departure to the U.S. From it he learned about the method of circumscribed ellipsoids which yielded a polynomial estimate for the number of steps in the solution of the problem of convex programming; in particular, for linear programming over the reals. It was Khachian who took the next step, essential for everything that was to follow, namely, he found the necessary computational complexity for the rationals (and it turned out to be polynomial).

As a result of all this the papers of Khachian and Nemirovskiĭ-Yudin became widely known. Their authors were honored with the prestigious Fulkerson award of the International Society for Mathematical Programming and the American Mathematical Society for 1982.

Peter Gács called L. Levin's attention to the paper [5] of his namesake A. Yu. Levin. Now we must go back somewhat in time. One of the reasons for the delay in the publication of the paper [5] was that A. Yu. Levin wanted to overcome the difficulties associated with the finding of the center of gravity of a polyhedron. (It later turned out that problem has indeed exponential complexity.) In [5] Levin advanced the idea that it was sometimes necessary to circumscribe simplexes about the A_n, and that this could be done without impairing the exponential decrease of their volumes. He gave no specific algorithms. Rather, it was an existence theorem for such algorithms. Beginning with this idea, Levin and his coauthor Boris Yamnitskiĭ showed in [10] that A. Yu. Levin's idea was correct in principle. The paper [10] contains a description of the method of "circumscribed simplexes" which coincides with the method of circumscribed ellipsoids, runs in polynomial time, and is in some respects superior to the method of ellipsoids.

CONCLUDING REMARKS. The events just described have resulted in an explosive development of methods of convex optimization in the eighties. There appeared a number of methods of sections (such as the method of inscribed ellipsoids of Tarasov-Khachian-Ehrlich, (see [19]) and others). Many methods were modified and perfected. Also, in 1984 Karmarkar proposed a polynomial method for the solution of problems of linear programming based on different ideas. This method turned out to offer many advantages and it resulted in a veritable flood of papers on algorithms dealing with convex programming in which the methods discussed above were modified, developed and perfected. The reader can find the relevant details in the monograph [11] of Nemirovskiĭ and Nesterov.

For additional information in English, see items [12]–[19].

ACKNOWLEDGMENT. The column editor wishes to thank Professor Michael J. Todd for his help.

REFERENCES

1. Dantzig, G. B. *Linear Programming and Extensions*, Princeton, Univ. Press, 1963.
2. Kantorovich, L. V. Mathematical methods in the organization and planning of production. Leningrad University, 1939. [Russian] For an English version see *Management Science*, Vol. 6, 1960, 366–422.
3. *History of Mathematical Programming* (A Collection of Personal Reminiscences). Ed. by J. K. Lenstra, F. H. G. Rinnooy Kan, A. Schrijver, North-Holland, 1991.
4. Yudin, D. B. and Gol'shtein, E. G. *Problems and methods of linear programming*. Moscow. Sovradio, 1961. [Russian]
5. Levin, A. Yu. On an algorithm for the minimization of convex functions. *DAN USSR* vol. 160, issue 6, 1965, pp. 1244–1247. [Russian]
6. Newman, D. J. Location of maximum on unimodal surfaces. *Journ. of the Assoc. for Computing Machinery* 12, 1965, pp. 395–398.
7. Kuzovkin, A. I. and Tikhomirov, V. M. On the number of computations needed for finding the minimum of a convex function. *Economics and mathematical methods*, 3:1, 1967, pp. 95–103. [Russian]
8. Yudin, D. B. and Nemirovski, A. S. Informational complexity and effective methods of solution of convex extremal problems. *Economics and mathematical methods* vol. 12, issue 1, 1976, pp. 357–369. [Russian]
9. Khachian, L. G. A polynomial algorithm in linear programming. *DAN USSR* vol. 244, issue 5, 1979, pp. 1093–1096 [Russian]. For an English version see *Soviet Math. Doklady* 20, 1979, 191–194.
10. Yamnitsky, B., Levin, L. An Old Linear Programming Algorithm Runs in Polynomial Time. In *23rd Annual Symposium on Foundations of Computer Science*, IEEE, New York, 1982, pp. 327–328.
11. Nemirovsky, A., Nesterov, Yu. *Interior-Point Polynomial Algorithms in Convex Programming*. SIAM Studies in Applied Math., 1994.
12. Nemirovsky, A. S., Yudin, D. B. *Problem Complexity and Method Efficiency in Optimization*. John Wiley and Sons, Chichester, 1983.
13. Grötschel, M., Lovász, L., Schrijver, A. *Geometric Algorithms and Combinatorial Optimization*. Springer-Verlag, 1988.
14. Schrijver, A. *Theory of Linear and Integer Programming*. John Wiley and Sons, 1986.
15. Shor, N. Z. *Minimization Methods for Nondifferentiable Functions*. Springer-Verlag, Berlin, 1985.
16. Bland, R. G., Goldfarb, D., Todd, M. J. The ellipsoid method: a survey. *Operations Research*, 29 (1981), 1039–1091.
17. Goldfarb, D., Todd, M. J. Linear programming in "Handbooks in Operations Research and Management Science," vol. 1, *Optimization* (G. L. Nemhauser, A. H. G. Rinnooy Kan and M. J. Todd, eds.), North Holland, Amsterdam, 1989, 73–170.
18. Shor, N. Z. Cut-off method with space extension in convex programming problems, *Cybernetics*, 13 (1977), 94–96.
19. Tarasov, S. P., Khachiyan, L. G., Erlikh, I. I. The method of inscribed ellipsoids. *Soviet Math. Doklady*, 37 (1988), 226–230.

A Few Expository Mini-Essays

Abe Shenitzer

Sometimes it is possible to describe a mathematical issue of great importance very briefly. This is exemplified by the essays below. (Of the six essays in this paper, two are quotes and two more are essentially quotes.)

(A) POLYNOMIAL EQUATIONS. Interest in the solution of polynomial equations has been a constant feature of mathematics from the time of the Babylonians to this day. In fact, it is safe to say that until the revolution wrought in mathematics by the ideas of Galois, algebra was synonymous with the solution of equations. The first substantial elaboration of Galois' ideas had to wait until 1870, when Jordan published his influential *Treatise*. Since then algebra has largely concerned itself with the study of structures such as groups, fields, rings, vector spaces, etc. Back to polynomial equations.

Why equations? Granted an interest in functions, it is natural to ask of any function f what is its range and what is the preimage $f^{-1}(a)$ of an element a in the codomain of f. The latter question asks for the solution of the equation $f(x) = a$. This can be normalized as $\phi(x) = f(x) - a = 0$. It is obtaining an explicit solution of this problem for a polynomial which has occupied mathematicians for some 4000 years, and which has been (almost) completely solved only in 1984 by the Japanese mathematician Hiroshi Umemura.

Umemura's achievement seems to be a well-kept secret. I found out about it by reading (parts of) a remarkable book-length essay (172 pages) titled *Mathematics and Physics* by Krzysztof Maurin [1]. Here is how Maurin leads up to Umemura's result:

Some 2000 years BC the Babylonians could solve certain quadratic equations (so too could the Chinese and the Indians). In the 16th century the Italians (del Ferro, Cardano) solved cubic and quartic equations by extraction of roots. All attempts to obtain similar solutions of the general quintic equation failed. (That they were bound to fail follows from the Abel-Ruffini theorem (1826) which established the impossibility of the solution of the general quintic equation by means of extraction of roots.) It was Galois who stated a group-theoretic condition for the solvability of an n-th order equation by means of extraction of roots. (The solvability of an algebraic equation is equivalent to the solvability of its Galois group.) As early as 1858 Hermite and Kronecker showed (independently) that the quintic equation could be solved by using an elliptic *modular function*. Since

$$(*) \qquad \sqrt[n]{a} = \exp\left(\frac{1}{n}\log a\right) = \exp\left(\frac{1}{n}\int_1^a x^{-1}\,dx\right),$$

extraction of roots involves integration of the function $1/x$ and the use of the exponential function. The idea of Hermite and Kronecker was the following: to solve a fifth-degree equation it is necessary to replace the exponential function exp in (∗) by another transcendental function (by an elliptic modular function), and the integral by elliptic integrals. Kronecker thought that the roots of an arbitrary algebraic equation could be found in a similar way. Umemura showed that Kronecker's conjecture was correct: Every algebraic equation has a root that can be expressed in terms of a modular function and hyperelliptic integrals....

(B) A CAPSULE HISTORY OF DYNAMICS. This is a quote from pp. 2–3 in: Strogatz, NONLINEAR DYNAMICS AND CHAOS, ©1994 by Addison-Wesley Publishing Company, Inc. Reprinted by permission of the Publisher.

Although dynamics is an interdisciplinary subject today, it was originally a branch of physics. The subject began in the mid-1600s, when Newton invented differential equations, discovered his laws of motion and universal gravitation, and combined them to explain Kepler's laws of planetary motion. Specifically, Newton solved the two-body problem—the problem of calculating the motion of the earth around the sun, given the inverse square law of gravitational attraction between them. Subsequent generations of mathematicians and physicists tried to extend Newton's analytical methods to the three-body problem (e.g., sun, earth, and moon) but curiously this problem turned out to be much more difficult to solve. After decades of effort, it was eventually realized that the three-body problem was essentially impossible to solve, in the sense of obtaining explicit formulas for the motions of the three bodies. At this point the situation seemed hopeless.

The breakthrough came with the work of Poincaré in the late 1800s. He introduced a new point of view that emphasized qualitative rather than quantitative questions. For example, instead of asking for the exact positions of the planets at all times, he asked: "Is the solar system stable forever, or will some planets eventually fly off to infinity?" [The latter may happen. See [2].] Poincaré developed a powerful *geometric* approach to analyzing such questions. That approach has flowered into the modern subject of dynamics, with applications reaching far beyond celestial mechanics. Poincaré was also the first person to glimpse the possibility of chaos, in which a deterministic system exhibits aperiodic behavior that depends sensitively on the initial conditions, thereby rendering long-term prediction impossible.

(C) THE ORIGIN OF "GEOMETRIC" TOPOLOGY. You can "do" analysis of singlevalued meromorphic functions in the plane, but "doing" analysis of multivalued meromorphic functions calls for finding suitable habitats for such functions. These habitats were introduced by Riemann—hence Riemann surfaces—in connection with his study of algebraic functions and their integrals, and are topologically spheres with handles. The number of handles of a Riemann surface is its genus. The preeminent role of this topological invariant of a Riemann surface in the characterization of its "meromorphic inhabitants" is described by Herman

Weyl in a lengthy passage in [3] which begins as follows:

> The classical example of the fruitfulness of the topological method is Riemann's theory of algebraic functions and their integrals. Viewed as a topological surface, a Riemann surface has just one characteristic, namely its connectivity number or genus p. For the sphere $p = 0$ and for the torus $p = 1$. How sensible it is to place topology ahead of function theory follows from the decisive role of the topological number p in function theory on a Riemann surface. I quote a few dazzling theorems: The number of linearly independent everywhere regular differentials on the surface is p. The total order (that is, the difference between the number of zeros and the number of poles) of a differential on the surface is $2p - 2$. If we prescribe more than p arbitrary points on the surface, then there exists just one single valued function on it that may have simple poles at these points but is otherwise regular; if the number of prescribed poles is exactly p, then, if the points are in general position, this is no longer true. The precise answer to this question is given by the Riemann-Roch theorem in which the Riemann surface enters only through the number p. If we consider all functions on the surface that are everywhere regular except for a single place P at which they have a pole, then its possible orders are all numbers $1, 2, 3, \ldots$ except for certain powers of p (the Weierstrass gap theorem). It is easy to give many more such examples. The genus p permeates the whole theory of functions on a Riemann surface. We encounter it at every step, and its role is direct, without complicated computations, understandable from its topological meaning (provided that we include, once and for all, the Thomson-Dirichlet principle as a fundamental function-theoretic principle).

In addition to his discovery and use of Riemann surfaces in function theory, Riemann—and independently of Riemann, Betti—discovered a class of topological invariants called Betti numbers. What are Betti numbers? Hilton and Pedersen [4] explain that

> With any topological space K we may associate certain abelian groups $H_r(K)$, $r = 0, 1, 2, \ldots$ called the homology groups of K, which, roughly speaking, count the r-dimensional 'holes' in K. If the space is n-dimensional then we only have homology groups up to dimension n. Then p_r is the *rank* of $H_r(K)$.

Betti numbers are intimately related to such topological household words as genus and Euler characteristic. Specifically, for a closed, orientable surface S_g the Betti numbers are $p_0 = 1, p_1 = 2g, p_2 = 1$. Since the Euler characteristic $\chi(S_g)$ of S_g is connected with its Betti numbers by the relation

$$\chi = p_0 - p_1 + p_2,$$

it follows that

$$\chi(S_g) = 2 - 2g.$$

After Riemann, "geometric" topology was magnificently advanced at the end of the century by Poincaré, of whose topological contributions Dieudonné says (in his article on Poincaré in [**11**]) that "until the discovery of the higher homotopy groups in 1933, the development of algebraic topology was entirely based on Poincaré's ideas and techniques."

(D) THE ORIGIN OF "GENERAL" TOPOLOGY. "General," or point-set, topology came into being as a result of Cantor's interest in Fourier series. I quote from

<small>Jaenich, TOPOLOGY, ©1984 Springer-Verlag New York, Inc. Reprinted by permission of the Publisher. (See pp. 3-4 of the book's introduction.)</small>

A... contribution of paramount importance to the emergence of point-set topology was... the work of Cantor. [An indirect indication of this is that the] dedication of Hausdorff's book [*Grundzuege der Mengenlehre*, 1914] reads: "To the creator of set theory *Georg Cantor* in grateful admiration"...

Cantor had shown in 1870 that two Fourier series that converge pointwise to the same limit function have the same coefficients. In 1871 he improved this theorem by proving that the coefficients have to be the same also when convergence and equality of the limits hold for all points outside a finite exception set A in $[0, 2\pi]$. In a paper of 1872 he dealt with the problem of determining for which *infinite* exception sets uniqueness still holds. An infinite subset of $[0, 2\pi]$ must of course have at least one cluster point:

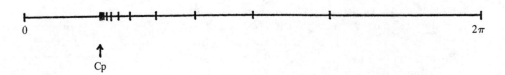

This is a very "innocent" example of an infinite subset of $[0, 2\pi]$. A somewhat "wilder" set is one whose cluster points themselves cluster around some point:

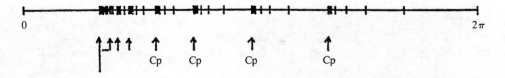

Cantor showed that if the sequence of subsets of $[0, 2\pi]$ defined inductively by $A^0 := A$ and $A^{n+1} := \{x \in [0, 2\pi] | x \text{ is a cluster point of } A\}$ breaks off after finitely many terms, that is if eventually we have $A^k = \emptyset$, then uniqueness *does* hold with A as the exception set. ...[While] the motivation for Cantor's investigation stemmed from classical analysis and ultimately from physics, [it led him] to the discovery of a new type of subset A in **R** which must have been thought to be quite exotic, especially when the sequence $A, A^1, A^2, \ldots$ took a long time to break off. Now the subsets of **R** moved to the fore as objects to be studied for their own sake, and, what is more, studied from what we would recognize today as a topological viewpoint. Cantor continued along this path when later, while investigating general point sets in **R** and **R**^{n}, he introduced the point-set topological approach on which Hausdorff could later base himself.

To flesh out the last sentence in this quote, let me add the following observation from [5] (p. 107):

> A not insignificant part of the nomenclature that is basic to general topology was introduced by Cantor in the course of his researches on point sets. Terms such as *perfect, dense, separable..., denumerable, continuous, closed, open*, are illustrative of the terms Cantor applied to point sets. Cantor obtained the essential theorems concerning the structure of these sets on the line, i.e. the topology of the line.

(E) FROM GEOMETRY TO GEOMETRIES. The first step in the transition from one geometry to many was the discovery of hyperbolic geometry. It became public knowledge in the late twenties and early thirties of the last century but it began to affect the thinking of large numbers of mathematicians only some forty years later. Its technical and intellectual significance for mathematics is immeasurable. One vital intellectual effect of its discovery was the realization, spelled out with all necessary precision by Gauss, that unlike arithmetic, the question of the geometry of physical space should be regarded as a question of physics.

In 1828 Gauss introduced an infinity of geometries, the so-called intrinsic geometries of surfaces in **R**^{3}, in which the geodesics played the role of straight lines. This was a tremendous step forward from a mathematical viewpoint but not a conceptual revolution comparable to the discovery of hyperbolic geometry.

In his famous address of 1854 Riemann generalized Gauss' intrinsic geometry of a surface by extending it to n dimensions. Riemannian geometries are metric geometries.

In another famous address, in the Erlangen Program of 1872, Klein introduced his notion of a geometry as the totality of invariants of the subsets of a set acted upon by a group. Of the gap between these two conceptions Cartan had this to say [6]:

> The principle of general relativity brought into physics and philosophy the antagonism between the two leading principles of geometry due to Riemann and Klein respectively. The space-time manifold of classical mechanics and of the principle of special relativity is of the Klein type, and the one associated with the principle of general relativity is Riemannian. The fact that almost all phenomena studied by science for many centuries could be equally well explained from either viewpoint was very significant and persistently called for a synthesis that would unify the two antagonistic principles.

The geometric ideas of Riemann and Klein were combined in the concept of "spaces with connections." A space with a connection is a Riemannian space with a group of motions (Euclidean, affine,...) grafted onto it. The group of motions is the "tool" needed to propel a vector parallel to itself along a curve in the Riemannian space.

Chronologically, the first step in this development was Levi-Civita's extension of the obvious notion of parallel transport of a vector along a curve in a plane to more general spaces. This was in 1917. In 1918, H. Weyl used a Riemannian space with a Euclidean connection in an attempt to unify electromagnetic and gravitational phenomena (see [6], [9], and [10]). "Weyl's ideas undoubtedly were the source from which E. Cartan, a few years later, developed his general theory of [spaces with] connections..." (see Dieudonne's article on H. Weyl in the Dictionary of Scientific Biography [11]).

A footnote on parallel transport. The remarks that follow give some indication of how one extends the notion of parallel transport from the plane to a more general space and mentions an unexpected application of this concept.

What it means to parallel-transport a vector along a straight line in the plane is suggested by Figure 1 below:

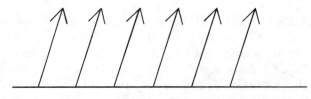

Figure 1

What is kept constant here is the angle between vector and straight line.

If we leave the plane for, say, a sphere and the straight line for a great circle G then, to obtain an analogue of Figure 1, we must draw tangent planes to the sphere at the points of G, tangent lines to G in the tangent planes, and draw vectors in the tangent planes that make the same angle with the appropriate tangent line.

It takes little to make this more general. "*Parallel translation of a vector tangent to a surface along a geodesic on this surface* is defined as follows: the point of origin of the vector moves along the geodesic, and the vector itself moves continuously so that its angle with the geodesic and its length remain constant."

It is obvious what to do if instead of a single geodesic arc we have a broken line consisting of several geodesic arcs. Finally, "*parallel translation of a vector along any smooth curve on a surface* is defined by a limiting procedure, in which the curve is approximated by broken lines consisting of geodesic arcs." (See Appendix 1: Riemannian [i.e. Gaussian] curvature, [7], pp. 301–317.)

What is the point of all this? Well, if you parallel-transport a vector around a triangle in a plane then, upon completion of the "tour," you end up with a vector that coincides with the initial vector. But when you parallel-transport a vector around a spherical triangle—see Figure 2—with angles α, β, and γ then the end-vector makes an angle of $\alpha + \beta + \gamma - \pi$ with the initial vector. Now comes the surprise (see p. 21 in [9]): $\alpha + \beta + \gamma - \pi = R^{-2}S$, where R is the radius of the sphere and S is the area of the triangle! In other words, *the angular change due to the parallel transport of a vector around a spherical triangle divided by its area is the (Gaussian) curvature of the sphere!* (If you are as use-obsessed as the writer of these lines, then you will enjoy the realization that this is, in principle, a prescription for finding out by means of a *local* test what kind of constant-curvature surface you are on!) Since anything related to the curvature of a manifold is important, parallel transport is important.

For more intuitive material on parallel transport see pp. 17–31 of [8]. For more precise material, see Ch. 8 of [9].

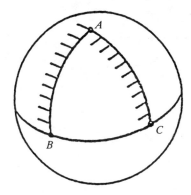

Figure 2

(F) FROM DANIEL BERNOULLI TO DISTRIBUTIONS. Extrapolating from physical experiments, Daniel Bernoulli made the brilliant guess that the shape of a vibrating string is describable by means of a trigonometric series.

Somewhat later, Euler and d'Alembert arrived independently at essentially the same solution of the equation of the vibrating string which did not involve trigonometric series, and was thus very different from that of Bernoulli. While the two disagreed on the admissible initial shapes of the string, they agreed that Bernoulli's guess made no sense.

Shortly thereafter, Fourier's work, while sorely lacking in rigor, vindicated Daniel Bernoulli's informed guess and ushered in the study of Fourier series as a new branch of analysis. As for the resolution of the argument between Euler and d'Alembert, that had to await the introduction of distributions and of generalized solutions of differential equations. (The latter idea is found in Hilbert's rhetorical question at the end of his Problem 20: "Has not every regular variation problem a solution, ...provided also, if need be, that the notion of a solution shall be suitably extended?")

REFERENCES

1. Krzysztof Maurin, *Mathematics and Physics* in: Mathematical Lexicon (Polish). Wiedza Powszechna, 1993. Umemura's article is titled *Resolution of Algebraic Equations by Theta Constants*. This is an appendix in David Mumford's book *Tata Lectures on Theta II*, Progress in Mathematics, Vol. 43, Birkhäuser, 1984.
2. D. G. Saari and Z. Xia, *Off to Infinity in Finite Time*. Notices of The AMS, May 1955, pp. 538–546.
3. H. Weyl, *Topology and Algebra as Two Roads of Mathematical Comprehension*. The Monthly, issues May and July–August, 1995.
4. P. Hilton and J. Pedersen, *The Euler Characteristic and Polya's Dream*. Amer. Math. Monthly 103 (1996), 121–131.
5. J. H. Manheim, *The genesis of point set topology*. Pergamon Press, 1964.
6. E. Cartan, *Group Theory and Geometry* (French). L'Enseignement mathématique, 1927, pp. 200–225.
7. V. I. Arnol'd, *Mathematical methods of classical mechanics*. Springer, 1978.
8. N. Ya. Vilenkin, *In Search of Infinity*. Birkhäuser (Boston), 1995.
9. J. A. Thorpe, *Elementary Topics in Differential Geometry*. Springer, 1979.
10. H. Weyl, *Space, Time, Matter*. Dover, 1922.
11. *Dictionary of Scientific Biography*. Published under the auspices of the American Council of Learned Societies, C. C. Gillispie, Editor in Chief. Charles Scribner's Sons, New York.

On the Emotional Assumptions Without Which One Could Not Effectively Investigate the Laws of Nature

Vl. P. Vizgin

Historico-Mathematical Investigations, issue 4, **39** (1999) 343–352
(Translated from the Russian by Abe Shenitzer)

1. INTRODUCTION. In many different texts by past great figures or by more recent leaders in theoretical and mathematical physics, from papers and textbooks to philosophical and methodological essays and recollections, we frequently come across elevated, emotionally charged expressions, and even quasi-religious terminology. E. L. Feĭnberg has often called attention to some of these expressions in connection with his study of intuitive judgments in science [1], [2], [3]. Such judgments are neither provable nor logically deducible. However, given their importance for theoreticians, they must be assigned an imperative character. In turn, this forces the scientists to employ expressions that are emotionally and even religiously colored. According to Feĭnberg, a relevant example is the thesis of the reality and knowability of the physical world, and, in particular, Einstein's credo: "The most incomprehensible thing about the universe is that it is comprehensible" [a popular paraphrase of the statement: "One may say 'the eternal mystery of the world is its comprehensibility. ...The fact that it is comprehensible is a miracle.'" [**21**, p. 292]. In the sequel we consider a large group of similar utterances concerning the mathematical nature of the physical world and the related notions of Einstein's "cosmic religion" and Wigner's so-called "empirical law of epistemology."

2. FROM THE PYTHAGOREAN "ALL THINGS ARE NUMBERS" TO THE "PREESTABLISHED HARMONY BETWEEN PHYSICS AND MATHEMATICS" AND "THE UNREASONABLE EFFECTIVENESS OF MATHEMATICS IN THE NATURAL SCIENCES". Heisenberg was right when he said of the Pythagorean school that "Here has been established the connection between religion and mathematics which ever since has exerted the strongest influence on human thought. The Pythagoreans seem to have been the first to realize the creative force inherent in mathematical formulations" [**4**, pp. 67–68]. The Pythagorean line of thought was continued by Plato in his conception of matter based on the doctrine about regular polyhedra.

Note by column editors. This paper contains many references to what may be called the Pythagorean-Platonic tradition. The author seems to be taking it for granted that the reader is familiar with this term. We think that Heisenberg's splendid explanation of what this tradition is about and of its role in modern physics is an indispensable prerequisite for this paper. The explanation in question is found in his book *Physics and Philosophy* [**4**, pp. 71–73].

The Pythagorean-Platonic tradition was developed in the doctrines of the medieval Scholastics who thought that God created the world rationally, on a mathematical basis. In the 17th century this tradition discovered an approach to the study of nature that amounted to the search for mathematical laws and structures that disclosed the essence of phenomena, and with it the designs of the Creator. The leading figures of the Modern Era—Kepler, Galileo, Descartes, Leibniz, Newton and others—emphasized in the most elevated terms the Pythagorean idea of the divine mathematicalness of the world. To quote Leibniz: "Cum Deus calculat, fit mundus" (As God calculates so the world is made) [5, p. 213].

In the second half of the 18th century and at the beginning of the 19th century God is gradually "eased out" of mathematical investigations [5, p. 214]. When French Enlightenment scholars, such as Lagrange, Laplace, Poisson, Fourier, Ampère and others, discussed the mathematical structure of nature they practically never resorted to quasireligious terminology or to references to God. And it is these French scholars who brought to light the mathematical structure of classical physics by showing that the fundamental laws of electricity and magnetism, of optics, and of heat phenomena can be described in the language of mathematical analysis: more specifically, in the language of the theory of second-order partial differential equations. This discovery became one of the key elements of the scientific revolution connected with the creation of classical physics [6].

We witness a rebirth of the Pythagorean-Platonic tradition, complete with rather frequent use of "elevated" and even quasireligious formulations, during the development of the mathematically sophisticated theories of relativity and quanta. Statements of the Kepler and Leibniz variety reoccur and are used both by mathematicians—Hilbert, Minkowski, F. Klein, and later Bourbaki and others—and by physicists—Sommerfeld, Heisenberg, Born, Dirac, Wigner, and of course Einstein, and others.

In his Paris lecture in 1900 Hilbert spoke of the "preestablished harmony" between mathematics and physics [7, p. 3]. Thirty years later, enriched by his participation in the development of general relativity and quantum mechanics, he again spoke of this "preestablished harmony," of which "general relativity and quantum mechanics" are the most magnificent and remarkable example [8].

In 1908 Minkowski concluded his Köln lecture on the four-dimensional formulation of special relativity with a reference to the "preestablished harmony between pure mathematics and physics" exemplified by this theory [9]. The occurrence of this "preestablished harmony" was noted not only by the eminent Göttingen mathematicians who made a significant contribution to the relativity and quantum revolution in the first third of the 20th century but also by the theoretical physicists —Einstein, Heisenberg, Born, Dirac, Wigner, and others—who founded these theories.

In 1930 Sommerfeld spoke very beautifully about this issue. He recalled the Platonic expression that God is a geometer and stressed that nature is a much better mathematician than we are [11]. In 1950 Heisenberg, one of his most eminent students, wrote in much the same vein:

> If we follow the Pythagorean line of thought [then] we may hope that the fundamental law of motion will turn out as a mathematically simple law.... It is difficult to give any good argument for this hope for simplicity—except the fact that it has hitherto always been possible to write the

fundamental equations in physics in simple mathematical forms. This fact fits in with the Pythagorean religion, and many physicists share their belief in this respect, but no convincing argument has yet been given to show that it must be so [4, pp. 72–73].

Here is an analogous assertion made in 1953 by Dirac, another founder of quantum theory:

It seems to be one of the fundamental features of nature that fundamental physical laws are described in terms of a mathematical theory of great beauty and power, needing quite a high standard of mathematics for one to understand it. You may wonder: Why is nature constructed along these lines? One can only answer that our present knowledge seems to show that nature is so constructed. We simply have to accept it. One could perhaps describe the situation by saying that God is a mathematician of a very high order, and He used very advanced mathematics in constructing the universe. [12, p. 53]

We note that the last phase in the passage just quoted is absent from the Russian translation of the Dirac article [13, p. 139]. What underlines the absurdity of the censor's arbitrary rule is the fact that, according to Heisenberg, Dirac was a "belligerent atheist" [14, p. 211].

This issue continues to be discussed in our time by mathematicians and theoretical physicists. Bourbaki said that physical theories fit into mathematical structures "as if this were the result of predetermination"; Arnol'd notes that the harmony between mathematics and physics "so amazed Newton that he regarded it as a proof of the existence of God"; Kobazarev and Manin speak of the "double life" or "double semantics" of the language of physical theories; Fadeev stresses the fundamental nature of the essentially empirical (and therefore baffling) fact of the mathematical nature of the physical world, and so on. For further relevant statements by those mentioned we refer the reader to Bourbaki [15], Arnol'd [16, p. 5], Manin [17, p. 4], Kobazarev and Manin [18, p. 176], and Faddeev [19, pp. 11–12]. In a lecture in 1959 Wigner used another fitting name for this fact: he referred to it as "the unreasonable effectiveness of mathematics in the physical sciences" (this was the title of the lecture; we discuss it in Section 4). [20, p. 222]

3. EINSTEIN'S "COSMIC RELIGION". In his writings devoted to philosophical-methodological and historical-scientific issues Einstein made abundant use of quasireligious terminology. It was he who introduced the expressions "cosmic religion" and "cosmic religious feeling" to denote the deep faith of physicists in the rational structure of the universe.

It seems that this expression appeared for the first time in his papers and speeches around the 1930s, although essentially similar ideas, a sketch of his theoretical-cognitive credo, is found in his address of 1918 delivered on the occasion of Planck's 60th birthday [21, pp. 224–227]. This sketch includes Einstein's model of the construction of a scientific theory (described later very laconically and intuitively in a letter to Solovine of May 7, 1952 [22]), and in it he speaks in an inspired way of the "supreme task of the physicist," of Leibniz' "pre-established harmony," and of the "state of mind" of the theoretician "akin to that of the religious worshiper or the lover." Because of the importance and "primary nature" of this text, we quote an excerpt from it:

The supreme task of the physicist is to arrive at those universal elementary laws from which the cosmos can be built up by pure deduction. There is no logical path to these laws; only intuition, resting on sympathetic understanding of experience, can reach them. In this methodological uncertainty, one might suppose that there were any number of systems of theoretical physics all equally well justified; and this opinion is no doubt correct, theoretically. But the development of physics has shown that at any given moment, out of all conceivable constructions, a single one has always proved itself decidedly superior to all the rest. Nobody who has really gone deeply into the matter will deny that in practice the world of phenomena uniquely determines the theoretical system, in spite of the fact that there is no logical bridge between phenomena and their theoretical principles; this is what Leibnitz described so happily as a "pre-established harmony."...

The longing to behold this pre-established harmony is the source of the inexhaustible patience and perseverance with which Planck has devoted himself... to the most general problems of our science, refusing to let himself be diverted to more grateful and more easily attained ends.... The state of mind which enables a man to do work of this kind is akin to that of the religious worshiper or the lover.... [21, pp. 226–227]

In 1930, in an essay on Kepler, Einstein returned to the thought of this harmony, but this time he spoke of the preestablished harmony between physical reality and mathematical structures: "Our admiration for this splendid man is accompanied by another feeling of admiration and reverence the object of which is no man but the mysterious harmony of nature into which we are born."[23, p. 265] He used the example of conic sections, realized in the orbits of celestial bodies, to explain the sense of this harmony, and summed up: "It seems that the human mind has first to construct forms independently before we can find them in things." [23, p. 266]

True, when Einstein, and Planck whom he honored, and other theoreticians talked of elements of religion and faith in scientific cognition, they frequently had in mind the faith in the real existence of the world and of nature, or the faith in the possibility of understanding them, faith in their rational structure, and so on.

We adduce a lengthy quote from Einstein's article of 1930 on religion and science in which, probably for the first time, he speaks of the meaning of "cosmic religious feeling," whose essence is a "deep conviction of the rationality of the universe":

> I maintain that the cosmic religious feeling is the strongest and noblest motive for scientific research. Only those who realize the immense efforts and, above all, the devotion without which pioneer work in theoretical science cannot be achieved are able to grasp the strength of the emotion out of which alone such work, remote as it is from the immediate realities of life, can issue. What a deep conviction of the rationality of the universe and what a yearning to understand, were it but a feeble reflection of the mind revealed in this world Kepler and Newton must have had to enable them to spend years of solitary labor in disentangling the principles of celestial mechanics! ... It is cosmic religious feeling that gives a man such strength. A contemporary has said, not unjustly, that in this materialistic age of ours the serious scientific workers are the only profoundly religious people."[24, pp. 39–40]

Einstein retained this conviction, which he restated on a number of occasions in the 1930's, 1940's, and 1950's, to the end of his life. In a letter to Solovine of 1 January 1950 he wrote briefly and expressively about, essentially, the same matter: "I cannot find a better expression than 'religion' to denote the faith in the rational nature of reality.... Where this feeling is absent, science degenerates into fruitless empiricism." [22]

At least during the period between 1920 and 1940, Einstein linked his conceptions of "rationality," the "rational construction of the universe," and the "rational nature of reality" with its "mathematicity," with the "preestablished harmony" between mathematics and physical reality. In 1933 he wrote:

> Our experience hitherto justifies us in believing that nature is the realization of the simplest conceivable mathematical ideas. I am convinced that we can discover by means of purely mathematical constructions the concepts and the laws connecting them with each other, which furnish the key to the understanding of natural phenomena. Experience may suggest the appropriate mathematical concepts, but they most certainly cannot be deduced from it. Experience remains, of course, the sole criterion of the physical utility of a mathematical construction. But the creative principle resides in mathematics. In a certain sense, therefore, I hold it true that pure thought can grasp reality, as the ancients dreamed." [25, p. 274]

Thus Einstein's cosmic religious feeling" is seen to be close to the "preestablished harmony" of Hilbert and Minkowski, the Pythagorean-Platonic divine mathematicity of the universe which Sommerfeld, Heisenberg, Dirac and others talked about, and to Wigner's "unreasonable effectiveness of mathematics." Incidentally, Wigner modified somewhat his conception of the unreasonable effectiveness of mathematics by ascribing to it the status of a fundamental epistemological law.

4. WIGNER'S "EMPIRICAL LAW OF EPISTEMOLOGY". Wigner felt "that the enormous usefulness of mathematics in the physical sciences is something bordering on the mysterious and that there is no rational explanation for it." [20, p. 223] He was no less baffled by the fact that "the mathematical formulation of the physicist's often crude experience leads in an uncanny number of cases to an amazingly accurate description of a large class of phenomena." [20, p. 230] He noted that while the mathematically formulated laws of nature are fantastically accurate, they hold for a strictly limited domain of variation of the relevant physical magnitudes. It was this combination of singular features of the interrelationship of physics and mathematics which Wigner proposed to call the "empirical law of epistemology." To quote Wigner: "The preceding three examples [the law of universal gravitation, quantum mechanics, and quantum electrodynamics, or the theory of the Lamb shift—V. V.], which could be multiplied almost indefinitely, should illustrate the appropriateness and accuracy of the mathematical formulation of the laws of nature in terms of concepts chosen for their manipulability, the 'laws of nature' being of almost fantastic accuracy but of strictly limited scope. I propose to refer to the observation which these examples illustrate as the empirical law of epistemology." [20, p. 233]

According to Wigner, this "law" refers to epistemology, and in this respect it is related to the methodological principles of physics. True, Wigner juxtaposed it only with the law of invariance of physical theories, which guarantee, in particular, the possibility of experimental confirmation of these theories. But the significance of his "law" lies elsewhere: "if the empirical law of epistemology were not correct, we would lack the encouragement and reassurance which are emotional necessities, without which the 'laws of nature' could not have been successfully explored." [20, p. 233] Unlike the laws of invariance of physical theories, which can be provided with a nontrivial theoretical (logical, philosophical) justification, the notion of a "preestablished harmony," or its Wigner form, cannot be so justified, but is nonetheless confirmed by the whole experience of the history of physics. That is why Wigner called his "law of epistemology" an empirical law and, following the advice of a colleague, assigned to it the status of "an article of faith of the theoretical physicist" for the obvious reason that it is not a "necessity of thought." [**20**, p. 233]

Wigner did not use quasireligious terminology. On the other hand, what was typical for him was the use of epithets such as "unreasonable," "marvelous," etc., in connection with his "law." He also stressed that many physicists and mathematicians were not surprised by the "preestablished harmony" and regarded it as an obvious "fact": "It is ... surprising how readily the wonderful gift contained in the empirical law of epistemology was taken for granted." [**20**, p. 233]

Wigner noted several difficulties connected with the "unreasonable effectiveness of mathematics" in physics: its empirical nature and the related indefiniteness of the bounds of its applicability; going outside the part of observed phenomena that we associate with initial conditions (i.e., "aspects of the world concerning which we do not believe in the existence of any accurate regularities"); the absence of a single mathematical structure that should manifest itself in the case of a complete "preestablished harmony"; the existence of patently false theories with a nontrivial mathematical structure and considerable applicability (e.g., Ptolemy's system and Bohr's early quantum theory). Nonetheless, Wigner holds that the difficult-to-explain achievements of the Pythagorean-Platonic conceptions are unprecedented:

> Let me end on a more cheerful note. The miracle of the appropriateness of the language of mathematics for the formulation of the laws of physics is a wonderful gift which we neither understand nor deserve. We should be grateful for it and hope that it will be valid in future research. [**20**, p. 237]

5. WIGNER'S "EMPIRICAL LAW OF EPISTEMOLOGY". The adduced and analyzed material shows the closeness, in fact the virtual equivalence, of such phenomena of theoretical cognition in modern physics (we could say in the physics of the 20th century) as the "preestablished harmony" between mathematics and physics, "the unreasonable effectiveness of mathematics in the natural sciences," Einstein's "cosmic religion," and Wigner's "empirical law of epistemology." Looked at as manifestations of a single fundamental law of scientific cognition, they share a few features that we can associate with that law:

(1) the unified law is obviously related to the Pythagorean-Platonic tradition of scientific cognition;
(2) it has an empirical character in the sense of being repeatedly confirmed by the evolution of theoretical physics;
(3) it has an empirical character in the sense of the absence of a convincing logical, theoretical, or philosophical justification of it;
(4) there is a halo of quasireligious terminology around it;
(5) it has been assigned the character of an epistemological imperative, and stress is laid on its extreme importance as a powerful emotional stimulus in the work of theoreticians.

The absence of a convincing logical justification of this law, as well as its rootedness in the historical evolution of physics and the concomitant faith of theoreticians in its effectiveness, made it necessary to elevate it to the rank of an imperative, to endow it with aesthetic and emotional attractiveness, to "sanctify" it;

hence its "investiture" in elevated forms and the quasireligious expressions used to describe it. There is an analogy between this situation and the interrelationship between ethics and religion. Moral principles, so vital for mankind and similarly unprovable, also had to be "sanctified." It is possible that all people have in mind when they speak to God is just the personification of traditional moral norms and values that support the life of their community." [26]

The "empirical law of epistemology," side by side with the methodological principles of physics, belongs to the few, but extremely important, tools used in the construction of new fundamental physical theories. The two tasks of the historian and the philosopher of science is to discover such epistemological imperatives and to try to provide their theoretical justification.

REFERENCES

1. E. L. Feĭnberg, *Intuition and Logic in Art and Science*, Moscow, 1992. (Russian)
2. _____, Science, art, and religion, in *Philosophical Issues* **7** (1997) 54–62. (Russian)
3. _____, *Art in the Science Dominated World, Science, Logic, and Art*. Translated from the Russian by J. A. Cooper, Gordon and Breach, New York, 1987.
4. W. Heisenberg, *Physics and Philosophy*, Harper, New York, 1962.
5. M. Kline, *Mathematics and the Search for Knowledge*, Oxford, New York, 1985.
6. V. P. Vizgin, Mathematics in classical physics, in *Physics of the 19th and 20th Centuries in the General Scientific and Sociocultural Contexts. Physics of the 19th century*, Moscow, 1995, pp. 6–72. (Russian)
7. D. Hilbert, Mathematical problems, in *Proceedings of Symposia in Pure Mathematics*, American Mathematical Society, v. XXVIII, Part 1, 1976, pp. 1–34.
8. _____, Naturkennen und Logik, in *Naturwissenschaften* (1930) 959–963. Also in Hilbert's collected works, vol. 3, pp. 378–387.
9. H. Minkowski, Raum und Zeit, *Phys. Zeitschr.* **20** (1909) 181–203.
10. L. Pyenson, *The Young Einstein: The Advent of Relativity*, Adam Hilger, Bristol and Boston, 1985.
11. A. Sommerfeld, Wege zur physikalischen Erkenntniss, *Scientia* **51** (1936) 181–187.
12. P. A. M. Dirac, The evolution of the physicist's picture of nature, *Scientific American* **208** No. 5 (1963) 45–53.
13. Russian version of [12].
14. W. Heisenberg, Der Teil und das Ganze, in *Gespräche im Umkreis der Atomphysik*, München, 1969. Translated by A. J. Pomerans. The English version is *Physics and Beyond; Encounters and Conversations*, Harper and Row, New York, 1972.
15. N. Bourbaki, The architecture of mathematics, *Amer. Math. Monthly* **57** (1950) 221–232.
16. V. I. Arnol'd, Introduction to the Russian edition of M. Atiyah, *The Geometry and Physics of Knots*, Moscow, 1995.
17. Yu. I. Manin, *Mathematics and Physics*, Moscow, 1979. (Russian)
18. I. Yu. Kobazarev and Yu. I. Manin, *Elementary Particles: Mathematics, Physics, and Philosophy*, Dordrecht, Boston-London, 1989.
19. L. D. Faddeev, Mathematical view of the evolution of physics, *Priroda* No. 5, 11–16. (Russian)
20. E. P. Wigner, The unreasonable effectiveness of mathematics in the physical sciences, in E. P. Wigner, *Symmetries and Reflections*, Indiana University Press, Bloomington & London, 1967, pp. 222–238.
21. A. Einstein, Principles of physics, in A. Einstein, *Ideas and Opinions*, Crown, New York, 1954, pp. 224–227.
22. _____, *Letters to Solovine*, Philosophical Library, New York, 1987.
23. _____, Johannes Kepler (1930), in A. Einstein, *Ideas and Opinions*, Crown, New York, 1954, pp. 262–266.
24. _____, Religion and science (1930), in A. Einstein, *Ideas and Opinions*, Crown, New York, 1954, pp. 36–40.
25. _____, On the method of theoretical physics (1933), in A. Einstein, *Ideas and Opinions*, Crown, New York, 1954, pp. 270–276.
26. F. A. Hayek, *The Fatal Conceit. The Errors of Socialism*, Chicago University Press, 1988.

The Significance of Mathematics: The Mathematicians' Share in the General Human Condition

Notes for a talk given on December 5, 1978 at
Fordham University

Wilhelm Magnus

1. INTRODUCTION AND HISTORICAL REMARKS. The topic I propose to discuss is a concern of the mathematician. However, it is not a topic of mathematics but of philosophy, at least if one agrees that philosophy is not itself a specialized science but a discipline that deals with the interaction of all human endeavors. Although I am a working mathematician with not more than an amateur's knowledge of philosophy, I nevertheless hope to be able to make at least some valid observations that will contribute to a better understanding of a rather complex situation.

Mathematics begins with an understanding of the abstract concept of a natural number (i.e., of the numbers 1, 2, 3,—ad inf.) and the ability to count indefinitely. Today, this understanding is practically universal and, in this sense, we may say that every human being is a mathematician. It is a curious fact that the mathematical component in the emergence of civilization is hardly ever mentioned by modern historians. I have found a reference to it only in Jacob Burckhardt. It was different in antiquity. In one of his plays, Aeschylus mentions "$\dot{\alpha}\rho\iota\theta\mu\acute{o}\nu, \ \ddot{\epsilon}\xi o\chi o\nu \ \sigma o\phi\iota\sigma\mu\acute{\alpha}\tau\omega\nu$" (Number, outstanding (concept) among the ingenious inventions). And Aristotle says that everything was created by God with the exception of the concept of number, which is man's invention.

Mentioning the name of Aristotle could be the starting point for a survey of the role of mathematics and of mathematical concepts as an object of philosophical investigations, including a history of epistemology. Being a mathematician and not a philosopher, I neither can nor will discuss these things. However, I should like to touch at least briefly on the work of three eminent philosophers who assigned to mathematics an extraordinary role in their systems. They are Plato, Leibniz, and Spinoza.

Plato considers knowledge of mathematics to be a prerequisite of citizenship. Specifically, he states that anyone who calls himself a civilized person should know that there exist incommensurable quantities in geometry. For example, it is impossible to find a unit of length such that both the side and the diagonal of a square are integral (= whole) multiples of this unit. This is indeed a surprising fact. It requires a sophisticated proof and it is something beyond the range of intuitive perception. But why should everybody know it? Plato wanted everybody to know that some facts, even surprising ones, are absolute certainties. To understand this need for certainty, one should read the plays of Aristophanes, which exhibit the emergence of nihilism in Plato's time. If we use Nietzsche's definition of nihilism as the doctrine: "Nothing is true. Everything is permitted", we find it fully illustrated in "The Clouds". In another play, "The Birds", we see the human race entering into an alliance with the birds in order to destroy the power of the gods. It should be remembered that these plays were performed in honor of a particular god, Dionysus. In still another play, "The Frogs", this very god receives a thrashing.

Plato tried to fight nihilism by exhibiting mathematics as a source of absolute truth and certainty. Today, we know that the truths and certainties of mathematics are relative rather than absolute, so we are not in a good position to fight nihilism in this way.

Leibniz was both a philosopher and an eminent mathematician. No one ever thought more highly of mathematics than he. According to Leibniz, mathematics is the science that tells us what is possible. As far as the physical world is concerned, i.e., that aspect of the world that Descartes called "res extensa", this statement contains at least some truth. But Leibniz goes further. According to him, God, the supreme mathematician, created our particular world by choosing of all possible worlds the one with the greatest plenitude and variety. In this sense, ours is "the best of all possible worlds."

The success of the exact sciences (which are based on the use of mathematics) has increased the range of our knowledge of the universe to a degree enormously beyond that available to Leibniz. Paradoxically, this has made many of us (including myself) more modest, because our more extensive knowledge has made us more aware of the range of our ignorance. We are more reluctant than Leibniz to make definite statements about the universe, and we certainly would not make a statement like that of Descartes who said: "Give me matter and motion, and I shall make the world once more."

Like Leibniz, Descartes was both a philosopher and an eminent mathematician. But mathematics does not play an explicit role in his philosophy although he is extremely important for the history of the exact sciences through his dichotomy of the world into "res extensa" and "res cogitans." We still are influenced by his perception of the world. But at least one philosopher made a heroic attempt to overcome this dualism, and thought he could do so by using not mathematics proper, but at least the methods of mathematics.

Spinoza proposed to derive definite philosophical truths from self-evident statements "more geometrico", i.e., in the manner of Euclid. Although there can be no doubt that Spinoza provided deep and important insights, this is not due to his method, which does not qualify as a mathematical argument. I shall illustrate with an example taken from the first page of his main work, the "Ethics". There we find the statement "By God, I understand Being absolutely infinite." What is "absolutely" infinite? Spinoza did not know of the discovery of Georg Cantor (published in 1895) according to which there are smaller and larger infinitudes. For instance, there are more points in a finite interval on a straight line than there are natural numbers $1, 2, 3 \ldots$ (ad inf). What is worse, there is an infinite sequence of infinitudes, each larger than the previous one. And assuming that there exists a largest infinitude containing all the previous ones would lead to a contradiction. Now we may be able to *live* with a contradiction, but we cannot *tolerate* it in a mathematical argument.

The lesson from this is simple enough. Before we start relying on mathematics we have to understand both its potential and its limitations.

2. WHAT IS MATHEMATICS? There exists a book by Courant and Robbins with this title. It tries to answer the question by giving examples. I shall try to give a general description first and then illustrate it with a few examples. Please note: The following remarks must not be taken for an attempt at giving an epistemological definition of mathematics. Their purpose is merely to provide an intuitive understanding of the nature of mathematics.

Mathematics deals with concepts subject to the rules of logic, in particular to the postulate of the excluded middle. There exists at least one set of concepts of this type, namely that of the natural numbers.

Comments. It is not true that all statements involve concepts that are subject to logic. If we have a green piece of cloth with a bluish tint we may be in doubt whether we should call the color green or blue-green, and we may even disagree about the name we wish to give to the color. Similarly, we cannot say that a person is either tall or not tall. Even if we give an artificial definition of tallness (say, 6 feet or more) we may run into trouble because no measurement is absolutely precise. (There is a good reason why we have something like 250000 laws in this country. The law uses strangely defined concepts and has to be more and more casuistic to make them fit reality.)

As far as I know, Nietzsche was the first to point out this fact. He claimed that only man-made concepts are subject to logic. With due respect for Aristotle I would take sides against him and Nietzsche and say with Kronecker (a 19th-century mathematician) that "God made the whole numbers. All the rest is the work of man." I hope that these remarks will suffice. I am not prepared to make statements about the "reality" of the natural numbers in philosophical (ontological) terms.

Mathematical research has two important and, I believe, unique characteristics: It involves an element of the infinite—being the only secular human activity to do so—and it produces an increasing wealth of problems with increasing abstraction.

Comments and examples. The element of the infinite in mathematics can be used to prove—in this case truly "more geometrico"—that the human mind is superior to any conceivable electronic computer. I cannot describe the arguments needed here without becoming rather technical. They are forever linked with the name of one of the greatest mathematicians of our time, Kurt Gödel (1906–1978). In an age where scientists as well as philosophers try to tell us that we are really nothing particular—a survival mechanism for our genes or, at best, a freakish and rather unpleasant animal that, after all, is just capable of doing some things a little better than the more pleasant chimpanzee—our mathematical abilities provide perhaps the simplest and strongest non-metaphysical argument for our special position in nature.

To illustrate these remarks I shall use two examples. The first one is a theorem of number theory, which can be stated as follows:

Every natural number N is the sum of the squares of at most four natural numbers. Unless $N + 1$ is divisible by 8, at most three squares suffice.

Obviously, no amount of direct calculations can prove this theorem because it involves an infinitude of numbers. The proof is neither easy nor obvious and was given (for the first part of the theorem) in the late eighteenth century by Lagrange. It clearly illustrates what I mean by "an element of the infinite." I shall need this theorem again later when discussing the motivation of the mathematician. But for now I need another example to illustrate my remarks about abstraction, and for this purpose I shall start with the *Koenigsberg Bridge Problem*:

In 1735, the great Swiss mathematician, Euler, came across a peculiar problem and described it as follows: "In the town of Koenigsberg there is an island called Kneiphof, with two branches of the river Pregel flowing around it. There are seven bridges crossing the two branches. The question is whether a person can plan a walk in such a way that he will cross each of these bridges once but not more than once.... On the basis of the above I formulated the following very general problem for myself: Given any configuration of the river and the branches into which it may divide, as well as any number of bridges, to determine whether or not it is possible to cross each bridge exactly once."

Figure 1 shows the layout of the seven bridges of Koenigsberg. Can one stroll across each of these bridges once but no more than once? If so, how?

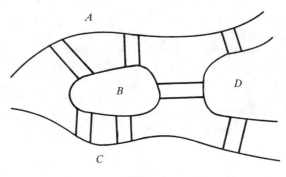

Figure 1

We begin to study this problem by throwing away unnecessary information. Since the shape and size of the islands and the countryside on the banks of the river do not matter at all, we contract each of the four areas labeled respectively A, B, C, and D to a single point. The width and shape of the bridges do not matter, either, so we replace each of them by a segment of a line or by a curve. Figure 2 shows the result of this process. This looks rather similar to a part of the subway maps exhibited in the trains in New York City, and we could rephrase the problem accordingly with stations and subway rides between them. However, we prefer to use the standard mathematical terminology. We shall call the whole figure a graph. The points A, B, C, D, are called *vertices*, and the connecting lines are called *edges*. Furthermore, we shall call the number of edges going through a vertex the *order* of the vertex. Obviously, the orders of A, B, C, D, are, respectively, 3, 5, 3, 3. Our problem now is: Can we find a path (i.e., a succession of edges, each having exactly one point in common with the previous one) such

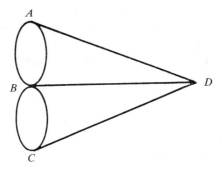

Figure 2

that this path goes through all the edges exactly once? Such a path is called an *Euler path*.

Any graph for which this question can be answered in the affirmative will be called an *Euler graph*. (This is an ad hoc notation and not common usage). Now we can show easily that our graph is *not* an Euler graph. The reason for this is the following theorem:

If a connected graph is an Euler graph then there are either no vertices of odd order or exactly two vertices of odd order.

To prove this, let us imagine that we erase any part of the path through which we have traveled in our attempt to travel through all edges exactly once. Whenever we enter a point and then leave it again, we will have to erase *two* edges through this point, reducing thereby its order by two. Therefore, if a point has odd order, then we may start our path at it, possibly pass through it several times, but then we cannot come back to it in the end, because its order would have to be even for this purpose. Therefore, if there is a point of odd order, we may try to start our path at it, but then we must end our journey at another point of odd order. (What is true for the starting point is also true for the terminal point, since we can reverse our journey). This proves our theorem, because all points other than the starting point and the terminal point must be of even order.

This is not a deep or difficult theorem, but Euler asked immediately the questions that every mathematician would ask in this situation: *Is the converse of our theorem true*? In other words: Suppose we have a connected graph in which all points are of even order. Can we start an Euler path anywhere and get back, in the end, to our starting point? And suppose we have a connected graph with exactly two points of odd order? Can we start an Euler path at one of the points of odd order and terminate it at the other one?

So far, we have established a first level of abstraction. It has led from a "finite" or "concrete" problem (the original Bridge Problem) to a general problem, and to a general theorem that covers an infinitude of "concrete" problems, namely, all possible graphs that we can draw. Now comes a second level of abstraction. We do not have to *draw* anything to define a graph. All we need is an "incidence table", which lists the points and number of edges joining any two of them. The incidence table for Figure 2 would then look as follows: The numbers in the first row simply mean that A is connected with no edge to itself, with two edges to B, with no edge to C and with one edge to D. The numbers in the other rows are similarly defined.

I hope that you will agree that incidence tables are more abstract than the graphs. But now abstraction raises immediately a new problem. We can design incidence tables without giving a graph.

Question: when does an incidence table define a graph that can be drawn in the plane so that no two edges cross each other? This is not always the case. The problem was solved in the 20th century by Kuratowski.

The reason why I brought up these things is this: In general, abstraction reduces the number of statements we can make and the number of questions we can ask. We can say more about one species of birds than about birds in general and more about birds than about the animal kingdom, etc. That the situation is different, at least in many cases, for mathematics is a curious and, as far as I know, unique occurrence.

3. WHAT CAN MATHEMATICS DO? The functions of mathematics may be described as an extension of some of the functions of language. Before I start explaining this I have to demolish a statement that I have learned or read frequently as a sort of platitude, namely: "Mathematics *is* a language." Now even platitudes can be true, but this one happens to be sheer nonsense. It *is* true that mathematics *uses* a special language, and the reason is that our everyday language uses concepts that are not subject to logic and therefore is not suitable for the formulation of mathematical arguments and results. But a loom is not a piece of cloth, and technical terms are not ideas.

In Genesis 2, God gives the privilege of *naming* all living beings to Adam, who represents the human race. To name things—and even non-material ones like feelings and sensations—is an act of abstraction. It has been pointed out by Hans Jonas that it is also an act of image-making, another important and specifically human privilege, a sort of secondary creativity. (The ability to make images has been used by Jonas for the purpose of determining man's "specific difference" in the animal kingdom.) The ability to name things is the basis of our ability to plan and provides us with a tremendous increase of our ability to communicate, which is essential for the emergence of a coherent human society.

Now I want to show that mathematics, too, has the function of image making and that this function gives us the ability to predict. I have to be rather sketchy here, in order to avoid technicalities as far as possible.

Surprisingly, mathematics provides us with abstract images of things that are not accessible to the direct perception of our senses. At the same time, the image can be made so precise or faithful that it allows us to know all aspects of the original that are of importance to us. However, I can give here only an example, which still is not really abstract. I hope it will give at least an idea of what all of this is about.

Let me start with the fact that we can draw a map of the globe on a flat piece of paper in such a way that the map can be used for navigation. This is an important achievement of mathematics. There are many ways to do this. One of the first was found by Gerhard Kremer, who is known under the latinized form of his name Gerhardus Mercator (1512–1594). The problem is a difficult one since it is impossible to make such a map without distorting distances.

This is still a rather elementary example of the image-making power of mathematics. A much more sophisticated and much more important example is the mathematical image of an atom with a nucleus and electrons. This mathematical image consists entirely of formulas. But these formulas permit us, at least under certain circumstances, to make predictions about the behavior of the atom. This is an enormous achievement, and it is but one example of the role of mathematics in physics, chemistry, and the branches of technology based on these sciences.

Here I have to put in a word of warning. We are always tempted to overestimate the power at our disposal. In the case of language, or, specifically, of "naming", the overestimation of our power appeared in the form of incantations or conjuring. (Mathematical objects, in particular the pentagram, have been abused for the same purpose.) In the case of mathematics, it appears in the more subtle form of applying mathematical deductions to situations where this is not justified. In both cases there is involved an element of cheating, trying to get something too easily. However, I cannot try here to describe the misuses of mathematics. It is a difficult topic, and it requires careful study. But I shall discuss briefly another aspect of the use of mathematics in physics and other exact sciences that was nicely formulated by Eugene Wigner as the title of a talk some years ago: "The unreasonable effectiveness of mathematics in the exact sciences."

Indeed, it has been a cause of astonishment for a long time that mathematics can be used at all to understand and even control the physical world. I believe that this astonishment is somewhat misplaced and that it is a consequence of Descartes' philosophy, which divides the world into "res extensa" and "res cogitans", and I should like to counter it with an aphorism by Lichtenberg (an 18th century physicist and essayist) who said: "I have found people who were astonished that cats have holes in their furs *exactly* at the places where the eyes are."

Obviously, the cat would not exist if it were otherwise. Similarly, we need order for our existence. It is therefore not so surprising that the world contains an element of order and that we have an organ to deal with it. Of course, this does not explain the extent of our ability to apply mathematics, nor does it explain the fact that the human race developed mathematics long before it became useful. So there is indeed a reason for some astonishment, which, however, should include other phenomena, such as our ability to appreciate and to create beauty and to do many other things that, at least originally, provided no visible help in the preservation of our species. I shall have to say more about these things in the next section. For now, I should like to mention a negative service that mathematics can perform. Mathematics can tell us that there are things we cannot do with the means at our disposal. For example, suppose we wish to seat the representatives, one for each, of the ca. 150 members of the United Nations at a conference table. We can not list the possible seating arrangements since their number would be greater than the number of electrons and protons in the known Universe. Of course, we are not particularly interested in such seating arrangements. But we might be interested in the arrangements of genetic material in chromosomes. I have no exact data at hand to discuss this problem, but the numbers are large, too.

4. THE PHENOMENON OF MATHEMATICS. We shall use the term "mathematics" in its strict sense: *The systematic derivation of theorems with the help of* *explicitly formulated arguments*. Some mathematical insights are intuitively clear, e.g., that a diameter divides a circle into two equal parts; Thales (ca. 624–548 B.C.) is supposed to have *proved* this. The fact that the side and diagonal of a square are incommensurable is not at all intuitively clear. Its discovery is ascribed to the Pythagorean school. A well-formulated proof of this and of related theorems appeared at the time of Plato and was due to his friend Theaetetos. Although Babylonian, Indian, and Chinese scholars developed a large body of mathematical knowledge, it is safe to say that mathematics in the strict sense is a creation of the Greeks. This does not mean the Athenians. With the exception of Theaetetos, none of the great Greek mathematicians lived in Athens. Euclid lived in Alexandria in Egypt. So did Apollonius. Archimedes lived in Syracuse in Sicily. Nothing like the systematic works of Euclid and Apollonius is known from other civilizations of the same or of earlier times.

What motivated these mathematicians? Not technology, not even astronomy, which, after all, was in its more sophisticated aspects not a "practical" matter at all. It is true that Archimedes developed technological applications of mathematics, in particular an instrument to compute the position of the planets. But the Romans who certainly needed and used high level technology never contributed anything to mathematics. In fact, the systematic use of mathematics for the development of technology (excluding astronomy) starts only in the 18th century.

The case for the development of mathematics was not usefulness. Earlier, we compared some functions of mathematics to some functions of language. The analogy goes even further. Language, too, is not merely an instrument of power or of usefulness. Nor is poetry. As far as mathematics is concerned, a good summary of its role appeared in an editorial (by Chandler and Edwards) in *The Mathematical Intelligencer*:

> It is a perennial problem for mathematicians to explain to the public at large what makes mathematics worthwhile if not its practicality. It is like explaining to someone who has never heard music what a lovely melody is . . . . Do let us try to teach the general public more of the sort of mathematics that they can use in everyday life, but let us not allow them to think—and certainly, let us not slip into thinking—that this is an essential quality of mathematics.
>
> There is a great cultural tradition to be preserved and enhanced. Each generation must learn the tradition anew. Let us take care not to educate a generation that will be deaf to the melodies that are the substance of our great mathematical culture.

In the past, some poets understood the beauty of mathematics. I already mentioned Aeschylus. Calderon speaks of "sublime mathematics" and Schiller calls it "divine". There are certainly more examples of this type, but they seem to become rare if not extinct in modern times. The reason for this is, of course, increasing inaccessibility of mathematics. Our latest products are available only to very few people. But the columns by Martin Gardner and occasional essays on mathematics in the *Scientific American* show that a much larger part of the population understands what mathematics is about. Some people are fascinated by Lagrange's theorem (mentioned earlier), but certainly not everybody is. However, little would be left of human civilization if we restrict it only to things that enjoy universal appreciation.

There is one more aspect of mathematics that, although well known, usually is mentioned as a mere curiosity. I believe it is more than that since it relates to the important idea of *evolution*. What are its uses? Is it a matter of pure chance or is it a response to a need, a change of conditions and environment or both? Consider the development of the exact sciences and its result, the human ability to dominate nature. In several cases, scientists found the mathematical tools they needed ready-made and available, sometimes formulated centuries earlier. The philosopher Whitehead mentions the conics, which had been thoroughly investigated by Apollonius in the third century B.C. and were available to Kepler in the 17th century A.D. The most surprising example I know of is the theory of probability. First of all, it is strange that even a situation of complete disorder, that of random events, should be subject to mathematical laws. Secondly, what provoked the study of probability was an almost universally despised human activity, namely, gambling. But one of the main contributors to the theory of probability was Pascal, who gave up mathematics because he thought that the only truly important thing in life was to work for the salvation of one's soul. And finally it turned out that the laws of probability are an essential ingredient of the laws of nature. This insight started in the 19th century with Boltzmann and culminated in our century with the development of quantum theory. Einstein could never overcome his intuitive objections against this development. He said: "God does not play dice." Niels Bohr's answer to that was: "We cannot tell God how to shape the universe".

Coming back to the causes of evolution: We can never refute those who say that everything is due to pure chance. At best, we may be able to embarrass them. But it is hard to see here a response to a need. The theory had been developed long before it was needed.

5. WHAT MAKES A MATHEMATICIAN? There exists a widespread resentment against mathematics. It is supposed to deal only with quantity (not true, since most of mathematics deals with structure and relations), or with computing (again not true, but I cannot explain that in a few words) and, on the whole, it is more worthy of a machine than of a human being. As an aid to science and technology, it does not provide values and is therefore dehumanizing. Even the claim of the mathematician to be concerned with truth is frequently answered by saying that mathematical statements are not true but merely correct. Nevertheless, it is undoubtedly true that the results of mathematics are found by human beings. Can anything be said about them?

The answer is: Not enough to enable us to recognize a mathematician if we meet one at a party. Nevertheless, there exist properties without which a mathematician cannot exist. One of them is, of course, a specific talent. But this is far from being enough. It must be supplemented by an interest in the matter, in fact by a fascination with the problems of the field. And the talent must be supported by persistence and by the willingness to spend the large amounts of time and energy needed to master a difficult craft. And the mathematician needs an exceptionally great ability to stand up under frustration. This is due to the fact (pointed out to me by a colleague) that ours is the only field with an all-or-nothing alternative. A painting or a piece of furniture may be more or less perfect. A theorem and a proof are either true or false. If either the proof or the theorem is false, we have absolutely nothing. Finally, we must be satisfied with the production of something intangible. I have found housepainting to be a gratifying supplement to mathematical research. At least one can see and touch what one has done.

It follows that the mathematician needs the support of a civilization that acknowledges as valuable the products of theory, of pure thought. Although we do not set a scale of values, we would not exist without such a scale. I can be brief here, since the arguments given by the philosopher Cantore for the humanistic significance of science apply, with small modifications, to mathematics as well.

Let me conclude by pointing out one advantage that the mathematician (and, with him, the representative of the exact sciences) has. Our thoughts are eminently communicable. Not, perhaps, from person to person. But certainly from nation to nation. Mathematicians understand each other no matter where they come from. Even across many centuries we understand each other. We may not see clearly what a particular expression in Euclid means. But we are confident that, could we talk with him, we would be able to clear up the matter quickly. Nothing is more international than the community of mathematicians.

Biographical Sketches of the Contributors

SIR MICHAEL ATIYAH (1929–) was educated at Victoria College, Cairo, the Manchester Grammar School, and Trinity College, Cambridge. He has been a Research Fellow at Trinity College, a fellow at Pembroke College, Cambridge, and University Lecturer. From 1963 to 1969 he was Savilian Professor of Geometry, Oxford University, and Fellow at New College. Between 1969 and 1972 he was a professor of mathematics at the Institute for Advanced Study at Princeton.

In 1990 Sir Michael was appointed Master of Trinity College at the University of Cambridge and has subsequently been appointed first director of the Isaac Newton Institute for Mathematical Sciences there. Also, in December 1990 he became President of the Royal Society. Earlier he had served as President of the London Mathematical Society and the Mathematical Association.

Sir Michael received the Fields Medal at the meeting of the International Congress of Mathematicians in Moscow in 1966. The Royal Society awarded him the Royal Medal in 1968 and the Copley Medal in 1988. In 1980 he received the De Morgan Medal from the London Mathematical Society. Five volumes of his collected works were published in 1988 and his book, *The Geometry and Physics of Knots*, appeared in 1990. He moved to the University of Edinburgh in 1997.

ISABELLA GRIGOR'EVNA BASHMAKOVA (1921–) completed her studies at Moscow State University in 1944 and has taught there since that time. She was awarded a doctorate in the physical and mathematical sciences in 1961 and was appointed professor in 1962. She is the author of fundamental works in the history of diophantine analysis, of ancient mathematics, and of Russian mathematics. She became a member of the International Academy of Historical Sciences in 1972.

SERGEI SERGEEVICH DEMIDOV studied at the Lomonosov Moscow State University in the Department of Mechanics and Mathematics, receiving his Doctor of Physical and Mathematical Sciences in 1968. His doctoral research was on the development of the theory of ordinary differential equations from Cauchy till the beginning of the 20th century. His research interests are in the history of mathematics and he is editor-in-chief of the journal *Historico-Mathematical Investigations*.

JOSEPH LEO DOOB (1910–) is a member of the National Academy of Sciences and of the American Academy of Arts and Sciences. He studied at Harvard and was a professor at the University of Illinois at Urbana-Champaign. Most of his works are in the theory of stochastic processes. He has developed a theory of stochastic functions and has made important contributions to the theory of Markov processes and the theory of martingales. He and Joseph Hunt have pointed out important connections between potential theory and the theory of Markov processes, and this has attracted the attention of many probabilists to potential theory.

Doob is a corresponding member of the Académie des Sciences, Paris. He was awarded the National Medal of Science in 1979 and the Steele Prize of the American Mathematical Society in 1984.

ISRAEL KLEINER is a professor in the Department of Mathematics and Statistics at York University. He received his Ph.D. at McGill University in ring theory. His interests are in the history of mathematics, in mathematics education (in a broad sense) and in their interface. He is one of very few mathematicians who have been awarded by the Mathematical Association of America the Carl B. Allendoerfer Award (twice), the Lester R. Ford Award, and the George Pólya Award for expository writing.

ERWIN KREYSZIG (1922–) received his Ph.D. at the Technical University in Darmstadt in 1949. He worked at a number of universities in Germany, the United States, and Canada. Since 1984 he has taught at Carleton University in Ottawa, Canada. He is the author of some 150 papers (of which about 60 deal with partial differential equations), and of a number of monographs and textbooks.

DETLEF LAUGWITZ (1932–2000) received his doctoral degree from the University of Göttingen in 1954. He taught at the Technical University in Darmstadt, Germany. His main research interests ranged from differential geometry and convexity to analysis and its history. For close to half a century (beginning in the 1950s) he worked on the creation of a genetic approach to a modern theory of infinitesimals (or nonstandard analysis). He is the author of several textbooks in these fields. In 1996 Birkhäuser published his book *Bernhard Riemann*, devoted to Riemann's mathematics, physics, and philosophy. An English translation of this book appeared in 1999.

NIKOLAĬ NIKOLAEVICH LUZIN (1883–1950) began his study of mathematics at Moscow University in 1901. While still a student, he went to Paris in 1905 and listened to lectures of Borel, Poincaré, Hadamard, and Darboux. Between 1910 and 1914 he studied mathematics in Göttingen and Paris. Upon his return to Moscow, he lectured on the theory of functions of a real variable and directed a special research seminar that gave rise to the Moscow school of the theory of functions. In 1915 he completed the paper "The integral and trigonometric series", for which he was awarded a doctorate in pure mathematics. He became a member of the Academy of Sciences of the USSR and directed the work of the Steklov Institute of the Academy devoted to the theory of functions. Among his students were Suslin, Lavrent'ev, Lyusternik, Bari, Kolmogorov, Shnirel'man, and Keldysh. Between 1916 and 1920 Luzin, Suslin, and P. S. Aleksandrov developed descriptive function theory. In 1930 Luzin published his *Lectures on Analytic Sets and their Applications*, in which he presented his own results and the results of his students in this area. In the 1930s Luzin continued his work on problems of descriptive set theory and on the foundations of mathematics. He also began to work in the area of applications of classical analysis and differential geometry.

Luzin was interested in the publication of good textbooks for technical schools. His version of Granville's textbook on the calculus was extremely popular in the 1920s and 1930s. In 1940 Luzin wrote a textbook on the theory of functions of a real variable. In the last years of his life he resumed research in differential geometry.

WILHELM MAGNUS (1907–1990) was born in Berlin and educated at the University of Frankfurt, receiving his Ph.D. there in 1931 under the supervision of Max Dehn. After appointments at the Rockefeller University and the Institute for Advanced Study he became a professor at Göttingen in 1945. He joined the Bateman project in 1948 and in 1950 he moved to New York University. He had more than 60 doctoral students. His collected papers were published by Springer-Verlag in 1984.

VICTOR W. MAREK received his Ph.D. from Warsaw University in 1968 and his D.Sc. in 1972. He worked at the Institute of Mathematics of Warsaw University between 1968 and 1983 and at the Institute of Mathematics of the Polish Academy of Sciences between 1973 and 1977. Between 1980 and 1982 he headed the section of the Institute of Mathematics of Warsaw University devoted to the study of the foundations of mathematics. Since 1983 he has been Professor of Computer Science at the University of Kentucky. He is the author of more than 120 papers in set theory, logic, and the logical foundations of computer science. In 1973 he received the Sierpiński Prize of the Polish Mathematical Society.

JAN MYCIELSKI (1932–) received his M.A. and Ph.D. from Wrocław University and worked there until 1968. He moved to the US in 1969. He is a professor of mathematics at the University of Colorado at Boulder. He is the author of some 150 papers in several areas, including logic and foundations, game theory, universal algebra, topology, geometry, and brain science. At present he works on topics in set theory, the topology of 3-manifolds, and in mathematical models of knowledge.

STEFAN MYKYTIUK is a graduate student of mathematics at York University in Toronto. He is completing his doctoral dissertation "On Hopf algebras of quasi-symmetric functions" under the supervision of Nantel Bergeron.

ZDZISŁAW POGODA studied at the Jagiellonian University in Cracow and obtained his doctorate in 1982. He works at the Institute of Mathematics of the Jagiellonian University. His main scientific interests are differential geometry and its applications, the history of mathematics (with emphasis on the 19th and early 20th centuries), and popularization of mathematics. He devotes a great deal of time to working with young people. In 1995 he, and his friend and frequent coauthor Krzysztof Ciesielski, obtained the Great Award of the Polish Mathematical Society for the popularization of mathematics. He and Ciesielski have written two popular-scientific books that explore various areas of mathematics.

MIKHAIL MIKHAILOVICH POSTNIKOV (1927–) completed his studies at Moscow State University in 1944. He was awarded a doctorate in the physical and mathematical sciences in 1953 and was appointed professor in 1954. He has worked in the Mathematical Institute of the Academy of Sciences of the USSR since 1950, and at Moscow State University since 1954. He is the author of the so-called "P-system" which enables one to compute the homotopy type of a space with given homotopy groups. He has written many monographs and textbooks.

GALINA S. SMIRNOVA received her Ph.D. in mathematics at Moscow State University under the supervision of I. G. Bashmakova. She has taught at MSU since 1987. Her research is devoted to the history of algebra, number theory, and differential geometry. She has made significant contributions to the study of the history of algebra in Western Europe in the 16th century.

LESZEK M. SOKOŁOWSKI studied at the Jagiellonian University in Cracow and obtained his doctorate in 1975. Since then he has been working at the Astronomical Observatory of that university. His present position is that of an associate professor of gravitational physics. His field of research includes relativistic cosmology, modern Kaluza-Klein theories, black-hole physics, and gravitational waves. He is also interested in the philosophical problems of physics.

JURIS STEPRĀNS obtained a B.Math. from the University of Waterloo in 1977 and a Ph.D. in mathematics from the University of Toronto in 1982. Since then he has been a member of the mathematics department of York University. His primary mathematical interest is the application of set theoretic techniques to other areas of mathematics.

VLADIMIR MIKHAILOVICH TIKHOMIROV (1934–) completed his studies at Moscow State University in 1957, receiving a doctorate in the physical and mathematical sciences in 1971. He has worked at MSU since 1960 and was appointed professor in 1975. He is the author of fundamental works on approximation theory and on optimal control, having established connections between exact solutions of problems in approximation theory and problems in topology and nonlinear analysis. He is editor-in-chief of *Mathematical Enlightenment*.

JACQUES LÉON TITS (1930–) was born in Belgium. He received his Ph.D. in 1950 from the University of Brussels. After appointments at Brussels and the University of Bonn, he moved to the Collège de France in 1973. He received the Henri Poincaré Prize in 1976 and the Wolf Prize in 1993. His work is primarily in group theory.

VLADIMIR PAVLOVICH VIZGIN was awarded his doctorate in 1967 from the Moscow State University. His research areas are the history and methodology of theoretical physics in the 19th and 20th centuries, as well as the social history of physics in Russia and the USSR and the history of the Soviet atomic project. He has published more than thirty papers.

HERMANN WEYL (1885–1955) obtained his doctoral degree under Hilbert in Göttingen in 1908. He worked in Göttingen in 1908–1913 and in 1930–1933 and at the Eidgenössische Technische Hochschule in Zürich in 1913–1930. He left Nazi Germany in 1933 for the Institute for Advanced Study at Princeton. His first papers dealt with trigonometric series, series of orthogonal functions, and almost periodic functions. He was the first to rigorize the parts of complex function theory based on the notion of a Riemann surface. In analysis he studied differential and integral equations and created the spectral theory of differential operators. He introduced the so-called Weyl sums, of great importance in additive number theory. Weyl and Peter proved the completeness of the system of irreducible representations of a compact group and studied the representations and characters of semisimple groups. Weyl's notion of a space with an affine connection plays a fundamental role in modern differential geometry. He obtained certain results in the theory of atomic spectra by using group-theoretical methods. He is regarded as an intuitionist in the philosophy of mathematics.

EDITORS

ABE SHENITZER was born in Warsaw in 1921. He received his Ph.D degree under Wilhelm Magnus from New York University in 1954 and has taught at York University since 1969. He is the translator of Russian and German books and articles, and has served as the translator of two MAA books published in the Dolciani Mathematical Exposition series: *Diophantus and Diophantine Equations*, by Isabella Bashmakova, and *The Beginnings and Evolution of Algebra*, by Isabella Bashmakova and Galina Smirnova. His abiding ambition is to present mathematics so that it is seen to be what it is—an art.

JOHN STILLWELL was born in Melbourne Australia, and received his Ph.D from MIT in 1970. He has taught at Monash University in Melbourne since 1970. He has given several invited addresses at conferences including the International Congress of Mathematicians in 1994, and the Joint Meetings of the MAA and AMS in 1998. Since 1980 he has made a career of writing books on subjects he failed to understand as a student—topology, noneuclidean geometry, and, most recently, algebra. His books are: *Classical Topology and Combinatorial Group Theory*, *Mathematics and its History*, *Elements of Algebra*, *Numbers and Geometry*, and *Geometry of Surfaces*. Almost invariably, it seems, the necessary understanding can be found by reading the masters and reconstructing the history of the subject in modern language. Number theory is the next topic he plans to attack in this manner.

NAME INDEX

Abel, N. H.   2, 99, 109, 102, 119, 137, 143
Ackerman, W.   237
d'Alembert, J.-B. le R.   18, 20, 273
al-Khazin   174
Alexandroff, O.   80
Ampère A. M.   276
Apollonius   85, 290, 291
Arakelov, S. J.   14
Archimedes   63, 88, 92, 189, 290
Aristotle   283
Arnol'd, V.I.   5, 277
Artin, E,   109, 117, 130, 162

Babbage, C.   227
Bachelier, L. J. B. A.   252
Bachet, C.   105, 113, 174
Baire, R.   28
Banach, S.   231
Bauer, H.   249
Beltrami, E.   189, 190, 192, 194, 221
Bernoulli, Daniel   18, 19, 273
Bernoulli, Jakob   71, 136, 137
Bernoulli, Johann   17, 71
Bernoulli, Nicholas   16
Bernshtein, S. N.   53
Bernstein, F   79, 256
Bernstein, S.   33
Bessel, F. W.   220
Betti, E.   76, 80, 269
Bézout, E.   214
Birkhoff, G. D.   80
Bishop, E.   228
Bohr, N.   280, 291
du Bois-Reymond, P.   78
Boltzmann, L.   291
Bolyai, J.   189, 220
Bolzano, B.   226
Bombelli, R.   92, 96
Boole, G.   226
Borel, A.   9
Borel, E.   28, 30–33, 38, 51–53, 250–253
Born, M.   276
Bourbaki, N.   5, 224, 276, 277
Broden, T.   27
Brouwer, L. E. J.   30, 198, 228, 229, 238
Burali-Forti, C.   223
Burckhardt, J.   283
Byushgens, S. S.   48

Cantor, G.   51, 67, 127, 197, 199, 225–231, 235, 270, 271, 284
Carathéodory, C.   71
Cardano, G.   267

Carleman, T. G. T.   33
Cartan, É   9, 112, 173, 179, 271, 272
Cartan, H.   7, 131
Cauchy, A.-L.   27, 30, 41, 52, 55, 58, 59, 65, 100, 101, 123, 127, 153, 158, 238
Cavalieri, B.   65
Cayley, A.   3, 101, 102, 127, 194
Cech, E.   198
Chaitin, G. J.   242
Chebotaryov, N. G.   182
Chevalier, A.   142
Chevalley, C.   131, 174
Choquet, G. A. A.   250
Christoffel, E. B.   59
Church, A.   227, 240
Clausen, T.   63, 185, 186
Clebsch, R. F. A.   101, 102
Cohen, P. J.   223, 225, 227, 233, 234, 244
Connes, A. 9, 14
Conway, J. H.   168
Costa ibn Luca, 93
Courant, R.   79
Coxeter, H. S. M.   173

Daniell, P. J. 252, 253
Dantzig, G. B.   259–261
Darboux, J. G.   63
Davis, M.   227, 241
Dedekind, R.   102, 103, 105, 109, 114–117, 120–125, 127, 128, 142, 162, 192, 221, 222, 225
Degen, C. F.   174, 176
Democritus   51
De Morgan, A.   126, 226
Denjoy, A. 33, 38
Desargues, G.   176–178
Descartes, R.   4, 83, 88, 93, 95, 276, 284, 289
De Witt, B.   195
Dickson, L. E.,   129, 174, 176
Dieudonné, J.   224
Dini, U.   75
Diophantus   84–85, 133, 134, 136, 138, 174–176
Dirac, P. A. M.   210, 276
Dirichlet, P. G. L.   25, 27, 30, 55, 56, 58–60, 76–79, 100, 102, 109, 114, 115, 121, 122, 124, 144, 149, 158, 159
Donaldson, S. K.   12, 13, 209
Doob, 255
Dorodnov, A. V.   181
Douglas, J.   79
Duhamel, J.-M.-C   41
Dyck, W.   127

Egorov, D. F.   40, 41, 48–50
Ehrenfest, P.   211
Einstein, A.   11, 275–279, 291
Eisenstein, F. G. M.   58, 59, 100, 102, 137
Euclid   84, 85, 88, 92, 105, 171, 219, 284, 290, 292
Eudoxus, 88
Euler, L.   13, 18, 20–23, 52, 72–75, 79, 89, 91, 105, 106, 108, 120, 171, 174, 175, 217, 269, 273
Eutocius   92

Faber, G.   252
Fadeev, L.   277
Fagnano, G. F.   143
Feller, W.   257
Fermat, P.   10, 65, 100, 105, 106, 108, 120, 121, 133, 134, 137, 141, 153, 174, 217
Ferrari, L.   186
del Ferro, S.   267
Fibonacci   174
Fischer, B.   168
Florenskiĭ, P. A.   53
Fourier, J.   23, 24, 26, 259, 276
Fraenkel, A.   117, 223, 230
Fréchet, M.   75, 251
Fredholm, I.   78
Freedman, M. H.   208, 209
Frege, G.   225, 226, 228, 229, 237
Frenkel, I. B.   167, 169
Freudenthal, H.   179
Friedman, H.   240
Frobenius, G. F.   112, 176

Gács, P.   264
Galileo Galilei   276
Galois, E.   3, 99, 101, 109, 115, 119, 120, 125, 126, 142, 143, 169, 174, 267
Gardner, M.   146
Garnier, R.   79
Gauss, C. F.   2, 13, 56, 58, 77, 99, 100, 102, 105, 113, 120–123, 125, 129, 143, 144, 152, 167, 175, 189, 190, 220, 271, 272
Gödel, K.   1, 67, 68, 222, 225, 227, 229, 233, 234, 237–244, 285
Goldbach, C.   25, 174, 243
Gol'shtein, E. G.   262, 263
Gompf, R.   209
Gordan, P. A.   102
Gorenstein, D.   169
Gosset, T.   176
Goursat, E. J.-B.   40, 41
Granville, W. E.   37, 41
Grassmann, H.   101
Graves, J.   176
Green, G.   76
Gregory, J.   71
Gregory, O. G.   126

Griess, R.   168
Grothendieck, A.   7, 8
Grünbaum, B.   262

Haar, A.   79
Hadamard, J.   28, 76
Hamilton, W. R.   3, 11, 101, 112, 175, 176, 221
Harish-Chandra   10
Harrington, L.   240
Hasse, H.   130
Hausdorff, F.   231, 270
Heisenberg, W.   275–277, 279
Hensel, K.   100, 116, 128–130
Henstock, R.   66
Hermann, J.   20
Hermite, C.   142–144, 146, 147, 247, 267
Heron   84–86
Hesse, L. O.   102
Hilbert, D.   1, 2, 5, 7, 8, 49, 60, 78, 79, 109, 114, 116, 117, 122, 124, 129–131, 155, 160, 177, 178, 192, 215, 225, 227–229, 237–241, 273, 276
Hippocrates   63, 181, 185
Hirzebruch, F. E. B.   8, 9
Hodge, W. V. D.   11
Hopf, H.   80
Huntington, E. V.   129
Hurwitz, A.   160, 174, 215
Hypatia   92, 93

Jacobi, C. G. J.   72–75, 92, 100–102, 122, 137, 143, 217
Jacobson, N.   131
Jones, V.   12, 13
Jordan, C. M. E.   40, 66, 101, 129, 174, 197, 267

Kac, M.   256
Kac, V.   167
Kantorovich, L. V.   259, 261
Karmarkar, N.   264
Kasparov, G. G.   9
Kelvin, see Thomson, W.
Kepler, J.   172, 173, 276, 278, 291
Khachian, L. G.   263, 264
Killing, W. K. J.   173
Kirchhoff, G. R.   76
Klee, V.   261
Kleene, S. C.   225
Klein, F.   9, 75, 102, 149, 150, 161, 189, 190, 194, 221, 271, 276
Kneser, A.   79
Kobazarev, I, Y.   277
Koebe, P.   60, 161
Kolmogorov, A. N.   242, 253–255
Koopmans, T. C.   259, 261
Korn, A.   79
Krasnosel'ski, M. A.   261, 262

NAME INDEX

Kronecker, L. 102, 114, 116, 120, 122–124, 128, 129, 142, 144, 146, 153, 162, 267, 285
Krull, W. A. L. H. 131
Krylov, A. N. 52
Kummer, E. 100, 102, 109, 114
Kuratowski, K. 237, 288
Kürschak, J. A. 130
Kurzweil, J. 66
Kuzovkin, A. I. 262

Lacroix, 40
Lagrange, J. L. 23, 72–74, 79, 119, 174, 276
Lakhtin, L. K. 40
Lambert, J. H. 189
Lamé, G. 120
Langlands, R. 10
Laplace, P. S. 248, 276
Lasker, E. 116, 117
Laurent, P. A. 40
Laver, R. 240
Lebesgue, H. 28, 53, 65–67, 79, 198, 199, 251–255
Leech, J. 168
Lefschetz, S. 159
Legendre, A.-M. 74, 143, 189
Leibniz, G. W. 4, 5, 65, 71, 72, 94, 197, 226, 236, 276, 277, 283, 284
Leonardo da Vinci 170
Leont'ev, A. F. 259
Lepowsky, J. 167, 169
Leray, J. 7
Levin, A. Y. 262–264
Levin, L. 263, 264
Lichtenberg, G. C. 289
Lie, S. 9, 171, 173, 174, 176, 178, 179
Liouville, J. 99
Lobachevski, N. I. 189, 220, 221
Lusztig, G. 10
Luzin, N. N. 33, 35
Lyusternik, L. A. 81

de Maupertuis, P. L. M. 72
Macauley, F. S. 116, 117
Manin, Y. 277
Markov, A. A. 203
Markov, A. A. (the elder) 256
Martin, D. A. 225, 235
Matiyasevich, Y. 227, 241
Maxwell, J. C. 4
Mazurkiewicz, S. 249
McKay, J. 147
McShane, E. J. 79
Medvedev, F. A. 38
Menger, K. 198, 199
Mercator, G. 288
Meurman, A. 169
Meusnier, J.-B.-M.-C 74

Milnor, J. 8, 208, 209
Minkowski, H. 149, 276
Minty, G. J. 261
von Mises, R. 249
Mittag-Leffler, M. G. 51
Mlodzeevskiĭ, B. K. 40, 42, 48
Molien, T. 112
Moore, E. H. 126, 129
Mordell, L. J. 217
Morse, M. 80
Moufang, R. 177
Mycielski, J. 235

Napier, J. 65
Nemirovski, A. 262–265
Neumann, C. 60, 78
von Neumann, J. 226, 231, 243, 259
Newman, D. J. 262, 263
Newton, I. 4, 5, 50, 65, 71, 73, 97, 107, 133–135, 236, 268, 276
Nietzche, F. 284, 285
Noether, E. 109, 117, 130, 162
Noether, M. 116, 155, 160

Olbers, H. W. M. 220
Ostrowski, A. 130

Painlevé, P. 32, 52
Pappus 176–178
Paris, J. 240
Pascal, B. 65, 291
Pasch, M. 127, 189
Peacock, G. 126
Peano, G. 127, 197, 221, 236
Pearson, E. 249
Peirce, C. S. 12, 226
Picard, E. 142
Planck, M. 247, 277
Plateau, J. A. F. 79
Plato 275, 283, 284, 290
Poincaré, H. 2, 5, 8, 31, 32, 52, 60, 78, 80, 91, 161, 162, 193, 197, 205, 206, 208, 209, 214, 215, 217, 221, 228, 247, 249, 268, 269
Poisson, S. D. 76, 276
Pólya, G. 150
Post, E. 227, 240
Privalov, I. I. 33
Proclus 219
Protter, M. H. 257
Ptolemy 84, 195, 196, 280

Quillen, D. 8

Radó, T. 79
Radon, J. K. A. 251, 254
Ramanujan, S. 146
Rego, E. 208

Riemann, G. F. B.   2, 4, 8, 9, 12, 55–61, 75–77,
    79, 80, 115, 138, 142, 149, 154–157, 159, 161,
    162, 189, 192, 268, 269, 271, 272
Riesz, F.   75, 76, 78
Robinson, A.   238
Robinson, J.   227, 241
Rosenfeld, B. A.   179
Rourke, C.   208
le Roux, J.   78
Ruffini, P.   119
Russell, B.   223, 225, 228–230, 237

Saccheri, G.   189
St. Vincent, G.   65
Saks, S.   254
Salmon, G.   102
Schering, E.   58
Schläfli, L.   172, 173
Schmidt, O. Y.   37
Schnirelman, L.   81
Schrier, O.   130
Schubert, H. C. H.   159
Schumacher, H. C.   220
Schwarz, H. A.   59, 60, 78, 79, 155
Schweikart, F. K.   189
Scott, D.   231, 235
Seiberg, N.   13
Serre, J.-P.   7
Shelah, S.   225
Shilov, G. E.   262
Shor, N. Z.   262–264
Simplicius   92
Skolem, T   223, 225, 228, 230
Smale, S.   208
Solomonoff, R. J.   242
Solovay, R.   225, 240
Solovine, M.   277, 278
Sommerfield, A.   276
Spinoza,   283, 284
Stallings, J.   208
von Staudt, K. G. C.   177
Steel, J. R.   235
Steinhaus, H.   235, 253
Steinitz, E.   129, 162
Stevin, S.   109
Steifel, E.   173
Stieltjes, T. J.   76, 247
de Sua, F.   223
Sylvester, J. J.   102

Takagi, T.   131
Tarski, A.   225, 231, 235, 238, 244
Taubes, C. H.   209

Taylor, B.   18, 19
Theaetetus   171, 290
Thales   290
Theon   92
Thomson, W.   77, 149, 158
Thurston, W. P.   14, 202
Tits, J.   179
Torricelli, E.   65
Tschirnhaus, E. W. G.   72
Turing, A. M.   225–227, 240–243

Ulam, S.   67, 68
Umemura, H.   267
Uryson, P.   198
Uspensky, J. V.   249

la Vallée-Poussin, C. J. G. N.   41, 259
Varignon, P.   19
Vasil'ev, N. A.   37
Veblen, O.   129
Viète, F.   83, 92–95, 97
Ville, J.   249
Vitali, G.   67
Volterra, V.   31, 75, 76, 78
Vygodskiĭ, M. Y.   35, 37

van der Waerden, B. L.   157, 160
Wallis, J.   65
Weber, H.   59, 103, 115, 116, 124, 125, 127–129,
    147
Wedderburn, J. H. M.   112, 117
Weierstrass, K.   2, 26, 29–32, 37, 38, 41, 42, 52,
    59, 72, 74, 75, 77–79, 102, 128, 154, 155, 157
Weyl, H.   163, 173, 176, 211, 221, 228, 268, 272
Whitehead,   228, 230, 237
Whitney, H.   199
Wiener, N.   231, 253
Wigner, E.   275–277, 279, 280, 289
Wiles, A.   10, 14, 109, 133
Witten, E.   9, 13, 195, 196
Woodin, W. H.   235

Yamnitzky, B.   264
Yudin, D. B.   261, 262

Zeeman, C.   207
Zermelo, E. F. F.   28, 29, 223–226, 230, 231
Zhegalkin, I. I.   41
Zeuthen, H. G.   50, 159
Zhukovskiĭ, N. E.   47
Zolotarev, E. I.   100, 102
Zorn, M.   176, 178

ACX6196

LIBRARY
LYNDON STATE COLLEGE
LYNDONVILLE, VT 05851